# SYSTEMS, AUTOMATION AND CONTROL

# SYSTEMS, AUTOMATION AND CONTROL

Ahmed Benzerga
Rabie Belkacemi
Hanna Bandarenka

Kruger Brentt
Publishers
2025

Kruger Brentt Publishers UK. LTD.
Company Number 9728962

Regd. Office: 68 St Margarets Road, Edgware, Middlesex HA8 9UU



ISBN: 978-1-78715-317-2



For information on all our publications visit our website at http://krugerbrentt.com/

# PREFACE

"Systems, Automation, and Control" delves into the dynamic and interconnected realms of systems theory, automation technologies, and control systems, offering readers a comprehensive exploration of the principles, methodologies, and applications that define these interdisciplinary fields. This book serves as an essential resource for students, engineers, researchers, and practitioners seeking to understand the fundamental concepts and emerging trends in systems engineering, automation, and control.

In today's increasingly complex and interconnected world, systems play a central role in virtually every aspect of human endeavor, from transportation and manufacturing to healthcare, energy, and communication systems. Automation technologies enable the efficient and reliable operation of these systems, while control systems provide the means to regulate and optimize their performance in real time.

The primary objective of this book is to provide readers with a comprehensive overview of systems theory, automation technologies, and control systems, covering key topics such as system modeling and simulation, sensor and actuator technologies, feedback control, optimization techniques, machine learning, and human-machine interaction. Each chapter offers insights, case studies, and practical examples to help readers understand the fundamental principles and applications of systems, automation, and control.

Contributions from leading experts in the field explore various aspects of systems engineering, automation technologies, and control systems, including industrial applications, cyber-physical systems, autonomous systems, robotics, and smart infrastructure. Special emphasis is placed on emerging trends such as Internet of Things (IoT), artificial intelligence, digital twins, and Industry 4.0, and their implications for the future of systems, automation, and control.

As we navigate the complexities of the digital age and the rapid pace of technological innovation, it is essential to adopt a holistic and interdisciplinary approach to systems, automation, and control. By integrating knowledge from diverse disciplines and embracing collaboration and innovation, we can address the grand challenges facing society and unlock new opportunities for improving efficiency, sustainability, and quality of life.

We extend our sincere gratitude to the contributors whose expertise and dedication have enriched this book. It is our hope that "Systems, Automation, and Control" will serve as a valuable resource and catalyst for dialogue, innovation, and collaboration in advancing the theory and practice of systems engineering, automation, and control.

**Ahmed Benzerga**
**Rabie Belkacemi**
**Hanna Bandarenka**

# CONTENTS

# 01 OPTIMIZATION OF ASSEMBLY LINE BALANCING MONOLITHIC AND BILATERAL STRUCTURES

## 1. INTRODUCTION: AN OVERVIEW

Henry Ford pioneered the implementation of the factory assembly line during the early 1900s. The concept was intended to create a streamlined and highly efficient manufacturing process for a certain product. The fundamental assembly line has a series of workstations arranged in a linear manner, with each station linked by a material handling equipment. The fundamental process of material flow along an assembly line commences with a component being introduced into the initial station at a pre-established rate of feeding. A station refers to any specific location along the assembly line where a particular task is carried out on the part. These jobs can be executed by automated systems, robots, and/or human operators. After the part arrives at a station, a specific duty is carried out on the part, and then the part is transferred to the next process. The duration required to finish a task at each operation is referred to as the process time (Sury, 1971). The cycle time of an assembly line is established in advance based on a desired output rate. The production rate is established to ensure the desired quantity of final product is manufactured within a specific timeframe (Baybars, 1986). In order for the assembly line to maintain a certain production rate, the sum of the processing times at each station must not exceed the station's cycle time (Fonseca et al, 2005). If the sum of the processing times within a station is less than the cycle time, idle time is said to be present at that station (Erel et al,1998). An essential aspect in the creation of an assembly line is the organization of tasks. The organization of this may be somewhat subjective, but it must be determined by the implicit principles established by the manufacturing sequence (Kao, 1976). In the production of any object, there are specific sequences of tasks that must be adhered to. The assembly line balance issue (ALBP) emerged alongside the creation of the assembly line. Helgeson et al (1961) were the pioneers in introducing the concept of the ALBP, whereas Salveson (1955) was the first to present the problem in its mathematical formulation.

Nevertheless, in the initial four decades of the assembly line's establishment, only experimental methods were employed to achieve equilibrium in the production lines (Erel et al., 1998). Since then, various techniques have been devised to address the various manifestations of the ALBP. Salveson (Salveson, 1955) was the first to make a mathematical attempt at solving the problem by formulating it as a linear program. Gutjahr and Nemhauser (1964) demonstrated that the ALBP issue belongs to the category of NP-hard combinatorial optimization problems. Consequently, problems of considerable magnitude do not have a guaranteed optimal solution. As a result, heuristic methods have emerged as the most often used approaches for tackling the problem. The author of this book chapter emphasizes the significance of estimating the end results and suggests updated measures for both single and two-sided assembly line balancing problems.

## 2. TWO-SIDED ASSEMBLY LINES

Two-sided assembly lines, as seen in Figure 1, are commonly used for manufacturing large-scale products, such as trucks and buses. Assembling these products differs in certain aspects from assembling tiny products. Certain assembly activities exhibit a preference for being carried out on either one of the two sides (Bartholdi, 1993).

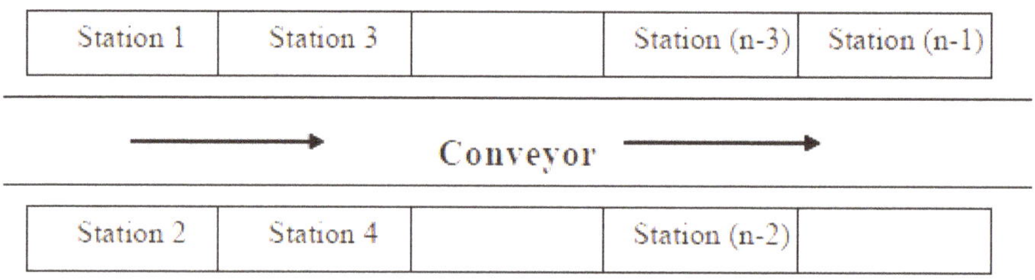


**Fig. 1. Two-sided assembly line structure**

Let's take, as an illustration, a truck assembly line. It is more convenient to install a gas tank, air filter, and toolbox on the left-hand side of the line, while mounting a battery, air tank, and muffler is better suited for the right-hand side. There is no specific requirement for the direction in which an axle, propeller shaft, and radiator are assembled. Therefore, they can be installed on any side of the line without any preference. The assessment of the optimal operation directions is crucial as it has a significant impact on the efficiency of the assembly line, especially when assigning duties, arranging facilities, and positioning tools and fixtures in a two-sided assembly line (Kim et al, 2001). According to Bartholdi (1993), a two-sided assembly line offers several significant benefits compared to a one-sided manufacturing line. The benefits of this approach are as follows: (1) it can decrease the length of the production line, resulting in a reduced need for workers, (2) it can therefore decrease the overall time required for production, (3) it can also take advantage of cost savings in tools and fixtures, as they can be shared between both sides of a connected station, and (4) it can minimize material handling, worker movement, and setup time, which would otherwise be difficult to eliminate. The benefits of using double-sided lines for manufacturing large-sized products provide a strong rationale for their utilization. The representation of a line balancing problem is typically depicted by a precedence diagram, as seen in Figure 2.

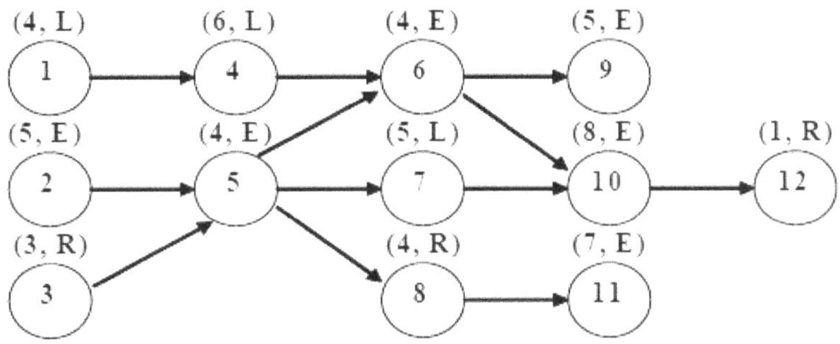


**Fig. 2. Precedence graph**

A circle symbolizes a task, whereas an arc connecting two tasks signifies the precedence relation between them. Every job is linked to a label (ti, d), where ti represents the duration of the task processing and d (which might be L, R, or E) indicates the desired direction of operation. The letters L and R indicate that the task should be allocated to a station on the left and right side, respectively. The task related to E can be executed on either side of the line. When balancing assembly lines, it is important to consider the unique characteristics of each line. If precedence relations are adequately regarded in a one-sided assembly line, all jobs assigned to a station can be executed continuously without any interruptions. Nevertheless, in a two-sided assembly line, certain duties allocated to a station may experience delays due to the tasks given to its counterpart (Bartholdi, 1993). Put simply, there are instances where idle time cannot be avoided, even when transitioning between activities given to the same station. Take, for instance, task j and its direct antecedent i. Assume that the variable j is allocated to a station, whereas the variable i is assigned to its corresponding companion station. job j is dependent on the completion of job i. Hence, when it comes to balancing a two-sided assembly line, it is crucial to take into account the finish time of activities, which is dependent on their sequence.

## 3. HEURISTIC APPROACHES FOR SOLVING THE ASSEMBLY LINE BALANCING PROBLEM

The heuristic approach relies on logical reasoning and practical wisdom rather than on quantitative demonstration. Heuristics do not provide a guarantee of finding the best possible answer, but they do yield satisfactory solutions that get close to the genuine optimal solution.

## 3.1. Heuristic methods for achieving balance in a single assembly line

The majority of the heuristic techniques documented in literature are specifically tailored to solve a single assembly line balancing problem. Furthermore, the majority of these systems rely on uncomplicated priority rules and produce just a limited number of viable options. Task-oriented processes select the most important task from a list of available tasks and assign it to the earliest station that can handle it. Within task-oriented procedures, two distinct techniques can be identified: immediate-update-first-fit (IUFF) and general-first-fit. The choice between these methods depends on whether the collection of available tasks is updated immediately after assigning a work or after all currently available tasks have been assigned. The immediate-update-first-fit method is more commonly utilized due to its higher level of flexibility. The primary concept underlying this heuristic is the allocation of jobs to stations based on their numerical score. There are multiple methods to ascertain the score for each activity. One might readily devise their own method for obtaining the score, although it is uncertain whether this approach would provide satisfactory results. In this section, we provide five distinct strategies that have been documented in the literature, along with the respective solutions they provide for our straightforward case. The methods are also incorporated into the Line Balancing program. Once the suitable score for each work is established, the execution of methods does not vary, and the necessary steps to achieve the answer are as follows:

STEP 1. Assign a numerical score $n(x)$ to each task $x$.

STEP 2. Update the set of available tasks (those whose immediate predecessors have been already assigned).

STEP 3. Among the available tasks, assign the task with the highest numerical score to the first station in which the capacity and precedence constraints will not be violated. Go to STEP 2.

The most popular heuristics which belongs to IUFF group are:

- IUFF-RPW Immediate Update First Fit – Ranked Positional Weight,
- IUFF-NOF Immediate Update First Fit – Number of Followers,
- IUFF-NOIF Immediate Update First Fit – Number of Immediate Followers, IUFF-NOP Immediate Update First Fit – Number of Predecessors,
- IUFF-WET Immediate Update First Fit – Work Element Time.

## 3.2 Method for balancing a two-sided assembly line using a heuristic approach

A task group is comprised of a specific task, referred to as task i, and all of the tasks that must be completed before task i can begin. A group is created for each work that has not been allocated. When balancing a two-sided assembly line, it is important to take into account the operation directions and sequence dependency of tasks, as well as the creation of new groups (Lee et al, 2001). During the formation of the initial groups IG(i), the operation direction is continuously monitored. It is prohibited for a group to include tasks with conflicting preferred operating directions. However, if each work in the first group is an E-task, the group can be assigned to either side. To establish the operational guidelines for these groupings, the rules (referred to as direction rules DR) are implemented:

- DR 1. Set the operation direction to the side where tasks can be started earlier.
- DR 2. The start time at both sides is the same, set the operation direction to the side where it's expected to carry out a less amount of tasks (total operation time of unassigned L or R tasks).

Typically, activities that arise from a "repeatability test" are considered initial tasks. However, there is an exception in the shape of the first iteration, where the method begins by searching for tasks (referred to as initial tasks IT) that are the first ones in the precedence relation. Following the initial step in the initial iteration, we obtain:

IG (1) = {1}, Time{IG (1)} = 2, Side{IG (1)} = 'L'

IG (2) = {2}, Time{IG (2)} = 5, Side{IG (2)} = 'E'

IG (3) = {3}, Time{IG (3)} = 3, Side{IG (3)} = 'R

where:

- Time{IG(i)} – total processing time of ith initial group, Side{IG(i)} – preference side of ith initial group.
- To those who are considered to be the first, the next tasks will be added, (these ones which fulfil precedence constraints).
- Whenever new tasks are inserted to the group i, the direction, cycle time and number of immediate predecessors are checked. If there are more predecessors than one, the creation of initial group j comes to the end.

First iteration – second step

IG (1) = {1, 4}, Time{IG (1)} = 4, Side{IG (1)} = 'L'

IG (2) = {2, 3, 5}, Time{IG (2)} = 12, Side{IG (2)} = 'R'

Whenever two or more initial groups are joined together, or when initial group is connected with those one coming from Final set – the "double task" is added to initial tasks needed for the next iteration. In the end of each iteration, created initial groups are copied to FS.

$$FS = \{ (1, 4); (2, 3, 5) \},$$

$$Side\{FS (1)\} = \text{'L'}, Side\{FS (2)\} = \text{'R'}$$

$$Time \{FS(2)\} = 12, Time \{FS(1)\} = 14,$$

$$IT = \{5\}.$$

In the second iteration, second step, we may notice that predecessor of last task coming from IG(1) is included in Final Set, FS(2). The situation results in connecting both groups under holding additional conditions:

$$Side\{IG(1)\} = Side\{FS(2)\},$$

$$Time + time < cycle.$$

After all, there is no more IT tasks, hence, preliminary process of creating final set is terminated.

The presented method for finding task groups is to be summarized in simplified algorithm form. Let U denote to be the set of un – assigned tasks yet and IGi be a task group consisting of task i and all its predecessors (excluded from U set).

STEP 1. If U = empty, go to step 5, otherwise, assign starting task from U.

STEP 2. Identify IGi . Check if it contains tasks with both left and right preference operation direction, then remove task i.

STEP 3. Assign operation direction Side{ IGi } of group IGi . If IGi has R-task (L-task ), set the operation direction to right (left). Otherwise, apply so called direction rules DR.

STEP 4. If the last task i in IGi is completed within cycle time, the IGi is added to Final set of candidates FS(i). Otherwise, exclude task i from IGi and go to step 1.

STEP 5. For every task group in FS(i), remove it from FS if it is contained within another task group of FS.

The resulting task groups become candidates for the mated-station

$$FS = \{(1,4), (2,3,5,8)\}.$$

The candidates are generated using processes outlined in the preceding section, which assert that they do not break precedence, operation direction limits, and also impose constraints on groups to be finished within a pre-determined cycle period. While all applicants may be allocated the same number of resources, only one group will ultimately be selected. In order to determine which group to select, the following rules that aid in decision-making will be outlined and clarified:

- AR 1. Choose the task group FS(i) that may start at the earliest time.
- AR 2. Choose the task group FS(i) that involves the minimum delay.
- AR 3. Choose the task group FS(i) that has the maximum processing time.

For the purpose of clarity, we shall conceptualize a mating station with distinct left and right sides, each previously assigned certain functions. To ensure a station is adequately balanced, the AR 1 method is used. An unbalanced station is defined as one that is likely to experience additional delays in future assignments. The decrease of the number of stations is not the exclusive objective. There are also indirect goals, such as the reduction of inevitable delay. This rule assigns more priority to the station with a lower allocation of tasks. In the event of a tie, the AR 2 algorithm is implemented, which selects the group with the lowest delay out of the options being examined. This rule may potentially lead to a tie. The final issue pertains to the correlation between labor and its related aspects within individual station group by choosing group of task with highest processing time. For the third rule the tie situation is impossible to obtain, because of random selection of tasks. The implementation of above rules is strict and easy except the second one. Shortly speaking, second rule is based on the test, which checks each task consecutively, coming from candidates group FS(i) – in order to see if one of its predecessors have already been allocated to station. If it has, the difference between starting time of considered task and finished time of its predecessor allocated to companion station is calculated. The result should be positive, otherwise time delay occurs.

Having rules for initial grouping and assigning tasks described in previous sections, we may proceed to formulate formal procedure of solving two – sided assembly line balancing problem (Kim et. al, 2005).

Let us denote companion stations as *j* and *j'*,

D(i) – the amount of delay,

Time(i) – total processing time (Time{FS(i)}), S(j) – start time at station j,

STEP 1. Set up j = 1, j' = j + 1, S(j) = S(j') = 0, U – the set of tasks to be assigned. STEP 2. Start procedure of group creating (3.2), which identifies

FS = {FS(1), FS(2), ..., FS(n)}. If FS = ∅, go to step 6.

STEP 3. For every FS(i), i = 1,2, ... , n – compute D(i) and Time (i). STEP 4. Identify one task group FS(i), using AR rules in Section 3.3

STEP 5. Assign FS(i) to a station j (j') according to its operation direction, and update S(j) = S(j) + Time(i) + D(i). U = U – {FS(i)}, and go to STEP 2.

STEP 6. If U ≠ ∅, set j = j' + 1, j' = j + 1, S(j) = S(j') = 0, and go to STEP 2, otherwise, stop the procedure.

## 4. METRICS FOR THE FINAL OUTCOMES OF THE ASSEMBLY LINE BALANCING PROBLEM

Some measures of solution quality have appeared in line balancing problem. Below are presented three of them (Scholl, 1999).

Line efficiency (LE) shows the percentage utilization of the line. It is expressed as ratio of total station time to the cycle time multiplied by the number of workstations:

where: K - total number of workstations, c - cycle time.

$$LE = \frac{\sum_{i=1}^{K} ST_i}{c \cdot K} \cdot 100\% \qquad (1)$$

Smoothness index (SI) describes relative smoothness for a given assembly line balance. Perfect balance is indicated by smoothness index 0. This index is calculated in the following manner:

$$SI = \sqrt{\sum_{i=1}^{K} (ST_{max} - ST_i)^2} \qquad (2)$$

where:

$ST_{max}$ - maximum station time (in most cases cycle time),

Time of the line (LT) describes the period of time which is need for the product to be completed on an assembly line:

$$LT = c \cdot (K-1) + T_K \qquad (3)$$

where:

c - cycle time,

K -total number of workstations,

$T_K$ – processing time of last station.

The final result estimation of two-sided assembly line balance needs some modification of existing measures (Grzechca, 2008).

Time of line for TALBP

$$LT = c \cdot (Km - 1) + Max\{t(S_K), t(S_{K-1})\} \qquad (4)$$

where:

$K_m$ – number of mated-stations

K – number of assigned single stations

$t(S_K)$ – processing time of the last single station

As far as smoothness index and line efficiency are concerned, its estimation, on contrary to LT, is performed without any change to original version. These criterions simply refer to each individual station, despite of parallel character of the method.

But for more detailed information about the balance of right or left side of the assembly line additional measures will be proposed:

$$SI_L = \sqrt{\sum_{i=1}^{K} (ST_{maxL} - ST_{iL})^2} \quad (5)$$

Where:

$SI_L$- smoothness index of the left side of two-sided line

$ST_{max}$ L- maximum of duration time of left allocated stations

$ST_{iL}$- duration time of i-th left allocated station

Smoothness index of the right side

$$SI_R = \sqrt{\sum_{i=1}^{K} (ST_{maxR} - ST_{iR})^2} \quad (6)$$

where:

$SI_R$- smoothness index of the right side of two-sided line,

$ST_{max}$ R- maximum of duration time of right allocated stations,

$ST_{iR}$- duration time of i-th right allocated station.

## 5. NUMERICAL EXAMPLES

An numerical example from Fig. 3. will be considered. The number of tasks, precedence graph and processing times are known and there are given in Table 1.The cycle time is 10.

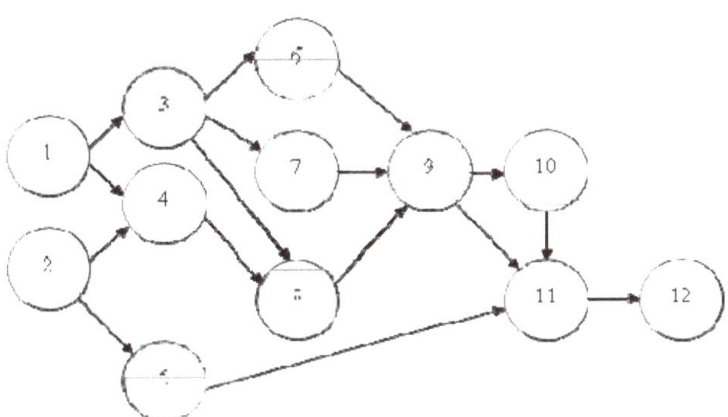


**Fig. 3. Precedence graph for single line**

**Table 1. Input data of numerical example – IUFF Ranked Positional Weight**

| Number of task | Processing Time | Weight | Positional Rank |
|---|---|---|---|
| 1 | 1 | 29 | 2 |
| 2 | 5 | 27 | 1 |
| 3 | 4 | 28 | 3 |
| 4 | 3 | 22 | 4 |
| 5 | 2 | 11 | 5 |
| 6 | 6 | 24 | 7 |
| 7 | 3 | 21 | 6 |

| Number of task | Processing Time | Weight | Positional Rank |
|---|---|---|---|
| 8 | 1 | 19 | 8 |
| 9 | 7 | 18 | 11 |
| 10 | 6 | 11 | 9 |
| 11 | 3 | 5 | 10 |
| 12 | 2 | 2 | 12 |

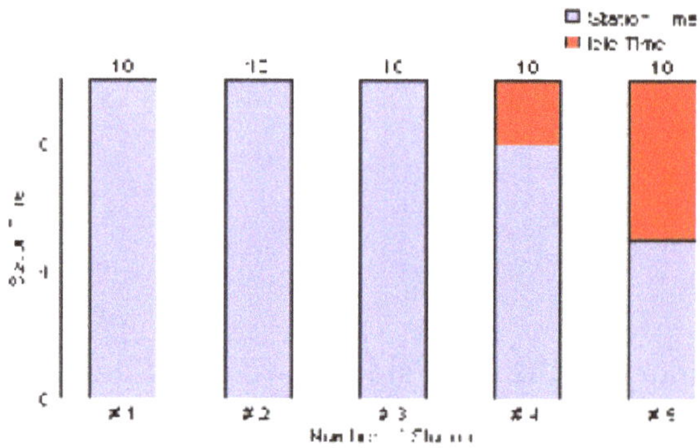

Fig. 4. Assembly line balance for IUFF-RPW and IUFF-NOF methods

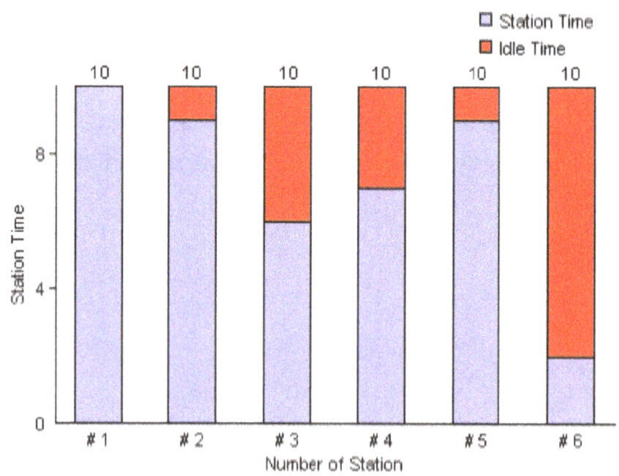


Fig. 5. Assembly line balance for IUFF-NOP and IUFF-NOIF methods

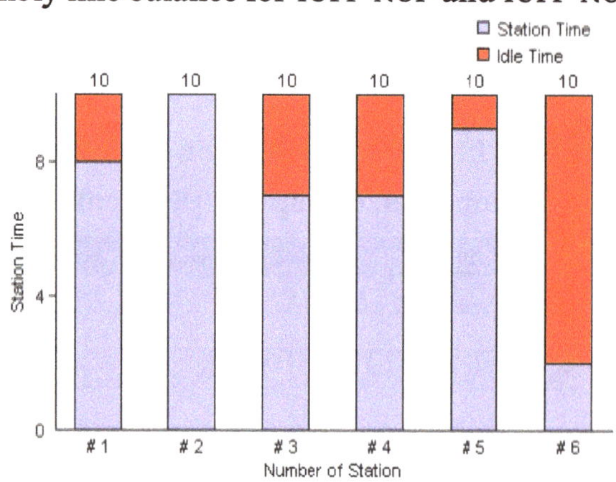


Fig. 6. Assembly line balance for IUFF-WET method

**Table 2. Results of balance for IUFF methods**

| Method | K | Balance | LE | SI | LT |
|---|---|---|---|---|---|
| IUFF-RPW | 5 | S1 – 1, 3, 2<br>S2 – 6, 4, 8<br>S3 – 7, 9<br>S4 – 10, 5<br>S5 – 11, 12 | 86% | 5,39 | 45 |
| IUFF-NOF | 5 | S1 – 1, 3, 2<br>S2 – 6, 4, 8<br>S3 – 7, 9<br>S4 – 10, 5<br>S5 – 11, 12 | 86% | 5,39 | 45 |
| IUFF-NOIF | 6 | S1 – 1, 2, 3<br>S2 – 5, 4, 7, 8<br>S3 – 6<br>S4 – 9<br>S5 – 10, 11<br>S6 – 12 | 71,67% | 9,53 | 52 |
| IUFF-NOP | 6 | S1 – 1, 2, 3<br>S2 – 5, 4, 7, 8<br>S3 – 6<br>S4 – 9<br>S5 – 10, 11<br>S6 – 12 | 71,67% | 9,53 | 52 |
| IUFF-WET | 6 | S1 – 2, 5, 1<br>S2 – 3, 6<br>S3 – 4, 7, 8<br>S4 – 9<br>S5 – 10, 11<br>S6 – 12 | 71,67% | 9,33 | 52 |

**Table 3. Input data of numeraical example – two-sided line from Fig. 2.**

| Number of task | Processing Time | Position (Constraints) |
|---|---|---|
| 1 | 4 | L |
| 2 | 5 | E |
| 3 | 3 | R |
| 4 | 6 | L |

| Number of task | Processing Time | Position (Constraints) |
|---|---|---|
| 5 | 4 | E |
| 6 | 4 | E |
| 7 | 5 | L |
| 8 | 4 | R |
| 9 | 5 | E |
| 10 | 8 | E |
| 11 | 7 | E |
| 12 | 1 | R |

The results of heuristic procedure for the example from Fig. 2 and cycle time c=16 are given in a Gantt chart – Fig. 7.

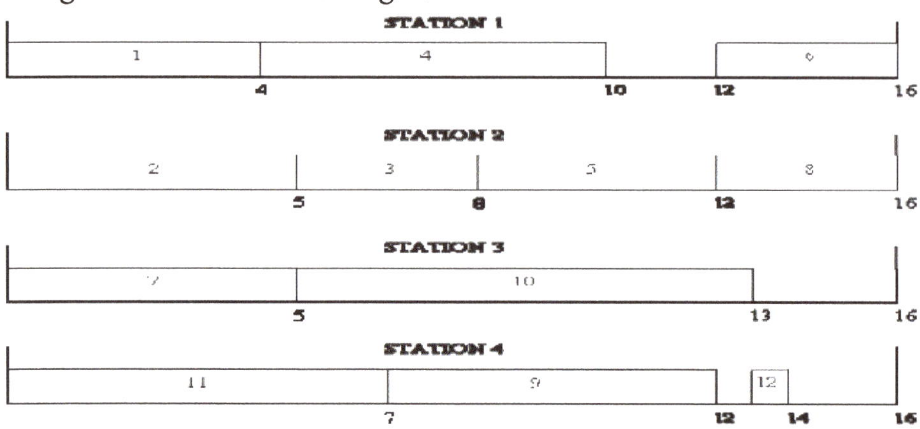


**Fig. 7. Gantt chart of assembly line balance of two-sided structure (Fig. 2.)**

Before presenting performance measures for current example, it would be like to stress difference in estimation of line time form, resulting from restrictions of parallel stations. In two – sided line method within one mated-station, tasks are intended to perform its operations at the same time, as it is shown in example in Fig. 7., where tasks 7, 11 respectively are processed simultaneously on single station 3 and 4, in contrary to one – sided heuristic methods. Hence, modification has to be introduced to that particular parameter which is the consequence of parallelism. Having two mated-stations from Fig. 7, the line time LT is not 3*16 + 13, as it was in original expression. We must treat those stations as two double ones (mated-stations), rather than individual ones Sk (4). As far as smoothness index and line efficiency are concerned, its estimation, on contrary to LT, is performed without any change to original version. These criterions simply refer to each individual station, despite of parallel character of the method. But for more detailed information about the balance of right or left side of the assembly line additional measures (5) and (6) was proposed (Grzechca, 2008).

**Table 4. Numerical results of balance of two-sided assembly line structure**

| Name | Value |
|---|---|
| LE | 84,38% |
| LT | 30 |
| SI | 4,69 |
| SIR | 2 |
| SIL | 3 |

## 6. CONCISE SUMMARY

In recent times, there has been a growing popularity of single and two-sided assembly lines. Hence, it is imperative to examine these structures employing various methodologies. This chapter presented a heuristic method. The single assembly line balancing problem frequently encounters challenges with the final station. Production engineers in firms find it impossible to accept a perfect solution that achieves 100% efficiency in workstations, except for the last one. Various heuristic methods enable the acquisition of distinct possible solutions, allowing for the selection of the best suitable outcome. The dual-sided assembly line arrangement is very responsive to variations in cycle time values. Frequently, the end outcome of a two-sided assembly line may exhibit an incomplete structure, with certain stations missing. Various metrics can be employed to compare the solutions, such as line time, line efficiency, and smoothness index. The author suggests the inclusion of two new measures: the smoothness index of the left side (SIL) and the smoothness index of the right side (SIR) for the two-sided assembly line arrangement. These measures provide additional insights into task allocation and the equilibrium on both sides.

## REFERENCES


1. Bartholdi, J.J. (1993). Balancing two-sided assembly lines: A case study, *International Journal of Production Research*, Vol. 31, No,10, pp. 2447-2461
2. Baybars, I. (1986). A survey of exact algorithms for simple assembly line balancing problem, *Management Science*, Vol. 32, No. 8, pp. 909-932
3. Erel, E., Sarin S.C. (1998). A survey of the assembly line balancing procedures, *Production Planning and Control*, Vol. 9, No. 5, pp. 414-434
4. Fonseca D.J., Guest C.L., Elam M., Karr C.L. (2005). A fuzzy logic approach to assembly line balancing, *Mathware & Soft Computing*, Vol. 12, pp. 57-74
5. Grzechca W. (2008) Two-sided assembly line. Estimation of final results. *Proceedings of the Fifth International Conference on Informatics in Control, Automation and Robotics ICINCO 2008, Final book of Abstracts and Proceedings*, Funchal, 11-15 May 2008, pp. 87-88, CD Version ISBN: 978-989-8111-35-7

6. Gutjahr, A.L., Neumhauser G.L. (1964). An algorithm for the balancing problem, *Management Science*, Vol. 11,No. 2, pp. 308-315

7. Helgeson W. B., Birnie D. P. (1961). Assembly line balancing using the ranked positional weighting technique, *Journal of Industrial Engineering*, Vol. 12, pp. 394-398

8. Kao, E.P.C. (1976). A preference order dynamic program for stochastic assembly line balancing, *Management Science*, Vol. 22, No. 10, pp. 1097-1104

9. Lee, T.O., Kim Y., Kim Y.K. (2001). Two-sided assembly line balancing to maximize work relatedness and slackness, *Computers & Industrial Engineering*, Vol. 40, No. 3, pp. 273-292

10. Salveson, M.E. (1955). The assembly line balancing problem, *Journal of Industrial Engineering*, Vol.6, No. 3. pp. 18-25

11. Scholl, A. (1999). *Balancing and sequencing of assembly line*, Physica- Verlag, ISBN 9783790811803, Heidelberg New-York

12. Sury, R.J. (1971). Aspects of assembly line balancing, *International Journal of Production Research*, Vol. 9, pp. 8-14

# 02 AN ARCHITECTURAL FRAMEWORK FOR COGNITIVE TECHNICAL SYSTEMS DESIGNED FOR ASSEMBLY TASKS IN A PRODUCTION ENVIRONMENT

## 1. INTRODUCTION: AN OVERVIEW

In recent years, many corporations have embraced production in low-wage countries due to the significant cost advantages it offers. In order to impede the acceleration of production relocation to countries with lower wages, it is imperative to devise novel approaches for production in nations with higher wages. Presently, the extensively mechanized sector demonstrates exceptional efficiency in manufacturing a limited variety of products in enormous quantities. Automotive giants such as GM, Daimler, and BMW serve as prime instances of this. Modifications in the production lines result in substantial financial investments in both equipment and personnel. In order to achieve the anticipated level of production, the entire production process is meticulously planned in advance. Conversely, there is a sector of the economy that focuses on producing personalized goods in limited quantities. Super sports vehicle manufacturers are exemplified by these examples. The product is constantly changing due to personalization. These various manufacturing methods can be categorized into two issues that enterprises in high wage countries must address.

The initial quandary lies in the conflict between value orientation and planning orientation. Value orientation suggests that the manufacturing process lacks significant pre-production planning. The manufacturing process is modified and refined during production. The opposite of value orientation is planning orientation. Prior to manufacturing, the entire process is carefully designed and optimized. The second quandary is in the choice between scale and scope. Scale refers to the process of mass production, in which the efficiency of manufacturing is enhanced due to the effects of scaling. In contrast, scope refers to the dominance of small batch quantities in manufacturing. The mainstream automobile industry is characterized by large-scale production and a highly organized approach, driven by the level of automation used. The producer of high-performance sports cars is implementing a focused strategy that prioritizes value. The polylemma of production technology is represented by two dilemmas, as shown in Figure 1 (Brecher et al, 2007). The primary objective of the cluster of excellence "Integrative Production Technology for High-Wage Countries" at RWTH Aachen University is to minimize these issues. The research goal is to produce a wide range of items in small quantities, while keeping costs competitive with mass production. This will be achieved by fully utilizing the advantages of value orientation and planning orientation. In order to achieve the desired outcome, four key research areas were identified. The four sections are referred to as "Individualized Production Systems", "Virtual Production Systems", "Hybrid Production Systems", and "Self-optimizing Production Systems". Self-optimizing production systems aim to implement value-oriented strategies by improving planning efficiency through the application of previously acquired information to new production situations.

The research hypothesis is that only technical systems which incorporate cognitive capabilities are capable of showing self-optimizing behavior (Heide 2006). In addition to that these Cognitive Technical Systems can reduce the planning efforts required to adapt to changes in the process chain (Brecher et al. 2007). In this chapter a software architecture for such a Cognitive Technical System will be described and a use case in the context of assembly processes will be presented. Section 2 will deal with the definition of the terms "Self optimization", "Cognition" and "Cognitive Technical System". Section 3 deals with related work in the context of Cognitive Technical Systems and the involved software architectures. The fourth section describes an excerpt of the functional as well as the non- functional requirements for a Cognitive Technical System. Afterwards the software architecture will be presented. The sixth section will introduce an assembly use case and the chapter closes with a final conclusion in section 7.

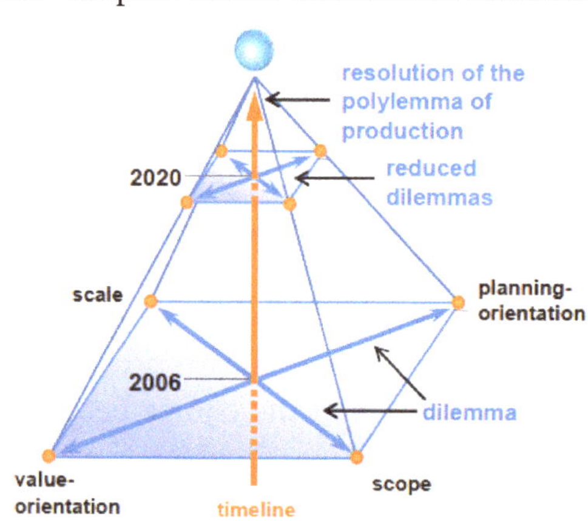


**Fig. 1. Polylemma of Production Technology**

# 2. DEFINITION OF TERMS

## 2.1 Self-Optimization

Self-optimization within the realm of artificial systems encompasses three collaborative actions. Initially, it is necessary to examine the present condition, followed by the determination of objectives. These objectives can be contradictory. In this scenario, the system must make a tradeoff between the objectives. The third step involves adapting the behavior of the system. A system can be considered a self-optimizing system if it has the ability to analyze and identify significant changes in the environment or within itself. It can then adjust its objectives in response to these changes, whether they come from the technical system's surroundings, the user, or the system itself. Additionally, the system can autonomously adapt its behavior by making changes to its parameters or structure in order to achieve its objectives (Gausemeier 2008). In order to adapt, the system must integrate cognitive capabilities to effectively understand the present circumstances and subsequently modify its behavior accordingly.

## 2.2 Cognition

Presently, the term "Cognition" is predominantly associated with human beings and lacks a clear definition in the fields of psychology, philosophy, and cognitive science. According to Matlin (2005), cognition in the field of psychology refers to the process of acquiring, storing, using, and transforming knowledge in humans. This entails a multitude of diverse methods. Zimbardo (2005) identifies perception, reasoning, remembering, thinking, decision-making, and learning as fundamental cognitive processes. These definitions are not readily transferrable to artificial systems. Strasser (2004) provides a first definition of "Cognition" within the context of artificial systems. She derived her term from Strube's work in 1998:

Cognition, which is a relatively new trait in evolution, enables the sensory and motor aspects of an organism to be connected in an adaptive and indirect manner. This flexible and indirect connection allows for the capacity to acquire knowledge (as seen by conditioning, a method that even works in simple animals like flatworms), and in more complex organisms, the ability to engage in thoughtful decision-making. From this, Strasser derives the definition that a system, whether it be technical or biological, must include a flexible connectivity between input and output, which in turn implies the capacity to acquire knowledge. The term "learning" refers to the basic capacity to adjust the output in order to maximize the predicted usefulness, based on the input received. These are the fundamental criteria that a system must include in order to be considered cognitive, according to her. When the term "Cognition" is used, this definition should be taken into consideration. This definition represents the minimum extent to which the phrase can be applied. Thus, this does not mean that a system that may be classified as "Cognitive" is capable of integrating cognitive processes at the level of human beings.

## 2.3 Cognitive Technical Systems

The phrase "Cognitive Technical System" can be defined based on the above definitions for "Self-optimization" and "Cognition". Furthermore, a detailed account of the intermediary stages in the development of a Cognitive Technical System capable of integrating advanced cognitive processes can be provided. In the purpose of this chapter, a Cognitive Technical System is defined as an artificial system that possesses cognitive capacities and can adapt to various environmental changes.

Fig. 2 shows the different steps towards a Cognitive Technical System capable of cognition on a higher level. As cognitive processes of a higher level the communication in natural language and adaption to the mental model of the operator can be named. Also more sophisticated planning abilities in unstructured, partly observable and nondeterministic environments can be accounted as cognitive processes on a higher level.

The present state of human-machine interaction within a manufacturing environment is depicted in the left portion of the image. Cognitive processes are exclusive to humans. The technological system, comprising an Interaction System and a Technological Application System, does not possess cognitive capabilities as defined by Strasser (2004). The interaction occurs in a traditional manner using a human-machine interface that is integrated into the Interaction System. The evaluation and optimization of the Technological Application System's output is only performed by the human operator. Therefore, it is only the human operator who has the ability to analyze and enhance the output of the Technological Application System by adjusting the parameters of the processes through the human-machine interface.

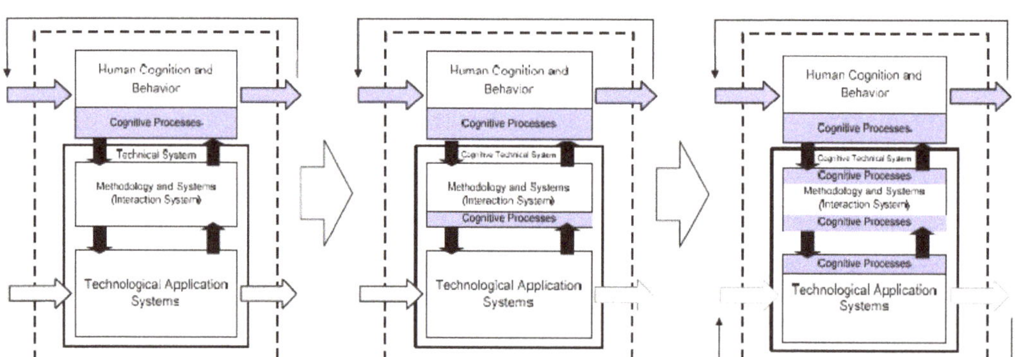


**Fig. 2. Steps in the development of Cognitive Technical Systems**

The intermediate level, located in the middle of the figure, involves integrating fundamental cognitive processes into the Interaction System. The technological system can now be classified as a Cognitive technological System. Cognitive processes encompass the mental activities of reasoning and decision making. The Cognitive Technical System must integrate a knowledge foundation from which decisions can be derived. The cognitive processes are integrated inside the Interaction System, which both communicates with and governs the Technological Application System. The right portion of the image displays an innovative Cognitive Technical System. This system integrates cognitive processes at all levels, resulting in human-machine interaction that is based on multimodal communication. Furthermore, the human-machine interface adjusts to the cognitive framework of the human operator as they engage in conversation. This significantly enhances the efficiency of the communication process. Thus, the human-machine interaction integrates cognitive capabilities. Furthermore, the Interaction System integrates cognitive processes in its communication with the Technological Application System, which also has cognitive skills. This implies that the level of communication can be elevated. The systems just communicate concepts of a specific nature, and the future tasks are generated autonomously by the system. The act of producing a report is an example of human communication that involves the interchange of ideas and concepts. The receiver of the task must carry out numerous steps that have not been specified. However, the outcome aligns with the "intentions" of the person who assigned the assignment.

Furthermore, such a system would include the capability to assess the outcome of the procedure, along with the corresponding parameters that contribute to it. This facilitates the manifestation of self-optimizing behavior. The evaluation process is shown as a feedback system originating from the Cognitive Technical System (located in the lower right part). Regarding manufacturing processes, a Cognitive Technical System has the ability to oversee and manage various components of the Technological Application System, such as robots and belt conveyors, in order to carry out diverse jobs. A variety of Cognitive Technical Systems can collaborate to manage a sequence of actions and optimize it as a unified entity, based on the specified optimization goals assigned to the systems. The system can receive task descriptions in a more abstract manner. One potential description could encompass the physical form of the object or certain attributes of the thing. The Cognitive Technical System utilizes this description to deduce the necessary actions for creating the intended product. The description provided at such an abstract level is significantly lacking in detail, which is similar to a work description given by one person to another (Hägele 2008). In order to solve the assignment, a substantial knowledge base is essential for obtaining the necessary missing information.

## 3. RELATED WORK

Given the extensive research conducted in several domains such as artificial intelligence, software engineering, electrical engineering, and others, this section does not aim to provide a comprehensive overview. Instead, it presents a curated collection of relevant studies, with a specific emphasis on software architectures. Possible application sectors for Cognitive Technical Systems include autonomous vehicle control, manufacturing environments, and service robotics. There are numerous additional fields that are not further assessed within this particular context. The DARPA grand challenges in 2005 and 2007 demonstrated the feasibility of controlling ground vehicles in a somewhat unstructured environment while adhering to traffic regulations (Montemerlo et al., 2006). The Cognitive Technical Systems employed a multi-layered software design that extensively utilized state machines (Urmson et al., 2007).

Several architectural techniques have been proposed for autonomous robots, including those by Karim (2006), Konolige (1998), Gat (1998), and others. These software architectures prioritize the integration of a cognitive component for the actual planning process with a responsive component for motion control (Putzer 2004). The research conducted by the cluster of excellence "Cognition for Technical Systems" (CoTeSys) focuses on developing software architectures for Cognitive Technical Systems in production contexts (Ding et al., 2008). The research centers on incorporating cognitive capabilities into safety controls for industrial control. The primary evaluation scenario in this context is the cooperation between humans and machines. None of the described methodologies specifically address the use of Cognitive Technical Systems in an assembly procedure.

## 4. REQUIREMENTS

### 4.1 Functional Requirements

This section outlines the necessary functional criteria for a Cognitive Technical System that is capable of performing effectively in a production environment. A functional requirement is a requirement that becomes evident during the functioning of the system. (Sommerville 2007).

A Cognitive Technical System must possess the ability to process various sensor inputs, such as visual, tactile, or electric sensors, and combine them to extract crucial information. Given this information, the Cognitive Technical System is required to analyze the data and choose the most appropriate course of action based on the present environmental conditions and the specified purpose. In order to implement the next action determined by the Cognitive Technical System, it is necessary to regulate external elements such as robot actuators or conveyor belts.

The Cognitive Technical System is required to engage with a human operator through a human-machine interaction. The specific design of the human-machine interface is not included in the functional requirements, as stated by Cockburn (2003), but it is outlined in the non-functional requirements. In order to make a decision based on the information received, the system must possess a knowledge base that encompasses the domain expertise. Additionally, the procedural information regarding the many actions it can utilize to modify its surroundings must be kept.

A Cognitive Technical System requires additional functional requirements due to the context of a production plant. It is necessary to ensure communication with machinery such as programmable logic controllers (PLC) and multi-axis robots using various protocols.

## 4.2 Non-Functional Requirements

Non-functional requirements are characteristics that can be used to evaluate the performance of a system, rather than its specific behaviors (Sommerville 2007). The non-functional requirements for the human-machine interaction are based on DIN ISO 9355-1 and can be categorized into 14 groups, which will not be elaborated on in this context.

The software architecture criteria are partially drawn from ISO 9126. The subsequent categories are deemed indispensable and will be expounded upon in further depth:

- Modularity, Extendibility, Flexibility
- Robustness and Reliability
- Response times
- Information- and Datamanagement
- External communication
- User Interaction

**Modularity, Extendibility, Flexibility:** The software architecture of a Cognitive Technical System designed for an assembly task in a production setting must fulfill the criteria of modularity, extendibility, and adaptability. Modularity, in this context, refers to the ability to interchange components without the need to redesign the entire system. This pertains to the user interface, the various controller components, and the components responsible for decision making. This requires the process of enclosing individual functionalities behind components and utilizing clearly defined interfaces to connect them. The software design should possess the quality of extensibility, allowing for seamless integration of new components with minimal effort. This also fulfills the need for adaptability.

**Robustness and Reliability:** In a production environment the requirements for the reliability and the robustness of a system are high. The technical system must have a high reliability because of the high costs of a possible production stop in case of a system failure. Because of this certain safety measures must be implemented in the Cognitive Technical System. This can be realized through redundancy of components or by fault tolerant code. This also ensures a high robustness.

**Response times:** Processes in a production setting are optimized to achieve maximum throughput. These additional limitations impose more restrictions on the software structure of such a system. The response time should be sufficiently short to promptly respond to abrupt alterations in the surroundings. The deliberative component of the Cognitive Technical System is unable to provide real-time decisions due to the vast volume of knowledge being processed. Hence, it is imperative to guarantee the overall response time of the system using a method that is independent of deliberate decision-making.

**Information and Datamanagement:** The information flow within the Cognitive Technical System is highly extensive. The sensory information must undergo processing and be directed to the relevant components. The software architecture must include an internal communication mechanism to transmit information to the components. Furthermore, it is important to guarantee the storage of data in several repositories to accommodate the substantial bandwidth and the vast volume of accumulated data.

**External communication:** The Cognitive Technical System must engage in communication with many entities inside a production environment. These entities can include physical objects such as robots and programmable logic controllers. Additionally, the corresponding interfaces must also support various bus protocols, such as CAN-Bus and Process Field Bus (PROFIBUS). Furthermore, it is imperative that these interfaces may be easily extended.

**User Interaction:** The Cognitive Technical System must ensure effective communication with the system user. The user's input must undergo processing, and the decisions made by the Cognitive Technical System need to be communicated to the user.

### 4.3 Conclusion

The system's functional and non-functional needs have a significant impact on the design of the software architecture. The software architecture is determined by the specific needs of a production environment, particularly the requirement for a system with a low response time. Moreover, reliability is a crucial prerequisite.

## 5. SOFTWARE ARCHITECTURE

### 5.1 Multilayer approach

In order to fulfill the functional and non-functional objectives, a software architecture for a Cognitive Technical System designed for assembly activities must include many components.

The system must be compatible with many degrees of abstractions. Consequently, the deliberative mechanism is unable to process the raw sensor data obtained from the Technological Application System. Hence, it is imperative to create a conceptual representation of the data that has been received. This requires a component that can gather the received information for the deliberative mechanism. In order to fulfill the need for a quick reaction time, it is necessary to include a control mechanism that can operate independently of the deliberative mechanism's answer. In addition, the Cognitive Technical System must possess the capability to oversee and manage the operations of industrial facilities, while also ensuring effective communication between humans and machines. The selected strategy was primarily motivated by the principles of modularity and reliability. In order to fulfill these criteria, a software architecture for the system was selected, employing a multilayer method (Gat 1998).

Figure 3 depicts the software architecture that is integrated into the human-machine interaction. The Cognitive Technical System comprises both the Technological Application System and the Interaction System. The software architecture divides the Interaction System into four layers that encompass the various mechanisms needed. The Presentation Layer includes the user interface and an interface for modifying the knowledge base. The Planning Layer is the cognitive layer where the final determination for the subsequent course of action is formulated. The Coordination Layer offers services to the Planning Layer, which can be called upon by the Planning Layer to initiate the execution of actions. The Reactive Layer is accountable for ensuring a swift response time of the entire system during emergency scenarios. The Knowledge Module encompasses the essential domain knowledge of the system.

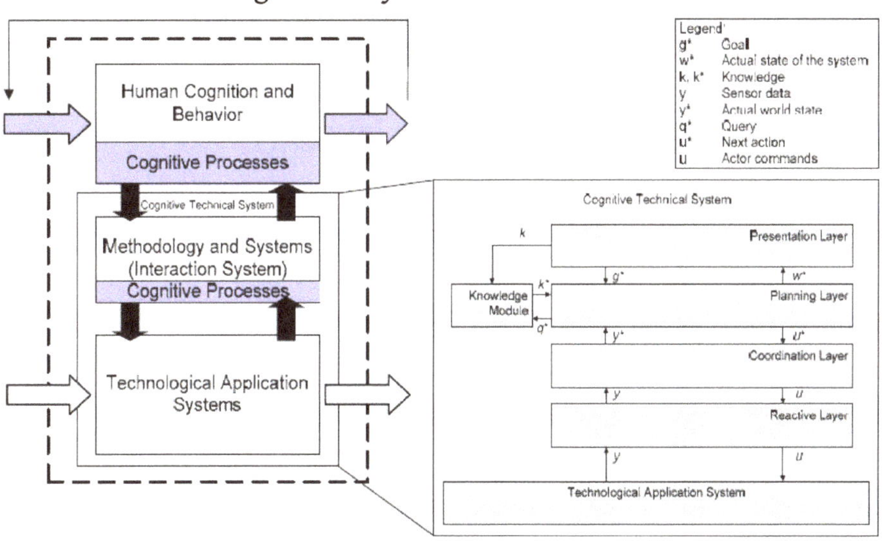


**Fig. 3. Software architecture embedded in the human machine interaction**

At the beginning the human operator gives the desired goal to the Cognitive Technical System via the Presentation Layer. This goal $g^*$ is then transferred to the Planning Layer where the next action $u^*$ is derived based on the actual world state $y^*$ and the desired goal $g^*$. The actual world state is based on the measured variables $y$ from the sensors in the Technological Application System which are transferred via the Reactive Layer. In the Coordination Layer $y$ is then aggregated to $y^*$. To derive $y^*$, the sensor data $y$ at a discrete time $t \in R \geq 0$ is taken into account. $y(t) \in Rny$ denotes the current vector of the current measured variables at time $t$. This vector is then transformed in the world state $y^*(t)$. This means that the base on which all decisions in the Planning Layer are made is the actual world state $y^*$ at a certain time $t$. Therefore the decision process must not take too long, because the state of the Technological Application System can have changed significantly in the meantime.

The next best action $u^*$ derived in the Planning Layer is sent back to the Coordination Layer, where the abstract description of the next best action $u^*$ is translated into a sequence of actor commands $u$, which are sent via the Reactive Layer to the Technological Application System. There, the sequence of commands is executed and the changed environmental state is measured again by the sensors. If the new measured variables $y$ of the Technological Application System indicate an emergency situation the Reactive Layer ensures a low response time. Then the sensor data is processed directly in the Reactive Layer and the according actor commands are executed.

Fig. 4 shows the software architecture in more detail. The different layers and their components will be described in more detail in the following section.

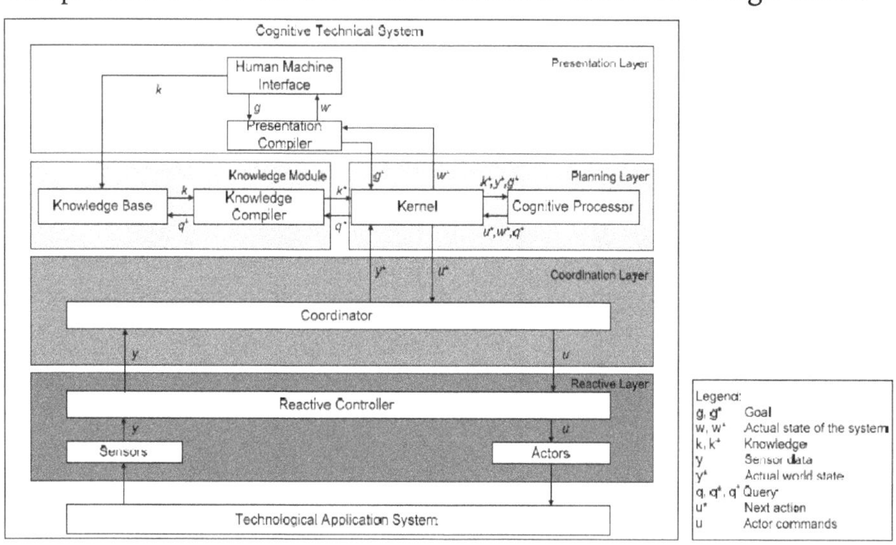


**Fig. 4. Software Architecture of the Cognitive Technical System with components based on a multilayer approach**

## 5.2 Presentation Layer

The Presentation Layer is accountable for facilitating the communication and engagement between the system and the user. The system includes a human-machine interface specifically intended to meet the unique demands of dealing with a technical system that has cognitive capacities.

The domain knowledge, denoted as k, is encoded using a predetermined representational formalism. One such option is the organization of the variable k within an ontology. The knowledge engineer precisely encodes the domain knowledge to suit the exact task that the system needs to perform. This is completed before the commencement of the system. While the system is in operation, a human operator engages with the system. This operator assigns a task description g to the Presentation Compiler component. For an assembly assignment, the description "g" can encompass both the shape of the pieces to be assembled and the details of the location and forms of these elements in the final assembly. This can be accomplished by utilizing a graphical representation, such as a CAD program, among other methods. The Presentation Compiler must translate the task description g into a goal state $g^*$ that can be understood by the Cognitive Processor of the Planning Layer.

The behavior of a cognitive system cannot be precisely predicted in advance due to the dynamic nature of the environment. Hence, it is imperative that the operator is continually aware of the current state of the system. The current state $w^*$ of the system is provided to the Presentation Compiler, where it is combined with other data to generate a machine feedback w that can be easily understood by humans. This feedback is then transmitted to the operator through the Human Machine Interface.

## 5.3 Planning Layer

The Planning Layer comprises the fundamental components that are accountable for decision-making. The device comprises the Kernel and the Cognitive Processor as its constituent elements. The Kernel allocates the signal flows within the Planning Layer. The Cognitive Processor calculates the optimal action $u^*$ by considering the desired goal state $g^*$ and the current state of the world $y^*$. In the event that the Cognitive Processor is unable to determine the most appropriate course of action, it has the capability to transmit a question $q^*$ to the Knowledge Module in order to obtain additional information.

The Kernel component subsequently triggers the execution of the action based on the action provided by the Cognitive Processor. If there is a need for additional information, the Kernel will search the Knowledge Base for actions that can be applied to the items in $y^*$. The information Base retrieves the information $k^*$ from the Knowledge Compiler based on the current processor being utilized.

The supplementary information is subsequently taken into account for calculating the subsequent optimal course of action. Figure 5 displays the activity diagram representing the operations of the Cognitive Processor. If the Cognitive Processor fails to locate an action and the Knowledge Base is unable to provide k*, the Cognitive Processor will request the human operator for the next action. Subsequently, the user has the option to either provide the further course of action or modify the current state of the environment. This implies that the user alters the physical environment without directly informing the system. The system subsequently identifies the new environmental condition by analyzing the measured variables y, evaluates the new state of the world y*, and determines the optimal course of action u* based on y*.

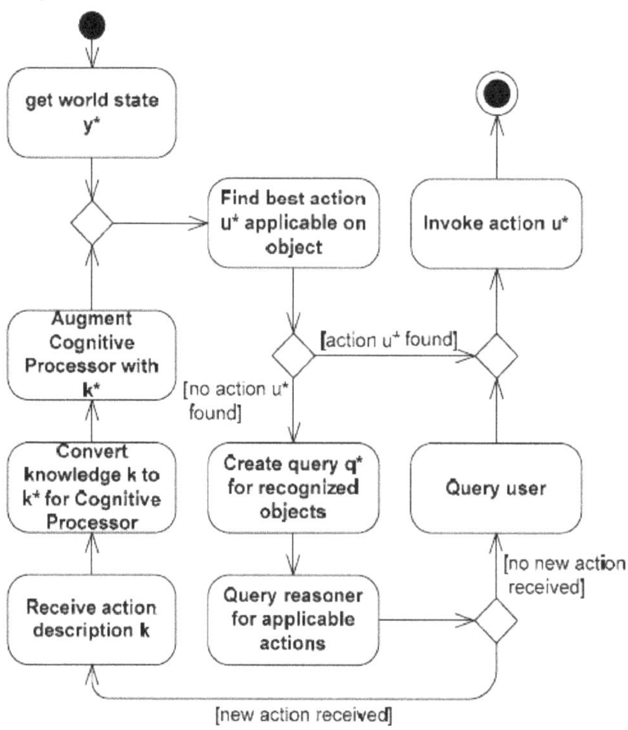


**Fig. 5. Activity diagram of possible actions of the Cognitive Processor**

Multiple designs have been devised to comprehend the human control behavior. The EPIC design integrates cognitive and perceptual processes with procedural task analysis (Keiras 2004). The several interconnected units, referred to as processors, function simultaneously. The Soar architecture is a cognitive framework that is rooted on the "unified theory of cognition" (Newell 1994). Its primary objective is to emulate general intelligence, as outlined by Laird in 1996. It represents behavior as the process of choosing and applying operators on a given state. A state denotes the present condition of knowledge and the process of problem-solving, whereas operators facilitate the movement of knowledge from one state to another. During execution, Soar attempts to execute a sequence of operators in a specific order to achieve a goal, as described by Laird (1996). Control in Soar pertains to the resolution of conflicts and is executed as a purposeful and knowledge-driven procedure. The ACT-R control mechanism is considered to be an automated procedure that employs an automatic conflict resolution technique (Johnson, 1998). The Soar architecture was selected as the Cognitive Processor among these architectures (Hauck 2008).

Soar is a manufacturing system that operates based on rules. Rules are triggered when they correspond to elements of the internal representation of the current y* and make changes to this representation. Through input and output links, Soar has the ability to communicate with its surroundings, such as retrieving a new world state or executing operations. Furthermore, there is ongoing research on integrating the Soar architecture with a conventional planning algorithm such as Fast Forward (Hoffmann 2001). This feature allows for the utilization of the full potential of Soar while also facilitating the rapid production of a solution plan for a given assignment.

## 5.4 Coordination Layer

The Coordination Layer serves as the operational layer of the Cognitive Technical System. It offers operational services to the Planning Layer. These services align with the actions that the Cognitive Processor can initiate. The Coordinator in the Coordination Layer receives the measured variables y from the Reactive Controller through the Reactive Layer and combines this information to generate the current world state y*. Additionally, the Coordinator component receives the upcoming action u* that needs to be executed. The abstract service called upon by u* is a series of actor instructions u. An elementary illustration is the act of fastening together two blocks using a stapler. The service move(blockA, blockB) is invoked when the positions of both blockA and blockB are known. This service involves bringing an actor, such as a robot, to the position of blockA, grasping it, transferring it to the position of blockB, and then releasing it. The value of 'u' is kept in the Coordinator component and will be executed using the parameters provided by 'u*'. Afterwards, the user is executed in the Technological Application System through the Reactive Layer. In this method, the Planning Layer is relieved from the specifics of the robot's actions, such as the precise coordinates of the block positions. This results in speedier decision-making due to a smaller problem space.

## 5.5 Reactive Layer

The Reactive Layer, namely the Reactive Controller component, is responsible for the system's low-level control. The vector of the measured variables y is observed for levels that suggest a potential emergency situation. The Reactive Controller then generates the appropriate actor commands, denoted as u. This guarantees minimal reaction times in the event of an emergency. The Reactive Controller is unable to guarantee a secure behavior for the entire system. If an incorrect sequence of actor commands is issued to the actors in the Technological Application System, the Reactive Controller will not evaluate the potential repercussions of this sequence on the Technological Application System based on its current state. The task must be completed by the Cognitive Processor.

## 5.6 Knowledge Module

The Knowledge Module houses the Knowledge Base, which comprises the essential domain knowledge required for the Cognitive Technical System to execute the intended task. The domain knowledge, represented by the variable k in the Knowledge Base, must be converted into a format that can be understood and processed by the Cognitive Processor. The process is carried out by a Knowledge Compiler, comprising of two components: The Reasoner and the Mediator. The Reasoner interrogates the information Base and acquires supplementary information k. Subsequently, this knowledge is converted into an intermediary format known as k' and then transmitted to the Mediator. The Mediator consolidates the information k' into the structure k* which is subsequently analyzed by the Cognitive Processor. Figure 6 depicts the pathways of the signal and the components that are involved. When the Cognitive Processor requests extra information q*, the Mediator translates q* into q' and the Reasoner uses the Knowledge Base to deduce the desired information.

In assembly tasks, the domain knowledge must encompass the players involved, which are regulated by the Cognitive Technical System. The Web Ontology Language (OWL) is the formalism employed for representing domain knowledge (Smith 2004). The original form is inadequate for storing the procedural knowledge that is utilized by the cognitive processor in the form of production rules. Thus, the Semantic Web Rule Language (SWRL), an extension to OWL, is employed along with a description formalism to directly describe procedural knowledge in the ontology (Horrocks et al., 2004).

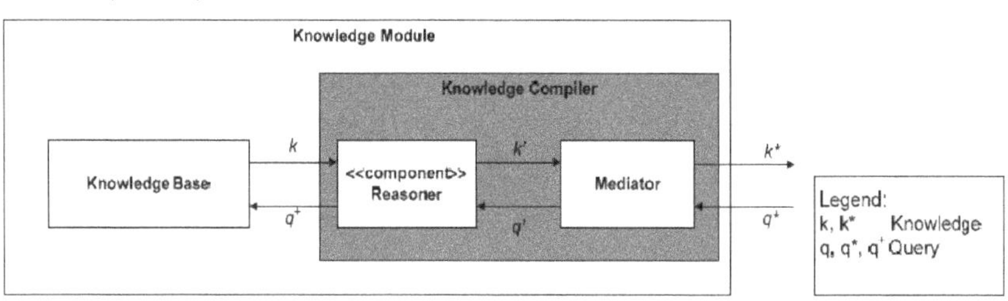


**Fig. 6. Component diagram of the Knowledge Module**

## 5.7 Conclusion

The multilayer technique guarantees the containment of various information abstractions within separate layers. The components of the Planning Layer function at the most abstract level of information. The Cognitive Processor executes the relevant service based on the next optimal course of action. The many services modify the environment without directly addressing the specific limitations imposed by the actors involved. The Coordination Layer comprises a series of actor commands that describe the services. The Reactive Layer then performs and manages these commands. As a result of this methodology, the system is capable of effectively handling a dynamic and ever-changing environment, and can adjust itself accordingly. The system is hybrid in two distinct ways. The system links a continuous flow of input signals to their discrete representation in states, incorporating both reactive and deliberative elements.

## 6. EXAMPLE: CONTROL OF AN ASSEMBLY CELL

Figure 7 displays the schematic arrangement of a robot cell that is under the control of the Cognitive Technical System. The system comprises of a pair of robots and a transportation mechanism that transports the components via a conveyor belt. The initial robot seizes the arriving components and places them onto the conveyor. An object recognition software uses a CCD camera to identify the colors and shapes of the parts. If the part is required for the current assembly, the second robot will grip the part and move it to either the assembly area if it is needed right away, or to the buffer region. If an article is no longer required, the conveyor will transfer it to the container for discarded items and it will be discharged. The second robot is outfitted with a robotic hand that has three fingers, enabling it to do intricate gripping tasks.

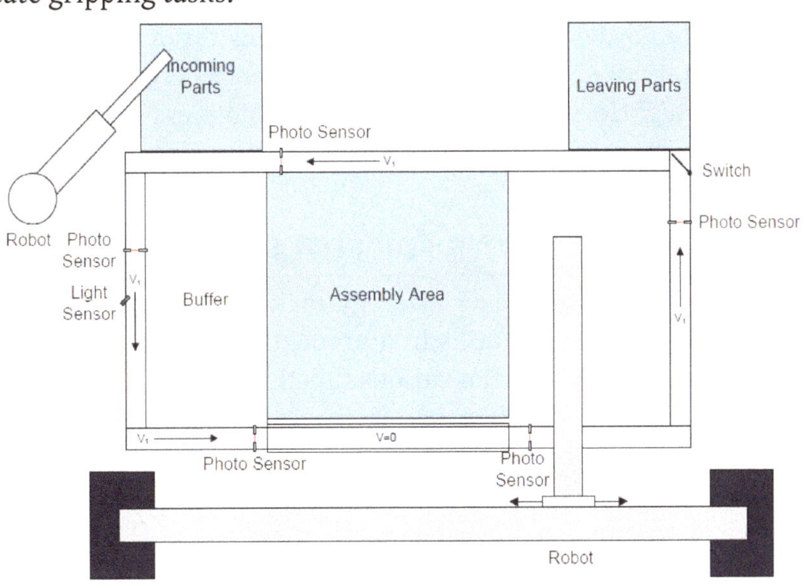


**Fig. 7. Schematic of the assembly cell used for the application of the Cognitive Technical System**

The initial assessments of the Cognitive Technical System will exclusively focus on components with basic shapes, such as blocks, spheres, and similar objects. This is essential since the identification of objects and the recognition of colors would be significantly delayed for intricate things. The system must be able to adjust to various conditions without the ability to prearrange the entire construction process. Thus, the distribution of the components is random. Furthermore, the current state of the world will be regularly examined to assess whether the internal representation in the Cognitive Technical System aligns with the ambient state.

Possible reasons for unexpected changes in the environmental state can be:

- Erroneous identification of a part
- Dropping or misplacement of a part by the robot
- Changes in the current assembly

Incorrect identification of a component can result in an inaccurate request for the entire assembly and have a substantial impact on the outcome of the assembly process. A part may be dropped if the three-finger robot hand incorrectly grasps an object or if the thing falls during the transfer operation. The final potential alteration in an environmental condition is the modification of the assembly. This is a situation in which the machine and the human collaborate. The system will detect the change using the monitored variables y.

Although it is not the primary focus of the current research, it must be taken into account for future applications.

Hence, the Cognitive Technical System must regularly monitor the current status of the world in order to avert any repercussions resulting from alterations in the environmental state. In order to assess the system's performance, the system will carry out a straightforward assembly task. The simplest form of geometry is represented by a tower of blocks, but this can be expanded to encompass more intricate geometries.

## 7. SUMMARY AND PROSPECTS FOR FUTURE RESEARCH

The implementation of a multilayer strategy for a Cognitive Technical System capable of performing assembly tasks in a production setting is a viable option. The software architecture fulfills the various functional and non-functional needs that a production environment demands from such a system. The present study is centered on the execution of the software structure and the simulation of environmental conditions. Subsequent tasks will involve establishing the link to the assembly cell and implementing the system on object outlines that are more intricate.

## REFERENCES

1. Brecher, C. et al. (2007). *Excellence in Production,* Apprimus Verlag, ISBN: 3940565008, Aachen
2. Cockburn, A. (2003). Writing effective use cases, Addison Wesley, ISBN: 9780201702255, London
3. Ding, H. et al. (2008). A Control Architecture for Safe Cognitive Systems, *10. Fachtagung Entwurf komplexer Automatisierungsysteme,* Magdeburg, April 2008,
4. Gat, E. (1998). On Three-Layer Architectures in *Artificial Intelligence and Mobile Robots*, Kortenkamp D., Bonnasso R., Murphy R., (Ed.), pp. 195-211, AAAI Press, ISBN: 9780262611374, Cambridge
5. Gausemeier, J. (2008). *Towards a Design Methodology for Self-optimizing Systems,* Springer Verlag, ISBN: 978-1-84628-004-7, London
6. Hauck, E.; Gramatke, A. & Henning,K. (2008). Cognitive technical systems in a production environment, *Proceeding of the 5th international Conference on Informatics in Control, Automation and Robotics,* pp. 108-113, ISBN: 9789898111326, Madeira, May 2008
7. Hägele, M. (2008). Industrial robotics, In: *Handbook of Robotics*, Siciliano, B., Khatib, O., (Eds.), pp. 963-986, Springer Verlag, ISBN: 9783540239574, London
8. Heide, A. & Henning, K.(2006). The cognitive car - A roadmap for research issues in the automotive sector, *Proceedings of the 9th IFAC Symposium on Automated Systems Based on Human Skill And Knowledge*, ISBN: 9783902661050, Nancy, May 2006,
9. Hoffmann J. & Nebel B. (2001). *The FF Planning System: Fast Plan Generation Through Heuristic Search.* Journal of Artificial Intelligence Research, Vol. 14, (2001), pp.253-302, ISSN:11076 – 9757
10. Horrocks, I. et al. (2004). *SWRL: A Semantic Web Rule Language Combining OWL and RuleML*, http://www.w3.org/Submission/2004/SUBM-SWRL-20040521/
11. Johnson, T.R. (1998). A comparison of ACT-R and SOAR. In: Schmid, U., Krems& J.,Wysotzki, F. (Eds.) *Mind modeling*, pp. 17–38, Papst, ISBN: 3933151252, Lengerich
12. Karim, S. et al. (2006). A Hybrid Architecture Combining Reactive, Plan Execution and Reactive Learning, *Proceedings of the 9th Biennial Pacific Rim International Conference on Artificial Intelligence (PRICAI)*, China, August 2006
13. Konolige, K. & Myers, K. (1998). The saphira architecture for autonomous mobile robots. In: Kortenkamp et al. (Eds.) *Artificial intelligence and mobile robots: case studies of successful robot systems*, pp.211-242, MIT Press, ISBN: 0262611376, Cambridge

14. Kieras, D. & Meyer, D. (2004). *EPIC Architecture – Principle of Operation*, Univ. of Michigan, Ann Arbor

15. Laird, J.E.; Lehman, J.F. & Rosenbloom P. (1996). A gentle introduction to Soar, an architecture for human cognition, In: *Invitation to Cognitive Science*, MIT Press, Boston

16. Matlin, M. W. (2005). *Cognition*, Wiley& Sons, ISBN: 0471427780, New York

17. Montemerlo, M. et al. (2006). Winning the DARPA Grand Challenge with an AIrobot, *Proceedings of the AAAI National Conference on Artificial Intelligence,* pp. 982-986, ISBN 9781577352815, Boston, July 2006, AAAI, Boston

18. Newell, A. (1994). *Unified theories of cognition*, Harvard University Press, ISBN: 9780674921016, Cambridge

19. Putzer, H. (2004). *Ein uniformer Architekturansatz* für *kognitive Systeme und seine Umsetzung in ein operatives Framework,* Dissertation, MünchenSmith, M. et al. (2004). *OWL Web Ontology Language Guide,* http://www.w3.org/TR/2004-/REC-owl-guide-20040210/*OWL Web Ontology Language Guide,* http://www.w3.org/TR/2004/REC-owl-guide-20040210

20. Sommerville, I. (2007). *Software Engineering*, Addison Wesley, ISBN: 9780201398151, London Strasser, A. (2004). *Kognition künstlicher Systeme*, Ontos Verlag, ISBN:393720296X, Frankfur

21. Strube, G. (1998). Modelling motivation and action control in cognitive systems. In *Mind modeling: a cognitive science approach to reasoning, learning and discovery*, Schmid, U., Krems, J., & Wysocki, F. (Eds.), pp 89-108, Pabst, ISBN 159326044X, Berlin

22. Urmson, C. et al. (2007). *Tartan Racing: A Multi-Modal Approach to the DARPA Urban Challenge*, Pittsburgh

23. Zimbardo, P. & Gerrig, R. (2005). *Psychology and Life*, Pearson, ISBN: 0205428673, Boston

# 03 THE KRIGING MODEL IS USED TO DEVELOP A TWO-STAGE APPROACH FOR MODELING THE EMISSION OF POLLUTANTS FROM DIESEL ENGINES

## 1. INTRODUCTION: AN OVERVIEW

The automotive industry must balance the objectives of creating vehicles with superior performance while also minimizing development time and costs. Manufacturers must prioritize the production of fuel-efficient automobiles that adhere to pollutant emissions regulations and meet customer expectations. In recent years, there has been a rapid increase in the complexity of engine reactions that we need to optimize and the number of parameters that need to be controlled throughout the design stage. To meet these objectives within an acceptable timeframe, organizations employ two methods: design of experiments (DOE) and modeling. DOE is used on one side, while modeling is used on the other side (Schimmerling et al., 1998). Design of tests (DOE) is a highly effective tool, but the automotive industry may face limitations in its utilization due to the high costs and lengthy duration of tests, especially when studying pollutant emissions.

Engine developers employ two primary methodologies to simulate engine behavior. The initial approach relies on chemical and physical models, employing a differential system. This technique is not the focus of this essay, as we do not have such models. In addition, these models, when they are accessible, are typically labor-intensive and not feasible for multi-objective optimization procedures. Moreover, they do not fully encompass all the patterns in the engine system as defined by the measured data, such as the Zeldovich model. This statement is especially true when there is a high number of control options and the engine reactions are intricate.

Statistical modeling, utilizing meticulously selected measured data of engine performance, in accordance with an experimental design, is a significant alternative technique.

Strategies utilizing the Lolimot (Castric et al., 2007) (Local Linear Model Tree) and Zeldovich mechanisms (Heywood, 1988) have been devised to forecast NOx emissions. In the initial scenario, the associated model may result in singular points, hence diminishing the accuracy of the outcomes. In the second scenario, the outcomes are insufficiently adequate.

The literature describes several statistical training approaches, including neural networks. This approach yields favorable outcomes, even when dealing with non-linear scenarios. However, it is not suitable for our situation as it necessitates a large number of trials to produce a meaningful estimation of its parameters, and we are severely constrained by the limited number of tests that the industrialist can conduct. The design of experiments procedures, as proposed by Cochran and Cox in 1957, were developed to address these types of difficulties. However, subsequent studies (Sacks et al. 1989; Bates et al. 1996; Koehler & Owen, 1996) indicate that polynomial models are not suitable for numerical experiments. For instance, a second-order response surface is insufficiently adaptable to represent a surface that has several extrema.

This chapter aims to describe the results we have obtained in predicting emissions of contaminants. These outcomes were attained without increasing the number of tests that the businessman can conduct. We utilize Kriging, a complex statistical model derived from the study of geostatistics. We employ this methodology, employing two distinct methodologies, to enhance the accuracy of forecasting NOx (nitrogen oxide) emissions and fuel usage.

During the initial phase, we directly determine the reaction by evaluating the controllable elements such as main injection timing and pressure in the common rail. This can be equated to a black box modeling. In the second phase, we present a novel methodology that enables us to forecast the reaction based on functional data. To be more specific, we assess the engine's performance based on indications such as pressure and cylinder temperature. The signals are derived from a combustion model. The primary benefit of the second technique is its ability to incorporate a comprehensive understanding of combustion. The absence of knowledge is frequently condemned in the context of black box models. The Kriging method is very suitable for the second strategy, which involves predicting engine reactions based on the state variables (signals) acquired from a physical model of combustion. This indicates that this approach is suitable for situations where there are several choice variables but a limited number of experiences. This is because the method relies on analyzing the similarity of the reaction of interest at several locations, in proportion to the distance between them. It is worth noting that software programs like R and Matlab include a toolbox that enables the utilization of this strategy. Regrettably, the latter are limited to fewer than 3 dimensions. The possibility of applying the technology to greater dimensions has been taken into account.

To implement this second approach, a model reduction is needed. This reduction will be made in two steps:

- Reducing the number of state variables from 10 to 2 by the study of correlations.
- Reducing the number of points in each signal using the theory of Fourier. Once the reduction is made, the Kriging can be applied to the model obtained.

This paper is organized as follows: In the second section, we describe the engine behaviour and recall the importance of controlling pollutant emissions. In the third section, the ordinary Kriging techniques are recalled. In the fourth section, two different approaches for modelling our problem are proposed. An efficient reduction model strategy is considered in order to apply the Kriging method. Finally, the Kriging method is applied to the reduced model. In the last section, numerical results are given followed by a short discussion.

## 2. ENGINE CALIBRATION

The engine calibration is a process which aims at defining the values of the engine control parameters. During the ten last years, the set of themes "engine calibration" took an important place in the process of the development of the internal combustion engines.

Indeed, under the impulse of the standards more and more severe, the car manufacturers are brought to integrate more and more elaborated technologies in the power unit.

Under these conditions, the principles of control used in the past (cartographic approach, buckles open...) are not enough sufficient. Indeed, the output variables (quantity injected, advances in lighting or in injection) were primarily given starting from two variables of entry (speed and load). Today, the use of new technologies in the conception of engines, in order to reduce the pollutant emissions, as for example EGR (exhaust Gas Recirculation) (Pierpont et al. 1995), multiply the number of parameters to control, as we can see it in Fig.1. This figure shows the exponential evolution of number of setting parameters due to the hardware complexity increase. This makes the cartographic approach impracticable. Moreover, this kind of approach does not take into account the dynamic of system.

The main drawback of this evolution is the increase of the difficulty to understand the engine behavior. To deals with all the parameters, we use a Kriging model which we define in the next section.

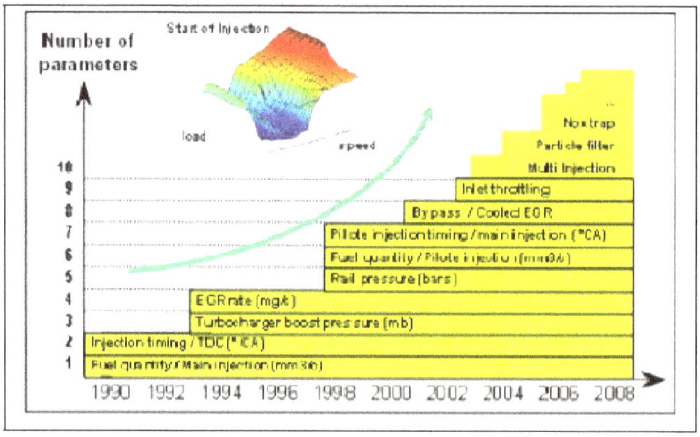


**Fig. 1. Parameters to tune a diesel engine function of technologies**

## 3. ORDINARY KRIGING TECHNIQUES

Kriging techniques are commonly employed for spatially interpolating soil characteristics (Krige, 1951; Matheron, 1963). Kriging is an algorithm for estimating values using linear least squares regression. This is an interpolation tool.The aim is to estimate the value of an unknown real function Z at point $x_0^*$, given the values of function Z at some other points xi ∈ Rd for each i = 1, ... , n .

Car manufacturers are increasingly compelled to incorporate more sophisticated technologies into the power unit due to stricter specifications.

In these circumstances, the control methods employed previously, such as the cartographic approach and buckles open, are insufficient. The output factors, such as the quantity injected and advancements in illumination or injection, were essentially determined based on two input variables: speed and load. Currently, the integration of modern technologies in engine design, such as exhaust Gas Recirculation (EGR), is being utilized to minimize pollutant emissions. This results in an increased number of factors that need to be monitored and controlled, as seen in Figure 1 (Pierpont et al., 1995). This graphic illustrates the exponential growth of the number of setting parameters as a result of the increasing complexity of the hardware. This renders the cartographic technique unfeasible. Furthermore, this technique fails to consider the system's dynamics.

An inherent limitation of this development is the heightened complexity in comprehending the engine's behavior. In order to account for all the variables, we employ a Kriging model, which will be explained in the following section.

### 3.1 Ordinary Kriging

The ordinary Kriging estimator $\hat{Z}(x_0^*)$ is defined by:

$$\hat{Z}(x_0^*) = \sum_{i=1}^{n} \lambda_i Z(x_i) \qquad (1)$$

Where n is the number of surrounding observations Z(xi) and λI is the weight of Z(xi). The weights should sum to unity in order to make the estimator unbiased:

$$\sum_{i=1}^{n} \lambda_i = 1 \qquad (2)$$

The weights are also determined such that the following Kriging variance is minimal under the constraint given by the equation 2:

$$var\left(\hat{Z}(x_0^*) - Z(x_0^*)\right)$$

This leads to a classical optimization problem with equality constraint. The Lagrange multiplier theory is used in order to work out this problem. This gives a linear system which must solved (Davis, 1986).

## 3.2 Variogram

The variogram is a function representing the spatial dependency. It is obtained from the stationarity definition. In fact, this stationarity hypothesis is an indispensable condition for the use of the Kriging method.

In the case of ordinary Kriging the expression of the variogram is obtained from the following definition of intrinsic stationarity:

1) $E(Z(x_i + h) - Z(x_i)) = 0 \ \forall\, i \in \{1, \dots, n\}$
2) $Var(Z(x_i + h) - Z(x_i)) = 2\gamma(h) \ \ \forall\, i \in \{1, \dots, n\}\ and\ h$

More precisely, the expression of the theoretical variogram, is deduced from the second condition of intrinsic stationarity. This condition means that the variation of a data set is only dependent on distance r between two locations, where the variables values are $Z(x_i+h)$ and Z(Xi) with $_r$ $|h|$.

Note that the variogram, γ (*h*), is a function of the separation between points *h* and not a function of the specific location ($x_i$, $x_i + h$). This mathematical definition is a useful abstraction, but not easy to apply to observed values.

Consider a set of n observed data: $(x_1, z_1)$, $(x_2, z_2)$... $(x_n, z_n)$, where xi is the location of observation i and Z1 is the associated observed value. There are $\frac{n(n-1)}{2}$ unique pairs of observations. For each of these pairs we can calculate the associated separation distance: $r=|x_i-x_j|$

To infer the variogram from observed data, we will then use the common formula for the experimental variogram (Cressie, 1993).

$$\hat{\gamma}(r) = \frac{1}{2N(r)} \sum_{N(r)} \left[Z(x_i) - Z(x_j)\right]^2 \qquad (3)$$

Where:

$$N(r) = \{(i,j) \quad such\ that\ |x_i - x_j| = r\ \}$$

N(*r*) is the pair number of Z(X*i*+*h*) and $Z(x_i)$: $\hat{\gamma}$ (r) is the experimental variogram.

## 3.4 Variogram Modeling

Equation 3 provides an estimation of the theoretical variogram using a limited amount of distances. Furthermore, it may not always constitute a genuine variogram. This implies that it is possible that it does not pertain to a negatively conditional function. Indeed, this condition is essential to guarantee the positive of the variance of a sum of random variables (Christakos, 1984). The experimental variogram is then represented by a function of negative conditional nature, encompassing all distances. This modeling enables the implementation of Kriging. A variogram model should be applied to fit the variogram. In order to determine the most suitable variogram model, one must choose from the different models available in the literature and adjust it to the experimental variogram (Arnaud & Emery, 2000). Consequently, it is necessary to estimate the parameters of the model. This modification can be accomplished visually, although it is typically performed via an estimating technique such as weighted least squares or maximum likelihood method. After selecting the variographic model and estimating its parameters, we calculate the weights λi that are used in equation (1) by solving the following system:

A Λ=B, with

$$A = \begin{bmatrix} \gamma(r_{11}) & \gamma(r_{12}) & \cdots & \gamma(r_{1n}) & 1 \\ \gamma(r_{21}) & \gamma(r_{22}) & \cdots & \gamma(r_{2n}) & 1 \\ \cdots & \cdots & \cdots & \cdots & \cdots \\ \gamma(r_{n1}) & \gamma(r_{n2}) & \cdots & \gamma(r_{nn}) & 1 \\ 1 & 1 & 1 & 1 & 0 \end{bmatrix} \quad (4)$$

$$\Lambda = [\lambda_1 \quad \lambda_2 \quad \ldots \quad \lambda_n \quad \lambda]^T$$

And

$$\Lambda = [\lambda_1 \quad \lambda_2 \quad \ldots \quad \lambda_n \quad \lambda]^T$$

$$B = [\gamma(r_{01}) \quad \gamma(r_{02}) \quad \ldots \quad \gamma(r_{0n}) \quad 1]^T$$

Where λ is the Lagrange multiplier.

γ (.) is the variogram model used for adjusting the experimental variogram.

$r_{ij}$ is the distance between the locations $x_i$ and $x_j$.

The variance of the estimate $s_i^2$ i.e. the square of the standard error at each point is obtained by the relationship:

$$s_i^2 = \Lambda^T . B$$

If we assume that the estimation errors are normally distributed around the true value, then the probability that the true value will be in $Z(x_i) \pm x_i$ is 68 %, while the probability that the true value will be in $Z(x_i) \pm 2\, s_i$ is 68 %, is 95 %, (Davis, 1986).

## 3.5 Kriging Emulator Validation

The ultimate measure of the accuracy of the fitted emulator model lies in its capacity to forecast the outcome when presented with untested factor values. To fully utilize the data for model fitting, the emulators are validated using leave-one-out cross validation. This procedure entails taking the previously fitted model and fitting it again to a subset of experimental data that was not previously used.

More precisely, for an experiment with d design factors $y = y_1 ,..., y_d$, the set of n experimental design points $X = x_1,...,x_n$, and responses Z=z1 ,..., $z_n$, contain the information used to build the Kriging model. A cross validation involves predicting at each design point in turn when that point is left out of the predictor equations. Let $\hat{Z}(x_i)$ be the estimate of the $Z(x_i)$ based on all the design points except $x_i$. The prediction error (the estimated root mean square error, RMSE) is then calculated as:

$$RMSE = \sqrt{\frac{1}{n}\sum_{i=1}^{n}\left[Z(x_i) - \hat{Z}(x_i)\right]^2} \quad (5)$$

An index of the accuracy of the emulator is made by expressing as a percentage of the range of the response Z,

$$RMSE = 100 \times \frac{RMSE}{\max(Z)-\min(Z)} \quad (6)$$

## 4. TWO APPROACHES TO MODEL ENGINE RESPONSES

This section introduces two-stage methodologies utilizing the Kriging technique to forecast NOx (nitrogen oxide) emissions and fuel usage.

During the initial phase, we directly determine the response by analyzing the controllable factors such as the main injection timing and the pressure in the common rail (black box).

In the second phase, we present a novel methodology that enables us to forecast the engine's reaction based on signals such as pressure and cylinder temperature (which are state variables of the combustion chamber).

### 4.1 First approach

The initial step involves constructing a Kriging model based on the controllable parameters. Therefore, the Kriging model was trained utilizing around 300 input/output data sets obtained through the utilization of the D-optimal design method. The training examples encompass engine speeds ranging from 1000 rpm to 5000 rpm, with intervals of 250 rpm, and load varying from 1 to 23 bar. The data was generated to encompass the whole range of the engine map's cycle point, with the purpose of creating a comprehensive Kriging emulator. Therefore, our model considers the engine speed as one of the control parameters.

| | | | |
|---|---|---|---|
| Prail : | rail pressure, | - Main: | Main injection quantity, |
| Mpil1: | pilot1 injection quantity, | - Pmain: | Main injection timing, |
| Mpil2: | pilot2 injection quantity, | - Ppil2: | pilot2 injection timing, |
| Ppil1: | pilot1 injection timing, | -VNT: | turbine vane position, |
| VEGR: | EGR valve position, | - Volet: | position component of admission, |

## 4.2 Second approach

### 4.2.1 Modelling

In our second approach, we present a novel method that enables us to forecast the reaction based on signals such as pressure and cylinder temperature (which are state variables of the combustion chamber).

To be more precise, we break down the problem of estimating the engine reactions into two separate subproblems (Fig.2):

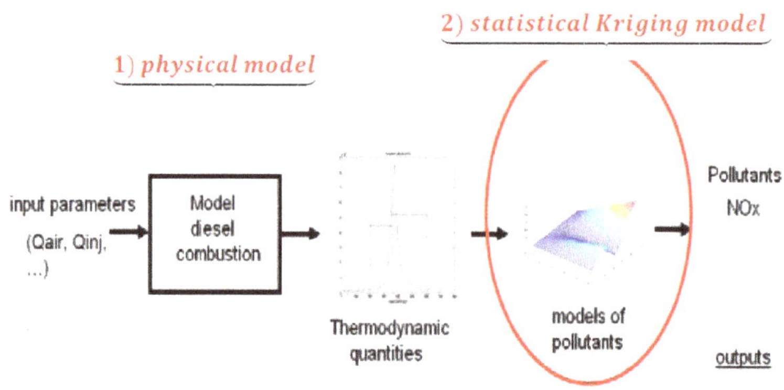


**Fig. 2. Coupling of the pollutants and consumption models with the combustion model.**

- The first step consists in simulating the various thermodynamic quantities from a physical model. In this work, we use the model developed by (Castric et al., 2007), which takes into account the input parameters. It leads to have a good representation of the experimental results. This model allows us to generate the following thermodynamic quantities:

The cylinder low pressure (the alone quantity that we can measure) The temperature in the cylinder,

The temperature of the fresh gas in the cylinder The temperature of the mixed gas in the cylinder, The temperature of the burned gas in the cylinder, The mass of the fresh gas in the cylinder,

The mass of the entrained gas in the cylinder, The mass of the burned gas in the cylinder, The turbulence in the motor,

The fuel vapor mass.

We precise, that each signal is represented by a vector of 1334 components.

- The second step consists in building a statistical Kriging model, from the 11 thermodynamics quantities generated by the model of combustion.

It is true that the advantage of this procedure is that, it allows us to include a physical knowledge of combustion. But this approach requires a great time of computing. Indeed to build Kriging from 11 signals, can pose a serious problem in memory capacity and the computing time can be considerable. Thus, to be able to implement this procedure, a reduction of the model is essential.

### 4.2.2 Model reduction

The data from the initial model can be directly utilized for Kriging. The second one does not conform to this. In the previous instance, it is necessary to decrease the amount of data.

The reduction procedure commences with an examination of the various correlations between the state variables and their related p-value. The selected criterion involves evaluating the p-value: if it is below 0.05, the correlation is considered statistically significant. This analysis enables us to limit our consideration to just two state variables: the pressure of the cylinder, denoted as P, and the temperature of the mixed gas within the cylinder, denoted as Te.

In the second stage, the quantity of components in the two remaining signals is diminished. This is achieved by employing the discrete Fourier transform. The Matlab function fft computes the discrete Fourier transform (DFT) of a vector using a fast Fourier transform (FFT) technique. After computing the coefficients, only a minimal number of them are kept. This enables us to replicate the original signal with a relative inaccuracy of approximately less than 0.02.

Reducing the number of points in each signal is equivalent to minimizing the number of Fourier coefficients used to represent that signal. The two signals indicating the cylinder pressure and the temperature of the mixed gas in the cylinder have been compressed to 40 Fourier coefficients. Each signal has been rebuilt using the 40 retained coefficients, resulting in a relative inaccuracy that is deemed acceptable. The table below displays the relative error incurred during the reconstruction of the two signals using the 40 specified coefficients:

$$\text{Relative error} = \frac{\|S - S_{rec}\|}{\|S\|}$$

S: is the experimental signal

$S_{rec}$ is the reconstruction of the signal S using the fast Fourier transformation.

$\|\,.\,\|$ is the Euclidian norm.

**Table 1. Relative error committed for the reconstruction of two signals.**

| Type of signal | relative error |
|---|---|
| the cylinder low pressure | 0.01 |
| the temperature of the mixed gas in the cylinder | 0.02 |

Figure 3 shows the experimental signals, resulting from the combustion and their reconstruction by using the fft Matlab function.

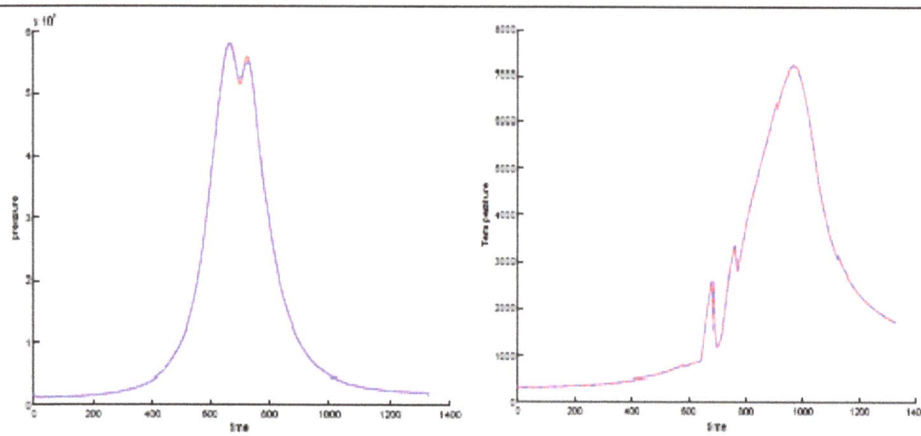

Pressure Temperature $T_e$

**Fig. 3 Rebuilding of the measured signals (red curve) by using the discrete Fourier transform (blue curve)**

Such reduction makes the Kriging possible. The considered entries of the model are:

$$\begin{cases} P_{cyl,i}(t) \rightarrow [c_1(P), \dots, c_n(P)]_i \\ T_{cyl,i}(t) \rightarrow [c_1(T), \dots, c_n(T)]_i \end{cases}$$

Where:

i is the index that corresponds to the ith operating point of engine. An operation point of the engine is defined by engine speed and engine torque $C_n(f)$ is the kept Fourier coefficient for the signal *f*, which corresponds to the ith operating point of engine.

## 5. APPLICATION FOR ESTIMATING ENGINE REACTIONS

In the preceding section, we have introduced and elucidated the two methodologies employed in this study to simulate the behavior of a diesel engine. Next, this section will focus on presenting the data produced by each approach, separately, for estimating each answer.

### 5.1 Quantitative outcomes obtained with the initial methodology

It is crucial to note that the creation and modeling of the experimental variogram play a pivotal role in the Kriging approach. Therefore, in this section, we will begin by presenting the selected model.

**Variogram fitting:** Variography modeling is a crucial and challenging phase in the development of a Kriging model. Consequently, multiple models were fine-tuned and subsequently compared. Choosing the superior graphical model was challenging. The process of cross validation simplifies the task. We can choose the option that minimizes the root mean square error.

The maintained model for NOx is a Gaussian model, represented by the equation:

$$\gamma(r) = c_0 + c\left(1 - exp\left(-\frac{r^2}{a^2}\right)\right) \quad as \quad r \geq 0 \tag{7}$$

The value of the model parameters was founded using the least square method. So, we obtain: $C_0$=10.929, c=1.829, a=309.559.

For the Consumption, the model used is an exponential model, given by the equation:

$$\gamma(r) = c_0 + c\left(1 - exp\left(-\frac{r}{a}\right)\right) \quad as \quad r \geq 0 \tag{8}$$

$c_0$=7.106, c=2.415, a=477.444.

This leads to: $C_0$=7.106, c=2.415, a=477.444.

Where:

r is the distance

$C_0$ is the Nugget effect

$C_0$+C is the sill correspond to the variance of Z(*x*)

$\sqrt{3a}$ and 3a are the range (the distance at which the variogram reaches the sill) for the Gaussian and exponential model respectively (Baillargeon et al., 2004).

Figures 4 shows the experimental variogram (red points), and Gaussian model (blue curve) corresponding to NOx response.

Figures 5 shows the experimental variogram (red points), and exponential model (blue curve) corresponding to consumption response.

The variogram is the tool which quantifies the spatial correlation of the response in of interest. It measures the variability of NOx and consumption as a function of distance. We notice that, when the distance reaches the range $535 \approx \sqrt{3a}$ (Fig.4) and 1300 ≈ 3a (Fig.5), the variation becomes stationary. This explains why we can have a similar behavior of consumption and NOx on two different operating points, thus with a pattern of different control parameters

The Cross-validation plots for the Kriging model are displayed in Figures 6 and 7. Figure 6 corresponds to the Gaussian variogram, while Figure 7 corresponds to the exponential variogram. The charts display the measured values, the Kriging estimated values, and 10% error bars.

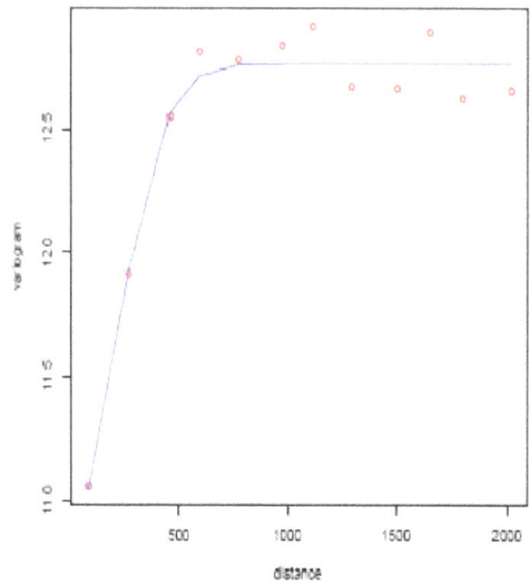


Fig. 4. Experimental and Gaussian model variogram in the case of NOx

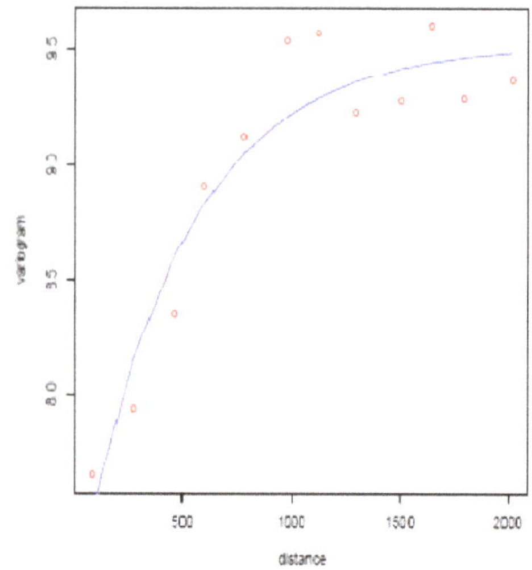


Fig. 5. Experimental and exponential model variogram in the case of consumption

**Fig. 4. and Fig. 5. Experimental and model variogram**

The predictive accuracy was comparable for both sets of validation data. The accuracy of both responses was satisfactory, being within a 10% margin for the majority of operating situations.

On the contrary, graph 7 displays several inaccurately calculated observations. The reason for this is that they are located at a considerable distance from the cluster of data points used for the adjustment. This inaccurate estimation is also attributed to the experimental design employed. The classical and optimum designs, namely the D-optimal designs, are not appropriate for Kriging due to their reliance on evaluating similarity across sites. The D-optimal design enables testing a limited number of levels for each variable and tends to yield points on the boundaries of the experimental field (Koehler & Owen, 1996). This ideal point distribution for fitting a polynomial model is unable to detect any anomalies within the experimental field, resulting in inaccurately estimated points. To tackle this issue, we suggest utilizing suitable designs for Kriging. Latin hypercubes, which fall under the category of 'space filling designs', are effective at distributing points evenly in space and are suitable for modeling using Kriging. This has been supported by Stein (1987) and McKay et al. (2000).

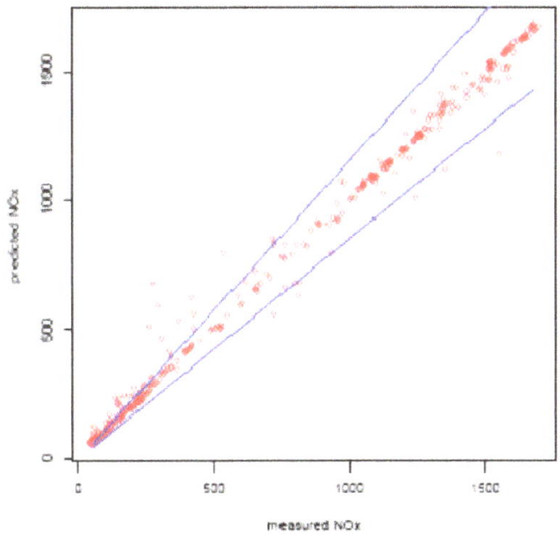


**Fig. 6. Measured and Kriging predicted NOx [ppm] with ± 10% error bands**

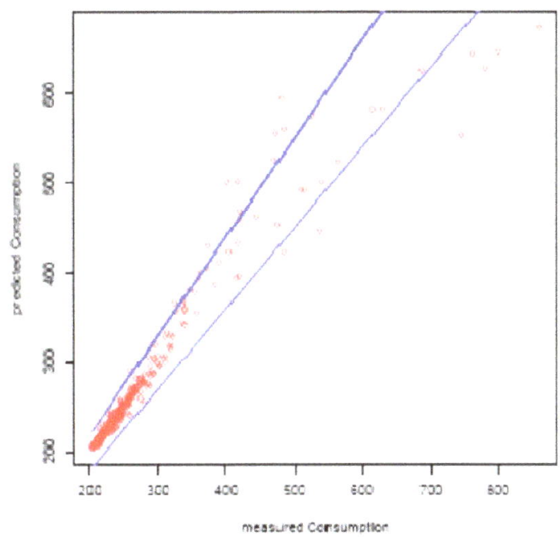


**Fig. 7. Measured and Kriging predicted consumption [g/kWh] with ± 10% error bands**

The emulator model is applied to each answer individually, and the root mean square error (RMSE) and percentage RMSE are recorded. The results are displayed in Table 2. The percentage root mean square error (RMSE) findings indicate that the model's %RMSE is below 7% of the range of the response data. This suggests that if the emulator is used to forecast the response at a novel input configuration, the prediction error is anticipated to be below 7% in comparison to the actual value.

**Table 2. Kriging RMSE end %RMSE for each response: first approach case**

| | NOx | Consumption |
|---|---|---|
| RMSE | 61.4 | 40.63 |
| %RMSE | 3.84 | 6.19 |

## 5.2 Numerical results using the second approach

Each answer is independently subjected to the emulator model, and the resulting root mean square error (RMSE) and percentage RMSE are documented. The outcomes are presented in Table 2. The findings of the percentage root mean square error (RMSE) reveal that the model's %RMSE is less than 7% of the range of the response data. If the emulator is utilized to predict the response at a new input configuration, it is expected that the prediction error will be less than 7% compared to the actual result.

**Variogram fitting:** The experimental variogram and the model which adjusts it for each response, were obtained by the same way that we have used in the first approach case.

For the NOx, the model used is a power model given by equation:

$$\gamma(r) = c_0 + cr^a \quad as \quad r \geq 0 \ and \ 0 \leq a < 2 \tag{9}$$

The value of the model parameters was founded using the least square method. So, c0=997.28, c=0.00018, a=1.52.

In this case the variogram does not show a sill. This means that the variance does not exist. For the consumption, the model used is an exponential model given by equation:

$$\gamma(r) = c_0 + c\left(1 - exp\left(-\frac{r}{a}\right)\right) \quad as \quad r \geq 0 \tag{10}$$

So co=5193, c=0.0327 10^5, a=5.953610^5

Where:

r is the distance.

co is the Nugget effect.

co + Cs the sill correspond to the variance of *Z(x).*

$\sqrt{3a}$ is the range (the distance at which the variogram reaches the sill) for the exponential model (Baillargeon et al., 2004).

Figures 8 shows the experimental variogram (red points), and power model (blue curve) corresponding to NOx response.

Figures 9 shows the experimental variogram (red points), and exponential model (blue curve) corresponding to consumption response.

We notice that when the distance reaches the range $535 \approx \sqrt{3a}$ (Fig. 9), the variation becomes stationary. In other term, this means that there is no correlation beyond the distance 3a. This explains that we have a similar behavior of consumption on two different operating points, thus with a pattern of different control parameters.

Let us notice that the model used here for the variogram of NOx, is of power type, contrary to what we had made in the first approach, where the Gaussian model was retained.

This explains that different engine configurations, lead to different behavior of the NOx. More details will be given in the section 6.

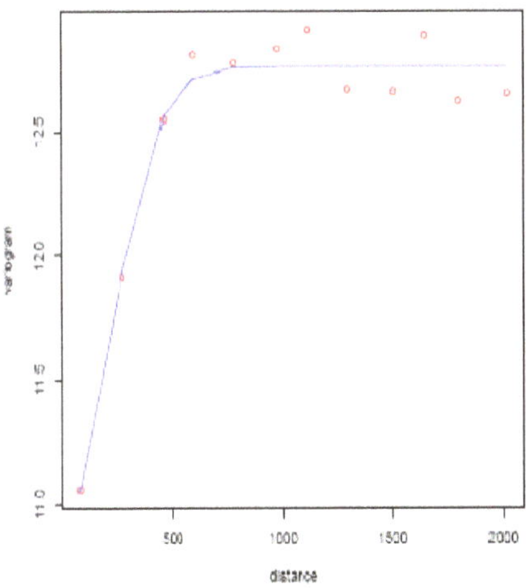


Fig. 4. Experimental and Gaussian model variogram in the case of NOx

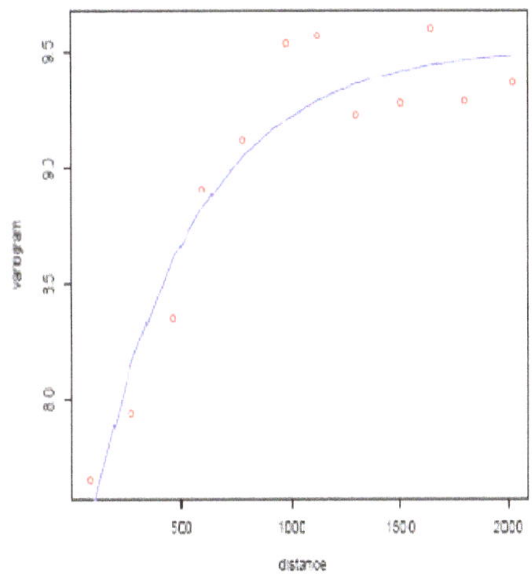


Fig. 5. Experimental and exponential model variogram in the case of consumption

**Fig. 4. and Fig. 5. Experimental and model variogram**

From our observation, the accuracy of the predictions is comparable for both the response and remains within a 10% range for most operating conditions. Upon observation, it is evident that the second strategy yields more accuracy in predicting both replies, as compared to the first approach. This enhancement greatly enhances the accuracy of consumption estimation.

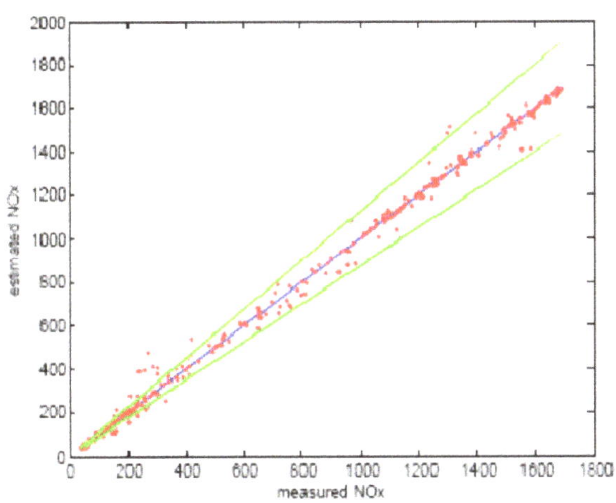


**Fig. 6. Measured and Kriging predicted NOx [ppm] with ± 10% error bands**

The improvement can be attributed to the incorporation of thermodynamic parameters, specifically pressure, in the second technique, which enables more accurate prediction of the two responses. By incorporating these quantities, it becomes possible to enhance the predictive capabilities for both answers. This information arises from the fact that these quantities are the state variables of our system and they describe the combustion behavior inside the combustion chamber.

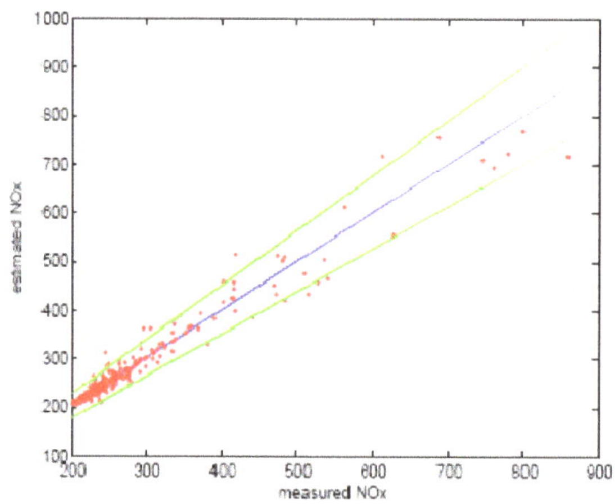


**Fig. 7. Measured and Kriging predicted consumption [g/kWh] with ± 10% error bands**

The emulator model is applied to each answer individually, and the root mean square error (RMSE) and percentage RMSE are documented. The results are displayed in Table 3. The percentage root mean square error (RMSE) findings indicate that the model's %RMSE is below 4% of the range of the response data. This suggests that if the emulator is used to forecast the response at a novel input configuration, the prediction error is likely to be below 4% in comparison to the actual value.

**Table 3. Kriging RMSE end %RMSE for each response: second approach case**

| | NOx | Consumption |
|---|---|---|
| RMSE | 40.51 | 19.99 |
| %RMSE | 2.45 | 3.04 |

## 6. COMPARISON AND DISCUSSION

In section 4, we previously discussed two distinct methodologies that rely on the Kriging model. In this section, we will attempt to compare these two methodologies and evaluate the benefits and drawbacks of each.

**Case of NOx:**

A legitimate question, which we could ask in the case of the estimate of NOx, is the following one:

Why do we obtain a variogram of power type in the second approach, while we had obtained a Gaussian variogram in the first approach, and the pressure is obtained with the same parameters of control?

In fact, the power variogram obtained in the second approach is a better representation of the true behavior of the emissions of NOx. Indeed, the interpretation of the power variogram suggests that the variability of the response increases with the distance between the points. This interpretation joins the opinion of the experts, who say that for two various engine configurations, the quantity of the corresponding NOx emissions will be also different.

Obtaining a Gaussian variogram in the first approach, is explained by the fact that the speed parameter of the engine take a raised values compared to the other control parameters. For example, if we take the first and the second line of the table 5, which correspond to two different engine speeds, we notice that the behavior of NOx is similar. However, the distance between these two points, is very tall (caused by the engine speed) which explains the sill on the variogram of the first approach.

Fortunately, this change in the behavior of variogram does not have an influence on the prediction of NOx. But the interpretation of the variogram in the first approach can lead us to make false conclusions. Indeed, in the case of the first approach, the variogram makes us believe that the quantity of the NOx emissions remains invariant when we consider very different configurations of control parameters. This does not reflect reality. In the case, where we wish to use the variogram, to understand how a response varies. We advise to check the values of the data, or to standardize the factors of the model.

**Table 5. Example of control parameters configuration**

| N | Prail | Main | Mpil1 | Mpil2 | Pmain | Ppil1 | Ppil2 | VNT | VEGR | Volet | NOx |
|---|---|---|---|---|---|---|---|---|---|---|---|
| 1000 | 407,7 | 5,9 | 1,0 | 1,0 | -4,4 | -18,7 | -11,2 | 79,9 | 36,0 | 75,9 | 67,0 |
| 2000 | 609,0 | 11,1 | 1,1 | 1,3 | -5,9 | -36,2 | -15,2 | 67,4 | 34,5 | 75,9 | 64,1 |

**Case of consumption:**

To manage to highlight the contribution of the second approach in the improvement of the prediction of consumption we consider another representation of the results in figure 12.

We note that for the first approach, the Kriging method could estimate with a good accuracy all the points which are close to the cloud used for the adjustment. The prediction of the points which are far from the cloud was bad (as it is explained in section 5.1).

The use of the second approach brought back an improvement for the estimate of these points. This gives a force of extrapolation to the Kriging method.

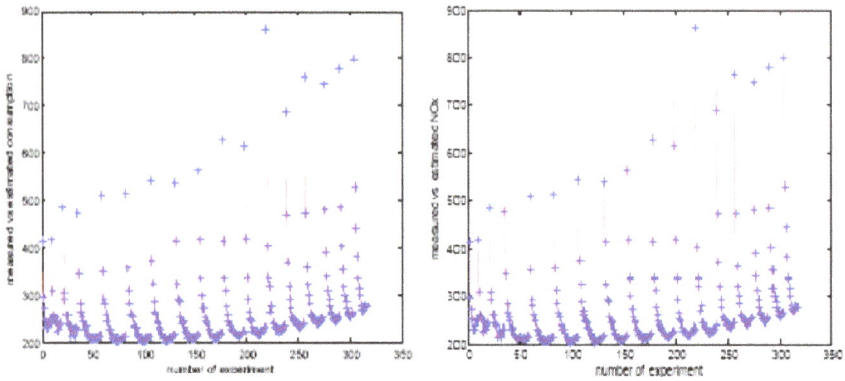


The first approach The second approach

**Fig. 8. Comparison of consumption estimation for the two case approaches. (the + points are the experimental data and the red line is the model )**

## 7. CONCISE SUMMARY

This study addresses the issue of engine calibration in situations when there is a significant number of control parameters. The procedure for resolving such problems typically consists of three consecutive stages: experimental design, statistical modeling, and optimization. This study focuses exclusively on the second step. We examine the significant impact of the experimental design on the accuracy of the Kriging model's prediction in relation to consumption response. The Kriging model was modified to enable the estimation of the response in situations involving larger dimensions. Two techniques were utilized to forecast the NOx emissions and fuel consumption of the two engines. The initial technique yields satisfactory outcomes. The outcomes were significantly enhanced in the second strategy, particularly in terms of consumption. We establish that the resultant model has the capability to forecast the diverse reactions of the engine. Generalizing for different diesel engine designs and applying it to real-time simulations is straightforward and feasible. In the future, this model will be combined with evolutionary algorithms to do multi-objective restricted optimization of calibration.

## REFERENCES

1. Arnaud, M.; Emery, X. (2000). Estimation et interpolation spatiale. *Hermes Science Publications*, Paris.

2. Bates, R.A.; Buck, R.J.; Riccomagno, E. ; Wynn, H.P. (1996). Experimental Design and Observation for large Systems. *J. R. Statist. Soc. B*, vol. 58, (1996) pp. 77-94.

3. Baillargeon, S.; Pouliot, J.; Rivest, L.P.; Fortin, V. ; Fitzback, J. interpolation statistique multivariable de données de précipitations dans un cadre de modélisation hydrologique, Colloque Géomatique 2004: un choix stratégique, Montréal (2004)

4. Castric, S.; Talon, V.; Cherfi, Z.; Boudaoud, N.; Schimmerling, N. P. A, (2007) Diesel engine com-bustion model for tuning process and a calibration method. IMSM07 The Third International Conference on Advances in Vehicul Control and Safety AVCS'07, Buenos Aires, Argentine (2007).
5. Castric, S. (2007) Readjusting methods for models and application for diesel emissions, PhD thesis, University of Technology of Compiègne, 2007.
6. Christakos, G. (1984). On the problem of permissible covariance and variogram models.*Water Resources Research*, 20(2):251-265.
7. Cochran, W. G.; Cox, G. M. (1957). Experimental Designs. Second edition. New York : Wiley. p 611.
8. Cressie, N. A. C. (1993) Statistics for spatial data. Wiley Series in Probability and Mathematical Statistics: Applied Probability and Statistics. John Wiley & Sons Inc., New York. Revised reprint of the 1991 edition. A Wiley-Interscience Publication.
9. Davis, J.C. Statistics and Data Analysis in Geology, second edition John Wiley and Sons. New York (1986).
10. Edwards,S.P.; A.D.P.; Michon, S.; Fournier, G. The optimization of common rail FIE equipped engines through the use of statistical experimental design, mathematical modelling and genetic algorithms, *S.A.E paper,* vol. 106, no3, (1997), pp. 505-523.
11. Goers, A.; Mosher, L.; Higgins, B. (2003). Calibration of an aftermarket EFI conversion system for increased performance and fuel economy with reduced emissions, *S.A.E. paper*, vol. 112, no3, March 2003, pp. 1390-1407, 2003-01-1051.
12. Heywood,J. (1988) Internal combustion engine fundamentals, London : Mac Graw-Hill (1988)
13. Koehler J.R.; Owen A.B.(1996) Computer Experiments. In Ghosh, S., Rao, C.R.,(Eds.), Handbook of Statistics, 13 : Designs and Analysis of Experiments, North- Holland, Amsterdam, p.261-308. (1996)
14. Krige, D.G. (1951) A statistical approach to some basic mine valuation problems on the Witwatersrand,*J. of Chem. Metal. and Mining Soc. of South Africa.* Vol. 52 pp 119-139 (1951).
15. McKay M.D., Beckman R.J., Conover W.J. Comparison of three methods for selecting values input variables in the analysis of output from a computer code, *Technometrics*, Vol. 42, no1, (February 2000) pp. 55 – 61, 239-245
16. Matheron, G. (1963) Principles of Geostatistics, *Economic Geology*, v. 58, no 8, (December 1963) pp. 1246-12688.

17. Pierpont D. A.; Montgomery D. T.; Reitz R. D. Reducing particulate and NOx using multiple injection and EGR in a D.I. diesel, *S.A.E paper*, vol. 104, no4 March(1995) , pp. 171- 183 950217.
18. Pilley, A.D.; A.J.B.; Robinson, D.; Mowll, D. (1994) Design of experiments for optimization of engines to meet future emissions target, International Symposium on Advanced Transportation Applications (1994).
19. Sacks J., Schiller S.B., Welch W.J. (1989) Designs for Computer Experiments. *Technometrics*, vol. 31,41-47.
20. Schimmerling, P.; J.C.S. ; Zaidi, A. (1998) *Use of design of experiments.* Lavoisier.
21. Stein, M. Large sample properties of simulations using Latin hypercube sampling,
22. *Technometrics*, vol. 29, no2, (1987) pp. 143-151, 0040-1706.

# 04 A METHOD FOR ACQUIRING A PROGRAMMABLE LOGIC CONTROLLER (PLC) SOFTWARE FROM A DISCRETE EVENT SYSTEM SPECIFICATION (DEVS) MODEL

## 1. INTRODUCTION: AN OVERVIEW

In order to thrive in the contemporary manufacturing landscape, a manufacturing company must possess the ability to adjust to the shortened lifespan of products within an ever-evolving market. Simulation is a valuable tool for manufacturers to adjust to the fast-paced market by designing and analyzing complex systems that are challenging to represent analytically or mathematically (Choi, 2000). Manufacturers employing simulation can decrease the time required to achieve a steady state in automated manufacturing processes by leveraging statistical analysis, identifying bottlenecks, and identifying scheduling errors. Manufacturers have utilized a range of simulation languages and tools, such as ARENA and AutoMod, for the purpose of simulating industrial systems. The majority of conventional simulation languages and software prioritize the depiction of autonomous entity movements between processes. This technique is usually known as a transaction-oriented paradigm. This study presents a proposed object-oriented strategy that utilizes a collection of object classes to accurately mimic the behavior of current system components. Object-oriented modeling (OOM) is a modeling paradigm that utilizes real-world items to create a design that is organized around these objects, independent of any specific programming language (Rumbaugh, 1991). Despite the well recognized effectiveness of OOM in modeling complex software systems, only a small number of researchers have attempted to utilize OOM in designing and simulating software models for industrial systems. Researchers have proposed several modeling methodologies within the Object-Oriented Modeling (OOM) paradigm, despite using different notations and expressing them in different ways. Choi et al. introduced the JR-net framework, which is built around the OOM paradigm of Rumbaugh et al. The framework consists of three sub-models: an object model, functional model, and dynamic model. Chen and Lu introduced an object-oriented modeling methodology for representing production systems using Petri-nets, entity relationship diagrams (ERD), and IDEF0 (Chen, 1994). The concept of a Virtual Factory (VF) is crucial in today's simulation environment. The effective implementation of the VF notion can be achieved by utilizing the Object-Oriented Modeling (OOM) paradigm (Onosato, 1993).

In a recent publication, Park (2005) introduced a 'three-phase-modeling framework' that may be used to develop a virtual model for an automated manufacturing system. This study utilizes the three-phase modeling framework to create a virtual model, and the Discrete Event System Specification (DEVS) (Zeigler, 1984) for process modeling. The suggested virtual model comprises four distinct categories of things. The virtual device model depicts the fixed arrangement of devices. This can be broken down into the shell and core components, which promote reusability and allow for the adaptation of various system configurations. The Transfer handler model manages a series of device-level commands that replicate the physical process of a transfer, ensuring the accuracy of the virtual model. The Flow controller model determines the firable transfers by utilizing decision factors that are determined by the State manager model. The State manager model and Flow controller model can be implemented as components of a Programmable Logic Controller (PLC). Once the process modeling is completed using the three-phase-modeling framework, these two models will serve as the control information for the conversion to PLC.

The chapter overall structure is as follows. Section 2 provides a concise overview of the Programmable Logic Controller (PLC), while Section 3 focuses on the Discrete Event System Specification (DEVS). The methodology for developing a manufacturing system model to generate PLC code is outlined in Section 4. In Section 5, a cell is provided as an illustration to examine the correlation between the PLC code and the DEVS model discussed in Section 6. Section 7 addresses the conclusion and debate.

## 2. PROGRAMMABLE LOGIC CONTROLLER(PLC)

A Programmable Logic Controller (PLC) is a computer specifically designed for industrial application, utilized to regulate and manage automated operations in manufacturing (Parr, 1999). A Programmable Logic Controller (PLC) is specifically engineered to handle various input and output configurations. It acquires data on the state of a process by utilizing sensing devices like limit sensors, proximity sensors, or signals from robots. The PLC then processes this information using its internal memory and executes appropriate logic. It subsequently activates the next command by utilizing actuators such as motors, solenoid valves, or command signals for the robots. A Programmable Logic Controller (PLC) carries out the control logic that is written using various programming languages. The International Electrotechnical Commission (IEC) released IEC 61131-3 with the aim of establishing standardized programming languages for Programmable Logic Controllers (PLCs). These languages include Ladder diagram, Sequential Function Chart, Structured Text, and Function Block Diagram (Maslar, 1996).

**Fig. 1. The PLC code in the form of Ladder diagram**

## 3. DISCRETE EVENT SYSTEM SPECIFICATION(DEVS)

DEVS formalism is introduced by Zeigler, which is a theoretic formalism and it supplies a means of modeling discrete event system in a modular, hierarchical way. With this DEVS formalism, we can perform modeling more easily and correctly by dividing large system into segment models and define the coupling between them. Formally, an atomic model **M** is specified by a 7-tuple:

$M = < X, S, Y, \delta int, \delta ext, \lambda, ta >$

$X$ : input events set;

$S$ : sequential states set;

$Y$ : output events set;

$\delta int : S \rightarrow S$ : internal transition function;

$\delta ext: Q \times X \rightarrow S$ : external transition function

$Q = \{ (s, e) | s \in S, 0 \le e \le ta(s)\}$ : total state of $M$; $\lambda: S \rightarrow Y$ : output function;

$ta : S \rightarrow Real$ : time advance function:

The second form of the model, called a coupled model, indicates how to couple several element models together to form a new and bigger model. Formally, a coupled model **DN** is defined as:

$DN = \langle X, Y, M, EIC, EOC, IC, SELECT \rangle$

$X$ : input events set;

$Y$ : output events set;

$M$: set of all component models in DEVS;

$EIC \in DN.IN \times M.IN$ : external input coupling relation;

$EOC \in M.OUT \times DN.OUT$ : external output coupling relation;

$IC \in M.OUT \times M.IN$ : internal coupling relation; SELECT : $2M - ø \rightarrow M$ : tie-breaking selector,

Where the extension .IN and .OUT represent the input ports set and the output ports set of each DEVS models.

## 4. METHODOLOGY FOR DEVELOPING A PRODUCTION SYSTEM MODEL FOR GENERATING PLC CODE

In order to create the automated process, the factory designers must take into account the whole process structure. Once the skeleton layout has been determined, the discrete event system software such as ARENA or AutoMod is used to model the process cycle time. This step encompasses the evaluation of co-working machines, focusing on factors such as process cycle time, production capability, physical validity, and efficiency. The software programs QUEST and IGRIP are utilized for the purpose of simulation and modeling (Breuss, 2005).

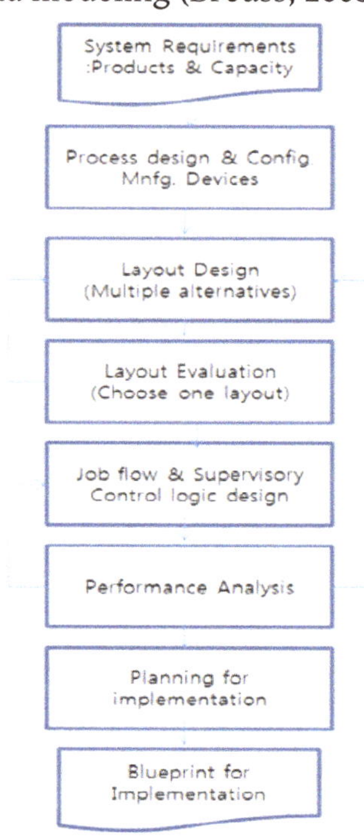


**Fig. 2. Automated factory construction procedure**

During the subsequent phase, the programming of the PLC code for logical operations is carried out independently of any information derived from earlier discrete event systems modeling. In order to effectively use process modeling and practical simulation for physical automated device movement, it is necessary to establish a connection between the high-level simulation of discrete event systems and the low-level physical process control logic. This work aims to establish a connection between two distinct simulation levels and provides a detailed explanation of the automated production of Programmable Logic Controller (PLC) code from the DEVS model. When constructing the DEVS model, the initial step is to represent the manufacturing system using the three-phase-modeling framework (Park, 2005). The framework outlines the modeling of a manufacturing system using four components: the Virtual device model, the Transfer handler model, the State manager model, and the Flow controller model, as depicted in Figure 3.

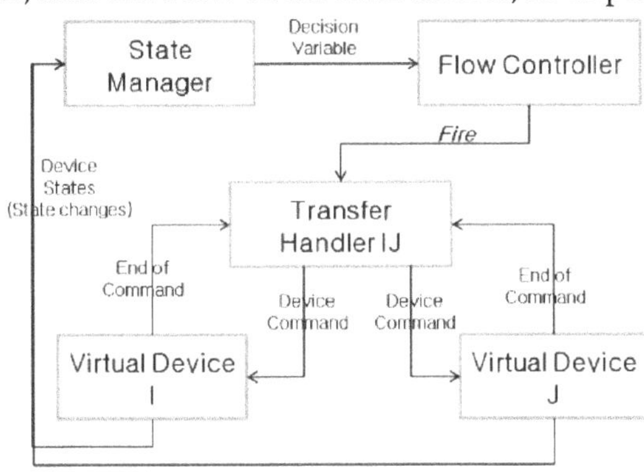


**Fig. 3. Outline of the virtual manufacturing model**

The Virtual device model displays the manufacturing devices. The device is equipped with an input port to receive the action signal and an output port to transmit the signal indicating the work done. The Transfer handler model manages the flow of parts and supporting resources (such as tools and pallets) between devices. This technique prioritized the examination of the physical process that facilitates the transfer, as opposed to more traditional approaches. Essentially, a transfer occurs through the coordination of device-level instructions between collaborating devices (sender and receiver devices). The State manager model gathers the state information from each device. Whenever there is a modification in the status of devices, it will automatically refresh the device statuses. Subsequently, this data will be transmitted to the Flow controller model as a determinative factor. Upon receiving the state information from the State manager model, the Flow controller model will determine the transfer that can be executed depending on the system state (decision variables).

This study utilizes the Discrete Event Systems Specification (DEVS) formalism to create the virtual manufacturing system model. DEVS allows for the specification of discrete event models in a hierarchical modular fashion. The formalism is extremely compatible with Object-Oriented Modeling (OOM) for simulation. In the DEVS formalism, it is necessary to define two types of sub-models: (1) the atomic model, which serves as the foundation for bigger models, and (2) the coupled model, which describes the hierarchical relationship between atomic models.

Once the DEVS model is created, the State manager atomic model, which is responsible for process monitoring, and the Flow controller atomic model, which handles actual control, can be substituted with the PLC component. Specifically, the control component for the manufacturing cell. The objective of this study is as follows.

## 5. THE DEVS MODEL REPRESENTS A BASIC CELL USING THE THREE-PHASE MODELING PARADIGM

In this Chapter, we will observe a small work cell example. The work cell is modeled according to the three-phase-modeling framework and converted to the DEVS model like mentioned above. Finally, we will compare the DEVS model and the PLC code to find some meaningful bridge.

Figure 4 shows the small cell example. At first, an entity is generated from the Stack, which will lay on the AGV machine in P1, then AGV senses this raw part and moves to the P2 for machining. When machine detects the part arrival by the AGV, the machine starts to operate.

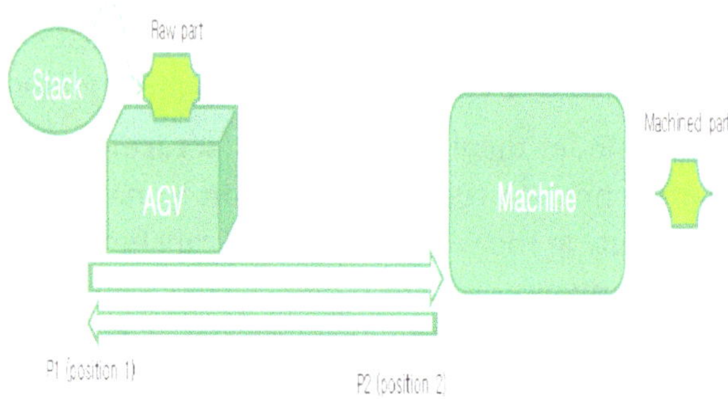


**Fig. 4. Example cell**

When we consider this example cell in terms of the three-phase-modeling framework, there are three virtual device models; the stack model, the AGV model and the machine model. The stack model generates the raw part entity and places it on the AGV for transfer. Until this point, the entity transfer process is between the stack and the AGV virtual device model as a result the transfer handler model is created between the stack the AGV model. Similarly, entity transferring between the AGV model and the Machine happens. This transfer handling model can be represented as *THam.* If there is any state change among the virtual devices, the changes are supposed to be reported to the State manager model. The State manager model maintains the decision variables in compliance with the reported state changes of the virtual devices and the Flow controller model will make a decision on firable transfer based on the decision variables. Figure 5 represents the constructed model about the example cell.

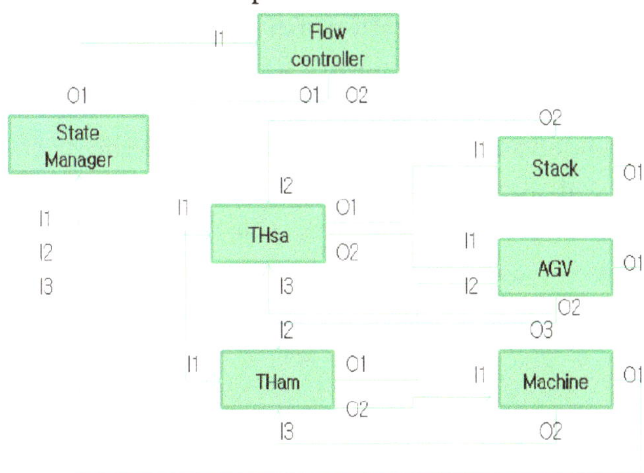


**Fig. 5. Modeling of the example cell in the Park's methodology**

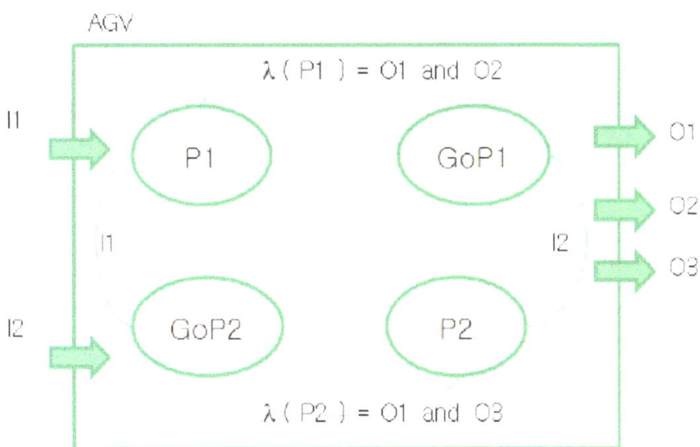


AGV atomic model

X = { I1, I2 }

Y = { O1, O2, O3 }

S = { P1, GoP2, P2, GoP1 }

$\delta_{int}$ :

$\delta_{int}$( GoP2, MT ) = P2,

$\delta_{int}$( GoP1, MT ) = P1

$\delta_{ext}$ :

$\delta_{ext}$( P1, I2 ) = GoP2, $\delta_{ext}$( P2, I1 ) = GoP1

λ ( P1 ) = O1 and O2, λ ( P2 ) = O1 and O3

, λ ( GoP2 ) = O1, λ ( GoP1 ) = O1

ta( MT) : Moving Time

**Fig. 6. DEVS model of the AGV**

Once the modeling by means of the three-phase-modeling framework is finished, second step is to convert the model to the DEVS formalism. In this example, every model is converted to the atomic model and entire cell will be the coupled model that is consist of all atomic models. Figure 6 is the converted DEVS model example of AGV. In the traditional implementation of discrete event system simulation using DEVS, DEVSIM++ is a simulation framework which realizes the DEVS formalism for modeling and related abstract simulator concepts for simulation, all in C++ (Kim, 1994). Through this open source frame, we can develop the discrete event system simulation engine easily. Once, both the DEVS implementation and the simulation with PLC control logic is done, we can achieve the overall physical control simulator for automated process.

## 6. THE RELATIONSHIP BETWEEN THE PLC CODE AND THE DEVS MODELS

For the auto generation of PLC code from the DEVS model, we need to examine the PLC code of example cell and the DEVS models, especially the State manager and the Flow controller model.

In the manufacturing unit, PLC collects the process state information through the sensors. These sensor signals are referenced to decide next command or operation. This task is done by the state manager model in the modeled frame. The State manager model detects every change in state of the virtual device and then updates the decision variables. Similar to PLC code, the Flow controller model is supposed to have running logic that is kind of combination of decision variables. As a result, PLC code from the DEVS model can be divided into two parts. One part is for updating the decision variable from the signal of input port in the State manager model. Another is for actual logic composed of decision variables to fulfill the intended process control.

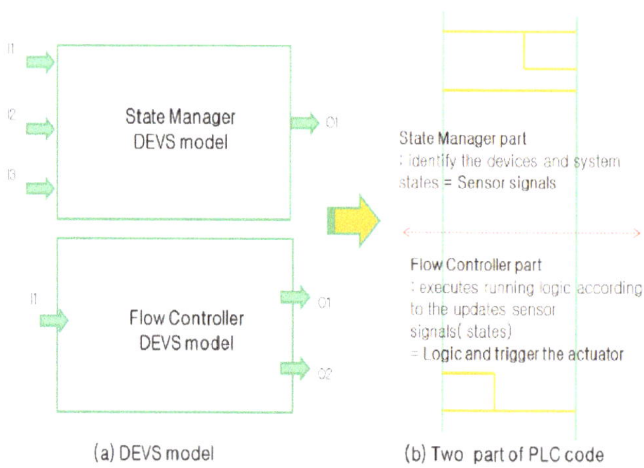


**Fig. 7. Two part of PLC code**

In the front part, the State manager model collects every state changes through the input port. The one input port of example cell has different kind of signal depend on the state. For example, the input port I2 is the signal from the AGV and it has 4 different kinds of state signals. With the same way, each input port of the State manager model has multiple input signals like shown in Table 1.

**Table 1. The States of Atomic models**

| | Atomic models | States | Input Signals |
|---|---|---|---|
| I1 | Stack | Idle, Release | STACK_IDLE STACK_RELEASE |
| I2 | AGV | P1, GoP2, P2, GoP1 | AGV_P1 AGV_GOP2 AGV_P2 AGV_GOP1 |
| I3 | Machine | Idle, Run | MACHINE_IDLE MACHINE_RUN |

The memory structure in the PLC code can be classified into three groups. The first group is input memory which consists of input signal names and the second group is the output memory consisting output signal names and the last is the internal memory which is used to maintain the signal information of input or output and for temporary numerical calculation.

The name of input signal can be determined with combination between the input port and its state name. In this way, we can give a name to all input signals.

As mentioned before, the flow controller model reads the decision variables to execute next command. Thus, we have to make decision variables representing the process state as the internal memory. As we did in the input variable for naming, we can give decision variables' name by putting the 'On' between the port name and the state name. Then, this decision variable shows the port's current state is active condition. Once decision variables are set, the Flow controller detects the firable output signals from the set variables. Figure 8 show the decision variables of each input of AGV model and moving condition. To the AGV, the possible condition to move from P1 to P2 is when the raw part is on the AGV, AGV's state is 'GoP2', and the machine state is 'Idle' at the same time.

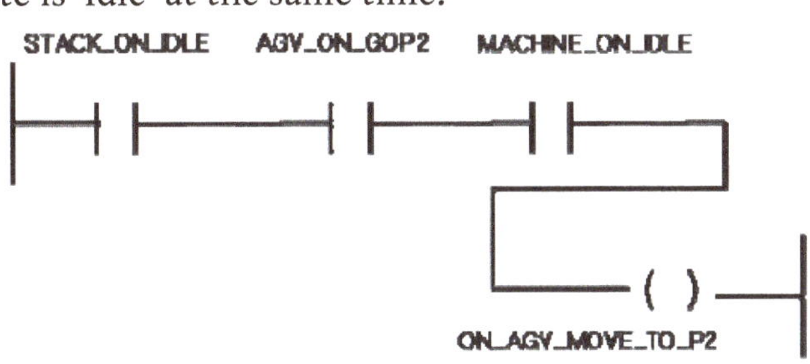


**Fig. 8. The triggering condition for AGV move**

Upon observation of the AGV model, it is evident that the operational state of additional devices can be inferred. Although the PLC code for the State manager model section can be automatically created using decision variables, the flow controller part can be unclear at times. The reason for this distinction is that the DEVS model, unlike the flow controller, is characterized by its abstract and high-level nature, while the PLC component is focused on a specific control region. While it is possible for a process system designer to create a DEVS model that incorporates a low level of Programmable Logic Controller (PLC), this is not typically how DEVS modeling is done. This aspect will either be a limitation or a deliberate decision made by the designer when it comes to the automatic production of PLC code. The DEVS modeling is specifically designed with the intention of generating PLC code for the Flow Controller model component. Figure 9 depicts the two components of the PLC code for the AGV, namely the State Manager and the Flow Controller model. The figure 10 displays the DEVS model of the Flow controller for automatic creation of PLC code in a small work cell.

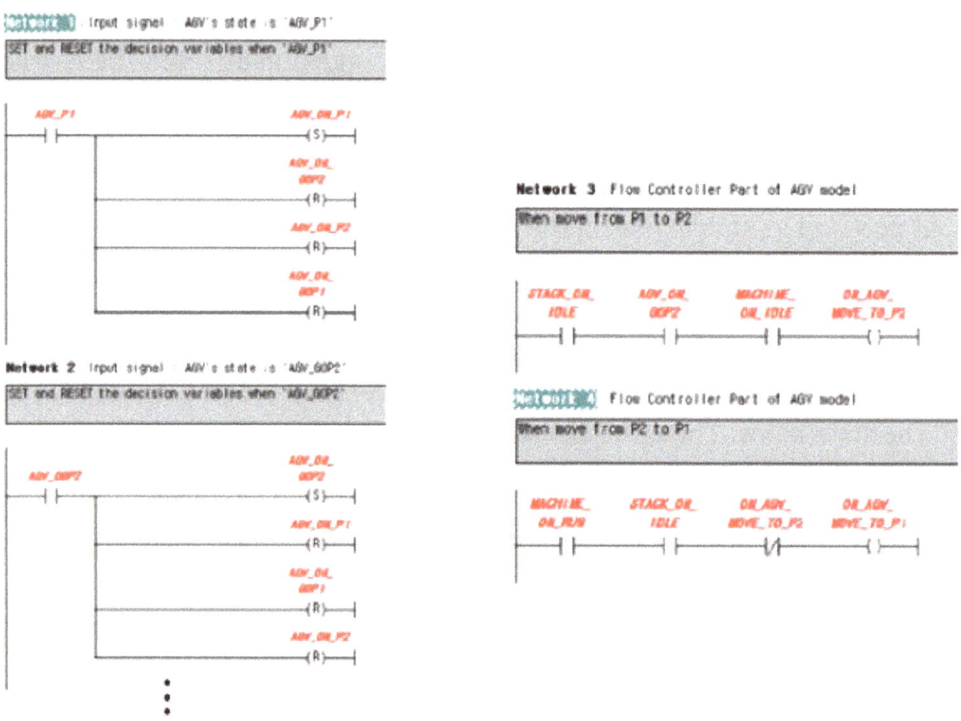


**Fig. 9. PLC code from the State Manager and the Flow Controller model**

## 7. ANALYSIS AND FINAL REMARKS

This document outlines the process for automatically generating PLC code from the DEVS model. The level control logic of PLC is a specialized and restricted field of engineering, but discrete event system modeling and simulation is extensively employed to quantify the process capability. The discrete event system simulation technique allows for the calculation of the process or overall cycle time and throughput.

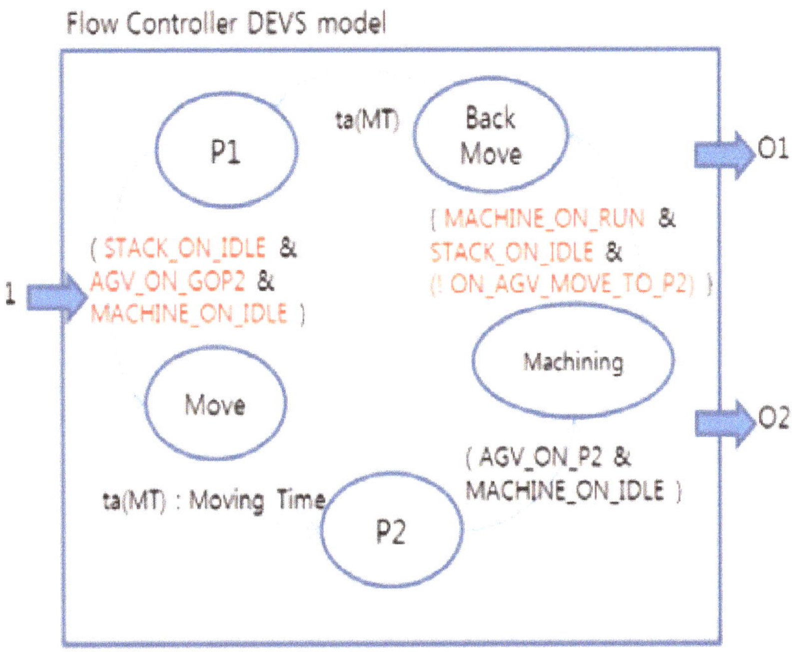


**Fig. 10. The Flow Controller DEVS model**

Nevertheless, there exists a significant disparity between the PLC code and the discrete event system simulation. This discrepancy results in the duplication of process analysis tasks for the PLC programmer and a delay in the implementation of an automated processing system in a manufacturing facility.

The proposed approach consists of three sequential steps. The initial stage is to create a model of the actual system using the three-phase modeling methodology. Subsequently, this model is transformed into the DEVS formalism. Out of the four types of models, the State Manager and Flow Controller models will be substituted by the PLC component.

The PLC code created by our approach may be divided into two sections: one originates from the state manager, and the other from the flow controller. The initial component is formed by combining the input signals and the decision variable. The latter portion is derived from the control component, which is a result of combining choice variables.

The later part of the generation process is not done with perfection due to the higher level of abstraction in the DEVS modeling compared to the PLC level. Nevertheless, this approach provides the comprehensive structure for generating PLC code from a DEVS model. In the foreseeable future, the aforementioned orientation will become the unavoidable trend for enhancing physical process simulation, reducing time in mass manufacturing, and improving the company's competitiveness.

## REFERENCES


1. B. K. Choi, B. H. Kim, 2000. Paper templates, In *Current Advances in Mechanical Design and Production Seventh Cairo University International MDP Conference.* New Trend in CIM: virtual manufacturing systems for next generation manufacturing.
2. J. Rumbaugh, M. Blaha, W. Premerlani. 1991. Paper templates, In *Prentice Hall Inc.* Object- Oriented Modeling and Design.
3. B. K. Choi, H. Kwan, T. Y. Park, 1996. Paper templates, In *The International journal of Flexible Manufacturing Systems.* Object-Oriendted graphical modelling of FMSs.
4. K. Y. Chen, S. S. Lu, 1997. Paper templates, In *International journal of Computer Integrated Manufacturing.* A Petri-net and entity-relationship diagram based object oriented design method for manufacturing systems control.
5. M. Onosato, K. Iwata, 1993. Paper templates, In *CIRP*. Development of a virtual manufacturing system by integrating product models and factory models.
6. Sang C. Park, 2005. Paper templates, In *Computers in Industry*. A methodology for creating a virtual model for a flexible manufacturing system.
7. B. P. Zeigler, 1984. Paper templates, In *Academic Press.* Multifacetted Modeling and Discrete Event Simulation.
8. E. A. Parr, 1999. *The book*, Programmable Controllers : An Engineer's Guide 3rd ed.
9. M. Maslar, 1996. Paper templates, In *IEEE Pulp and Paper Industry Technical Conference.* PLC standard programming language: IEC61131-3
10. F. Breuss, W. Roeger, 2005. Paper templates, In Journal of Policy Modeling. The SGP fiscal rule in the case of sluggish growth: Simulations with the QUEST
11. T. G. Kim, 1994. *The Book.* DEVS++ User's Manual

# 05 AN INFRASTRUCTURE FOR EMULATING DOMESTIC CONTROL NETWORKS

## 1. INTRODUCTION: AN OVERVIEW

Trasgu[1] is a design environment for control networks that is suitable for digital home environments and other locations. The deployment of services offered by information society technologies, particularly control networks, is expanding in various sectors such as businesses, buildings, and residences. Control networks offer a wide array of protocols and technologies. The array of control technologies that render the applications feasible is varied, with each adhering to its own set of criteria. Various control systems, such as X10 (Fuster & Azorín, 2005), KNX/EIB (Haenselmann et al., 2007), LonWorks (Ming et al., 2007), CAN (Jung et al., 2005), Zigbee (Pan & Tseng, 2007), and others, are used. Additionally, there are proprietary technologies like Bticino, Vantage, X2D, and more. Despite efforts to standardize, the process of designing and implementing control facilities remains intricate. Each technology has inherent restrictions. Among these, we discover the proprietary configuration tool. This program is proprietary and enables the creation and configuration of networks. The primary tool for designing and configuring control networks is proprietary software, which is supplied by the vendor. The standard automation network technologies utilize the European Installation Bus Tool Software (ETS) for Konnex (KNX) and the LonMaker Integration Tool (LonMaker) for Lonworks. Both tools provide the same function, but they differ in terms of their design and configuration methods. A design created using one tool is not compatible with another tool, and often cannot be connected. This has implications for the increase in both time and expense when the design requires the integration of multiple technologies in order to provide a more advanced service. The selection of technology is contingent upon the specific needs and preferences of the user, as not all technologies may adequately address the requirements at hand. There may be projects that cannot be solved using just one technology, so we will need to integrate multiple technologies. Integrating them into a shared system can often be challenging in several scenarios. Regardless of this, it is necessary to establish communication with them in order to offer superior services, even if they are part of distinct networks such as control networks, information networks, multimedia networks, and security networks. Residential gateways utilize middleware and discovery protocols to simplify the integration process.

Residential gateway middleware refers to software that enables communication between various technologies, namely control technologies. The commonly used middleware in the field of Information and Communication Technology (ICT) are CORBA, J2EE, and .Net. J2EE and .NET.The internet is the most widely used network. The first one is built on Java, while the second one is built on several Microsoft programming languages such as C#, Java#, and Visual Basic.Internet,...

Discovery protocols enable devices to connect to networks and negotiate services. Some examples of these protocols include Universal Plug and Play (UPnP) (Rhee et al., 2004), Jini Network Technology (Sun, 1999), Home Audio/ Video Interoperability (Havi), Open Service Gateway initiative (OSGi) (Kawamura & Maeomichi, 2004), and Services and Network for Domotics Applications (SENDA) (Moya & López, 2002)... UPnP is a decentralized open architecture that operates on a network and is technology-agnostic. The objective is to establish a seamless link between many devices using different technologies. Jini provides the necessary framework to connect and integrate services in a distributed system. The software is developed using the Java programming language. Jini's lack of success can be attributed to the limited number of devices that support it. Havi is an architectural design focused on electrodomestic appliances. The goal is to establish a service set that facilitates the development of interoperable and distributed networks in a home environment. OSGi provides an open and scalable architecture for executing services securely. The objective is to connect devices in a heterogeneous network at the application layer in order to compose services. The SENDA framework is a device networking system that utilizes the CORBA protocol. SENDAs objective aligns with the philosophy of OSGi, however with the intention of enhancing it. Several technologies exist for integrating control networks, but none of them have achieved widespread success in all situations, although a few have managed to capture a small market share.

Aside from integration, there is another issue present in current control networks. After designing a control network based on user requirements, it is necessary to proceed with the network installation in order to verify its proper functioning. This aspect leads to significant temporal and economic expenses throughout the installation process, as the designer must address any detected faults in the actual facilities. This issue can be prevented by conducting a simulation task after the design phase and before the installation of the network. However, the tools used for controlling networks do not support the execution of simulations. They successfully implement low-level validations. ETS conducts validations by assigning physical and group addresses to the devices. The ETS has a tool called the EIB Interworking Test Tool, also known as EITT. It specializes in device analysis and also provides the capability to simulate network protocols. Despite performing basic validations, none of these tools are capable of simulating the behavior of the control network. As a result, designers are unable to validate the proper working of a system until the control network has been constructed and installed. Simulation offers benefits in the design of control systems. Simulation is a valuable tool for techno-scientists in the development process as it enables the early detection of faults during the design phase (Denning, 1989). It has the capability to authenticate and confirm the control network design (Balci, 1998) in order to minimize expenses. Simulation plays an equivalent role in the design phase as testing does in the implementation phase, as both involve verifying the work that has been done. Simulation is seen as a test case, according to Norton and Suppe (2001), in which evaluations are conducted at a more abstract level. In addition to the professional tools, there are several control network simulators available in the literature, such as the VINT project (Breslau, 2000), DOMOSIM (Bravo et al., 2000, 2006), VISIR (González et al., 2001), and Conte (2007). The VINT project introduces a unified simulator that encompasses an extensive collection of network models. The system consists of a simulator foundation and a visualization tool. This simulation tool is specifically designed for studying TCP/IP network protocols, including TCP behavior, multicast transport, multimedia transmission, protocol response to topology changes, and application-level protocols.

DOMOSIM and VISIR are primarily designed for educational purposes, namely to teach the process of facility design. The drawback lies in the inability to integrate with professional design tools. The study conducted by Conte (2007) examines home automation systems using a simulation/emulation environment. The home automation system is designed to model agents, with each agent's action being characterized by its cost, length, and quality. The agent's behavior is represented by a collection of states and a set of rules. The simulator is a software environment where agents are implemented using LabView and LabWindows CVI. These agents are executed simultaneously. In order to conduct a simulation, the user must establish the parameters of the virtual environment and then initiate the execution process. This simulation environment is robust, but, it lacks integration with the designing environment. As a result, we need to initially design the control network using a separate tool and then subsequently redesign it within the simulation environment.

Control networks entail significant temporal and economic expenses. The high cost is mostly attributed to three factors: the need to integrate several technologies, the use of inadequate design processes, and the reliance on experimental design validation. The objective of this chapter is to establish an environment that enables the management of network topologies in the digital home. The goal is to ensure that these architectures are applicable to any technology, with a particular focus on network simulation tasks.

## 2. MODELLING CONTROL SYSTEMS

Modelling systems primarily rely on two types of techniques: bottom-up methodologies and top-down methodologies (Sommerville, 2004).

The technology and methodologies utilized for the modeling of home automated systems are limited in their level of development. They rely on the utilization of extremely basic technologies (Muñoz et al., 2004). The wide range of control technologies can make the process of developing control networks using bottom-up techniques challenging and result in systems that do not meet the necessary requirements.

The top-down approaches are primarily defined by their abstract nature. They insist that innovative ideas are generated prior to any deliberation. Ingenuities are conceived without any constraints imposed by technology. These techniques align seamlessly with our design philosophy of utilizing implementation technologies that are not reliant on certain platforms or systems.

Model Driven Architecture (MDA) and Services Oriented Architectures (SOA) are two distinct implementation philosophies, as described by Mellor et al. (2004) and Newcomer & Lomow (2005) respectively. On one hand, MDA enables the conversion of conceptual system models into code through automated generation, resulting in the creation of efficient and speedy middleware solutions. On the other hand, SOA offers a systematic approach and structure for documenting business capabilities and enabling the consolidation and integration efforts. The network nodes provide their resources to other participants in the network as autonomous services that have access to a defined protocol. Unlike object-oriented architectures, service-oriented architectures (SOAs) consist of application services that are loosely connected and highly interoperable. The communication between these services relies on a formal specification that is platform-independent and programming language-agnostic. The interface specification encapsulates the specific characteristics of an implementation, allowing it to be independent of the technology, programming language, or developer.

Designing control networks using this methodology necessitates addressing the following inquiries: What specific features do I intend to provide? For the specified functionalities of safety, comfort, and automation, we are required to offer the following services: intruder detection, access control, alarms, lighting, and temperature control.

What is the expected behavior of services? That is, how to establish a connection between them. Regarding the aforementioned examples, it is necessary to specify the desired method of temperature regulation or the desired interactions with other devices. Intrusion detection and alarm systems.

Which technologies will be utilized to implement the system? After determining the specific features and behavior of the network, we must then evaluate the various accessible technologies in order to select the most suitable option for this particular case.

We present a model that is founded on the three aforementioned questions. The model comprises three layers: functional, structural, and technical.

The functional layer is the highest level of abstraction. It explains the features and capabilities of the installation process. It circumvents the process of considering the methodology and execution of this task, as well as the application of technology. Our task is to convert customer needs into the specific functionality that the installation will offer. The control installation, CI, in this layer is defined by a collection of services, Si, that are requested by users.

$$CI = \{S_1, S_2, \ldots, S_n\} \quad (1)$$

Each service, *Si*, needs to satisfy a set of tasks, *ti:*

$$S_i = \{t_{i1}, t_{i2}, \ldots, t_{in}\} \quad (2)$$

The next level of abstraction is called structural. It is focused on the structure and behaviour of the generic devices. Since the structural layer, the control installation, *CI*, is composed of a set of generic resources, *Rs*, and a wide range of connections, *C*, which are established between resources:

$$CI = CI(Rs, C), \quad (3)$$
$$Rs = \{Rs_1, Rs_2, \ldots, Rs_n\},$$
$$C = \{C_1, C_2, \ldots, C_m\}$$

A resource represents the entity that may be physical or logical and provides some tasks, *tij*. At this level, both the resources and connections are independent of technology, because it only represents the installation structure. Resources are represented as a set of tasks, *tij*, offered to other entities:

$$RS_i = RS_i\{t_{i1}, t_{i2}, \ldots, t_{im}\} \quad (4)$$

Connections are represented as associations between two resources. They are formed by an input resource, *Rsi*, and an output resource, *Rso*:

$$C_i = C_i(Rs_i, Rs_o) \quad (5)$$

The lower abstraction layer is technological. It is an instance of structural layer using a specific technology. It takes into account more realistic aspects of the implementation. Resources viewed from the technological layer are defined by the set of tasks, *Ti*, and by his set of physical characteristics, *CAi*, like position, size, weight, etc:

$$R_i = R_i(T_i, CA_i), \quad (6)$$
$$CA_i = \{ca_{i1}, ca_{i2}, \ldots, ca_{ip}\}$$

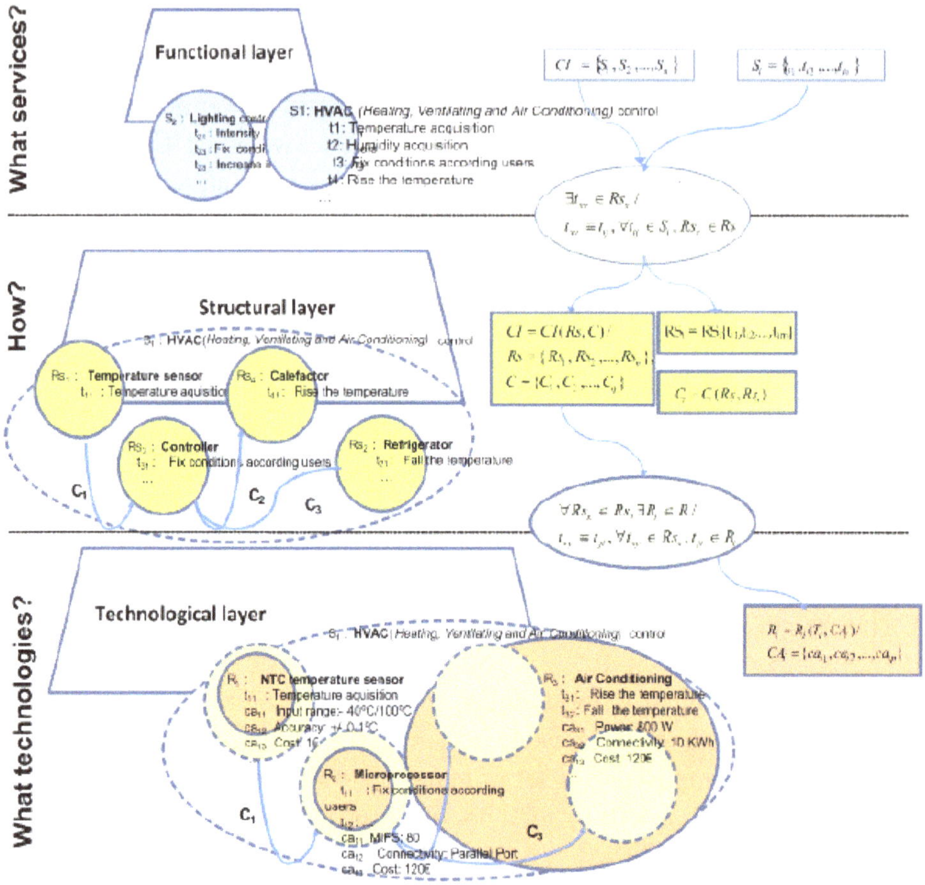


**Fig. 1. Correspondence between layers and model equations**

At this layer, resources, *Ri*, tasks, *Ti*, and characteristics, *CAi*, are determined by implementation technologies. This representation reflects reality faithfully.

As the level of abstraction is reduced, it is necessary to make transitions from functional to structural and, finally to technological level. To achieve the first transition, services should have all their tasks paired with some of tasks provided by resources of the technological level. Therefore task, *txy*, of any resource, *Rsx*, should be equivalent to task, tij, required by service *Si*. This must be satisfied for all tasks, *tij*, required by the service, *Si*:

$$\exists\ t_{xy} \in Rs_x \ /\ t_{xy} \equiv t_{ij},\ \forall\ t_{ij} \in S_i,\ Rs_x \in Rs \qquad (7)$$

Structural to technological level transition needs to match generic resources *Rs* with technological resources *Ri*. Therefore all tasks, *txy*, at all generic resources, *Rsx*, should be matched with task, *tij*, of technological resources *Ri*. This matching must be between equivalent tasks:

$$\forall\ Rs_x \in Rs,\ \exists R_i \in R \ /\ t_{xy} \equiv t_{ji},\ \forall t_{xy} \in Rs_x,\ t_{ji} \in R_i \qquad (8)$$

Once these transitions have been properly executed, the control network will be delineated into three distinct levels of abstraction: functional, structural, and technological. Figure 1 illustrates the relationship between layers and model equations, using a Heating Ventilating and Air Conditioning (HVAC) service as an example. The system determines the functional level of a series of tasks, such as temperature acquisition, humidity acquisition, adjusting conditions based on user preferences, increasing the temperature, decreasing the temperature, and so on. The structural level is determined by the resources that carry out the preceding activities, such as the temperature sensor, which performs temperature acquisition, the controller, which enforces user-defined conditions, the heater, which increases the temperature, and the refrigerator, which decreases the temperature. Ultimately, the technological level determines how technology utilizes the resources specified at the structural level. In this instance, an NTC temperature sensor has been selected as the temperature sensor. The temperature acquisition task can be configured based on its input range, precision, and cost. The microprocessor was selected as the controller. The fixing conditions of the system are determined according on the user's work and are specified by factors such as MIPS (Million Instructions Per Second), connection, and cost. The air-conditioning system performs two functions: increasing and decreasing the temperature. Its characteristics include power, connectivity, cost, and other factors.

## 3. TRASGU: DESIGN ENVIRONMENT

Trasgu is a prototyping environment that is built upon a pre-established model. The usage of it is applicable across three distinct levels of abstraction: functional, structural, and technological. The installation process can be characterized by a higher level of abstraction and the selection of specific implementation details in the final stages. The system offers a design that is strong and resistant, as well as a decrease in the amount of time needed for development. This is achieved by ensuring that the transitions from the most abstract level to the most detailed level are gradual and uncomplicated. It enables the completion of all necessary duties during the whole life cycle of a control network. The responsibilities mentioned will be performed by the individuals involved in a control installation, namely designers, integrators, property developers, and installation engineers.

The model possesses the capacity to build functionalities that are provided by the environment. The environment offers the following capabilities: autonomous design of control installations using any technology; conducting simulations to ensure that our design aligns with user specifications; creating an architecture that aligns with the chosen technology for implementation; generating a budget for each technology to determine its cost; generating various types of reports, including one that displays the wiring installation. Currently, the duties carried out by trasgu include designing control networks and doing simulation tasks. Figure 2 illustrates the potential users for each specific aspect of the environment, for example. The prescriber will be responsible for building the control network and doing simulations to validate early judgments. The programmers, on the other hand, will be responsible for modeling the architecture using implementation technology and conducting simulations. However, promoters will have the responsibility of managing finances, while installers will be responsible for creating installation reports.

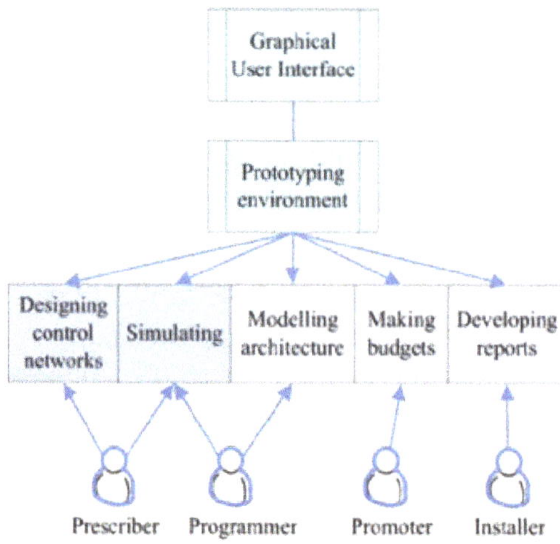


**Fig. 2. Environment features and their possible users**

The initial objective is to design the control network. This task involves the incorporation of additional resources into a building plan. Resources have been categorized into three distinct types: input, as defined by Equation (4); services, as defined by Equation (2); and output, as defined by Equation (4). Initially, it is necessary to incorporate services into the building floor. Next, we must incorporate the input and output resources and establish connections with their respective services. Every resource is linked to one or more services that govern their network behavior.

The choice to describe the control network based on services was made because this representation accurately demonstrates that several services might utilize the same input and output resources. Figure 3 depicts a demonstration of three distinct services: security, lighting, and gas detection. The presence sensor and binary light are utilized by both security and illumination services. The gas detection service utilizes a gas sensor and an audio alarm. Lastly, the acoustic alarm is communicated with the security service.

The simulation challenge is around the design of the control network. The simulation settings can be specified to model the control network. This task verifies that the design meets the specified requirements.

The architecture modelling work involves creating a representation of the building that can be implemented in an actual installation based on the design. The architecture is built on middleware, as described by Valdivieso et al. (2007) and Fuster et al. (2005). This middleware facilitates interoperability across diverse control technologies and protocols.

The budget development task enables the generation of budgets based on the technological design. It will provide a cost estimate for the customer. This task necessitates automated communication with the devices vendor for the purpose of database maintenance.

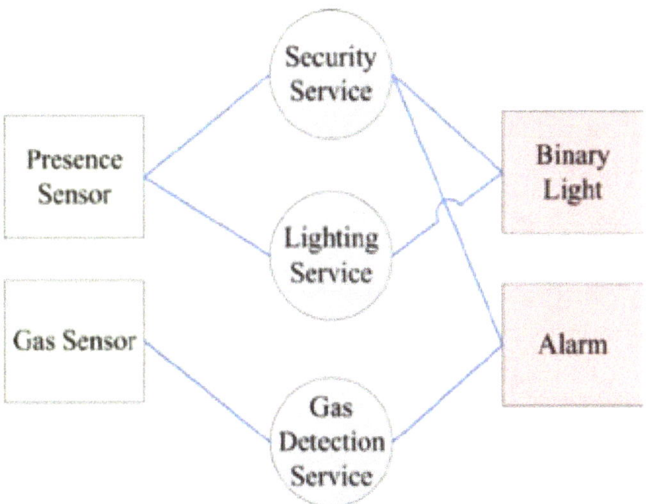


**Fig. 3. Representation based on services**

The work of report development enables the generation of several types of reports. Each installation feature is specifically focused on aspects such as wired connections, device placement, device configuration, and services. Environment features can be accessed and utilized at various levels of abstraction. Figure 4 illustrates the connections between environmental characteristics and the abstraction layer in which they are implemented. The installation design can be utilized across functional, structural, and technical aspects. Installation simulation can be utilized across functional, structural, and technical domains. Architecture modelling exclusively encompasses the structural and technological aspects. The technology aspect of this system is evident in its implementation on actual devices, while the structure aspect is characterized by the potential for shared architecture layers. Creating budgets is primarily beneficial from a technology standpoint as they are intended for customers. Reports can be generated from the functional, structural, and technological layers to cater to various purposes. The consumer may require a service report, a developer may require a structural report, and an installation engineer may require a wiring report.

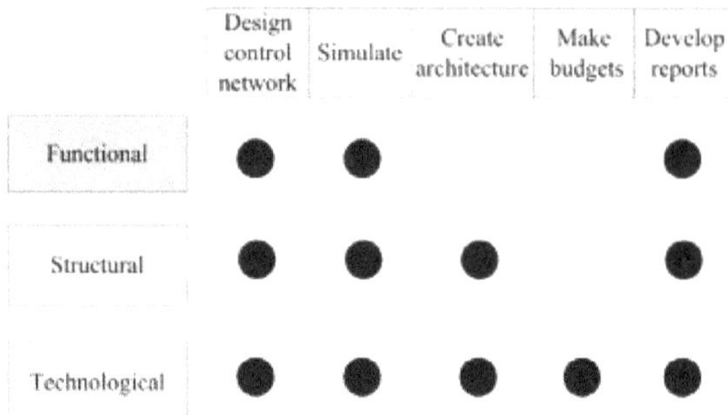


**Fig. 4. Abstraction layers supported by each feature**

## 3.1 Resources definition

Functional level resources are categorized based on their input and output. Resources are classified at a functional level to define subcategories within input and output resources.

Input resources can be further categorized into continuous and discrete sensors. Each family sensor is linked to a singular magnitude for measurement. Continuous sensors are devices that measure continuous variables such as temperature, humidity, illumination, and so on. Discrete sensors are designed to detect specific events such as gas leaks, presence of objects, or damaged windows. These sensors detect the occurrence of an event.

The output resources are categorized into two distinct groups: continuous actuators and discrete actuators. Continuous actuators are distinguished by their ability to provide feedback from continuous sensors, whereas discrete actuators do not provide feedback from sensors. The initial category is exemplified by an HVAC equipment that adjusts the previously measured magnitude using a continuous sensor. The second type is an auditory alarm, such as one that does not trigger any event. Continuous actuators engage with and alter the magnitude detected by sensors. Continuous actuators are defined by their ability to directly modify a specific magnitude and indirectly modify other magnitudes that can be altered by the actuator. For example, an HVAC system directly modifies temperature and indirectly affects humidity. Another element to be elucidated is the capability of certain actuators to alter the physical composition of the area in which they are situated. For instance, a blind engine or window engine can affect the construction of a wall, enabling or inhibiting the entry of light and adjusting the temperature by raising or lowering it.

## 3.2 Implementation

The environment has been developed using the Java programming language and utilizes an information system constructed using XML. The XML files contain data pertaining to the defining of resources that can be utilized during installation, as well as information regarding the current network. This information is derived from abstraction layers that are defined at the model level. The XML files contain a limited selection of the features available in the market. The environment has the capability to enhance the information system by including more resources through XML files. Adding new resources is a straightforward process since it only requires defining the XML files that specify the new resource and its interactions with magnitudes.

The prototyping environment presents four different areas, see figure 5. On the top of the window, we can see the services that can be added to the network. On the left side there are all the generic resources that can be used in the design. In the middle, there is the building floor map. In this area services and generic resources can be added for designing the control network. On the right side, there is hierarchical list of all the resources added in the network classified by resource type.

The control network design starts up opening a map of the building where user can drag and drop resources and services in order to design the network. The next action is adding services, input and output resources into the map, after doing that, user has to connect resources and services in order to establish relationships. Moreover, user has to configure parameters of input and output resources and, finally, user must determine the behaviour of each service through logical operators between input and output resources. It is not possible link inputs and outputs resources directly because it is necessary a service that link them. Figure 5 shows a control network designed at the environment. The control network contains two services: HVAC and technical security. HVAC is formed by a temperature sensor and an air-conditioning and heating. The technical security is composed by a gas sensor and an acoustic alarm. Although this design does not share any resource, it is possible share resources into some services as we have show at figure 3.

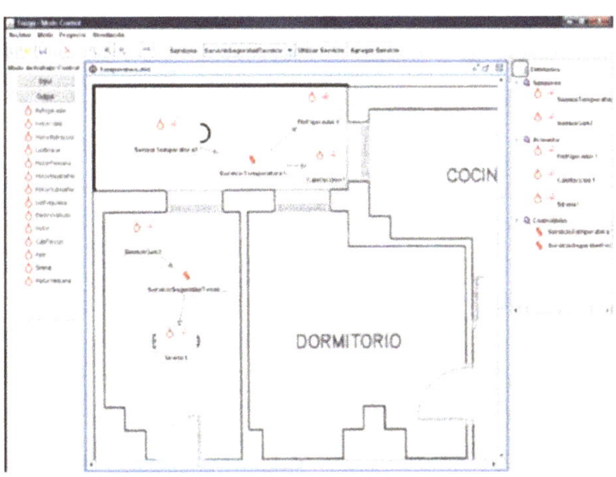


**Fig. 5. Environment aspect and control network design. It contains HVAC and technical security services**

Each service defines their behaviour based on some functions. These functions receive inputs from input resources and provide outputs to output resources. Functions are composed by logical, arithmetical or comparing operators. This specification contains inputs, and outputs that allows interconnect them using operators. Figure 6 shows services definition window. On the middle of the figure we can see inputs resources on the left and output resources on the right. Operators are situated on the left side of the window and it could be drag and drop into the previous area in order to link inputs and outputs. Operators should be configured in order to define the behaviour of the service. This figure contains the HVAC service behaviour. Refrigerator will be turned on when temperature rises above 25 Celsius degrees and it will be turned off when temperature falls under 25 Celsius degrees. Heating will be turn on when temperature reaches 15 Celsius degrees and it will be turn off when temperature rises above 15 Celsius degrees. According to this parameters temperature will be between 15 and 25 Celsius degrees.

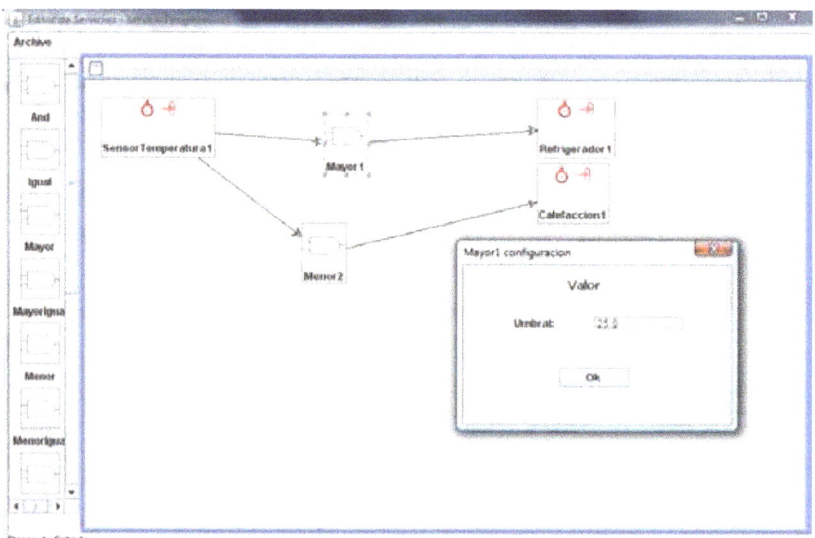


**Fig. 6. HVAC performance is defined by logical operators**

## 4. SIMULATION

The environment is specifically built to execute simulations across many levels of abstraction. The environment is simulated using equations (1) and (2) for the functional layer, equations (3) and (4) for the structural layer, and equation (5) for the technological layer. The functional simulation is tasked with emulating the behavior of a generic control network. By implementing this validation, we can ensure the accuracy of the general device configuration and connections. The structural simulation incorporates the functional aspects while introducing additional characteristics such as the accurate positioning of resources and compliance with installation regulations. The technological simulation assesses the actual performance that the built control network will deliver.

The functional layer is being described and simulated. The outcome of this simulation is influenced by two specific factors: the duration of the experiment and the rate at which events occur. The initial value represents the interval that we aim to simulate. The second factor is the frequency at which we shall extract values from each resource.

The simulation challenge requires a flag to differentiate between continuous and discrete sensors. Event generators are responsible for executing this action. We have developed two types of event generators: a discrete event generator and a continuous event generator. The first type of data collection involves discrete sensors that provide data based on probability percentages. Each discrete sensor in the network has a set probability %. The second category pertains to input data for continuous sensors and is determined by the average value and variability over the course of the simulation period. Generating a sinusoidal waveform that accurately depicts continuous variables such as temperature, humidity, etc...

The results are displayed in two perspectives. The first diagram depicts the layout of a building, with the control network and resource values represented separately. The values of resources are dynamically changing at a specified frequency to see the progression of the control network in simulation time. This perspective is depicted in figure 7.

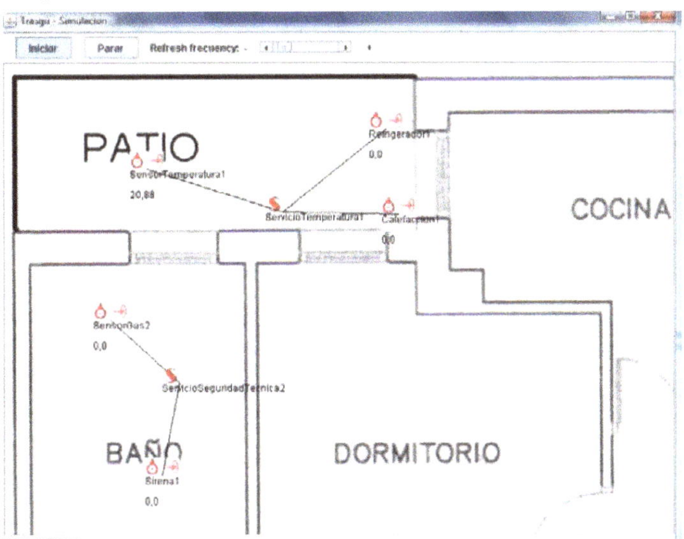


**Fig. 7. Simulation results view showing values at the frequency defined at the top of the window**

The second view represents a graph. A graph is showed for each resource in the previous view. The graph represents the resource behaviour in simulation time. There are three types of graphs.

The first graph type shows services. This graph shows all signals of the resources linked in the service, representing all the information in one graph, although sometimes data range are different so we have to view graphs individually at each resource. Thinking in our HVAC service example, the graph shows temperature sensor, air-conditioning and heating values. The second graph type represents sensors and shows event generators values and values given by the sensor. In HVAC example, the temperature sensor graph shows external temperature and the temperature sensor measured, which might have a little error. The third graph is reserved for actuators, and shows actuator values drawn as a single signal representing the actuator performance. In HVAC example, the air-conditioning and heating will be show in two graphs, each one shows a signal indicating when the air-conditioning is on or off. Figure 8 shows three graphs, temperature sensor at left, air-conditioning at right on top and heating at right on bottom. Temperature sensor graph shows the external temperature in blue, which is generated as a variation event, and sensor temperature in red. The red signal has peaks because the sensor has been defined with measure error. Moreover the red signal is ever between 25 and 15 Celsius degrees because the air-conditioning starts

when temperature rises and the heating starts when temperature falls. Air-conditioning and heating graph shows when they start to work.

**Fig. 8. Simulation results graph of temperature sensor, air-conditioning and heating**

## 4.1 Implementation

The process of simulation is executed by an external module. The implementation is done outside to allow for easy introduction of future changes in the simulation module, without affecting the environment. The choice to design it in Matlab / Simulink was made due to its ability to provide great precision and efficiency.

The interaction between the simulation module and the environment has been facilitated using a Java package known as jMatLink (Müller & Waller, 1999). It has the responsibility of transmitting commands to Matlab. Mastery of Matlab instructions is essential for doing this activity. We have created a software library called jSCA. Its purpose is to encapsulate Matlab commands in order to facilitate communication. The functionalities of jSCA are realized using methods that enable the creation of Matlab instances, the creation, configuration, and removal of blocks in Simulink, the configuration of simulation parameters, and the retrieval of simulation results. This library surpasses jMatLink in terms of quality and performance. The environment utilizes the jSCA library, which in turn invokes the jMatlink library.

The figure 9 illustrates the architecture that facilitates communication between a Java program and Simulink. The initial three layers are situated within the internal environment, while the latter two are positioned in the exterior world. The initial layer consists of the individual's prototyping environment. The jSCA library is the second option. The third option is the jMatlink library. The fourth component is the Matlab engine. The daemon actively listens for commands sent by jMatlink. The Matlab Engine serves as an intermediary between jMatLink and Simulink. We initiate Simulink and execute commands via the Matlab Engine.

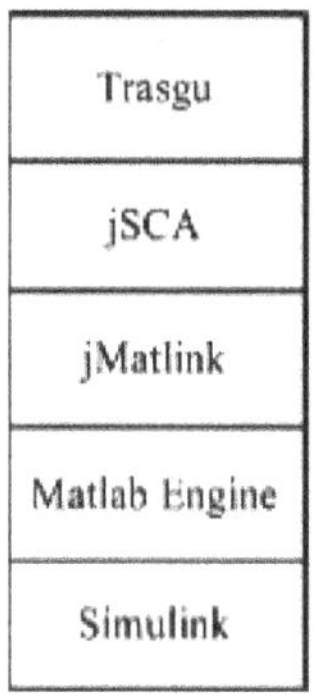


**Fig. 9. Architecture that achieves communication among Java application and Simulink**

The sequence diagram illustrates the interaction between the user and the previously described architecture layers. This figure illustrates the chronological order of the primary telephone conversations. Figure 10 depicts the sequence diagram that aligns with the Unified Modeling Language (UML), a specification language. The temporal sequences of calls commence when the user formulates the control network. The subsequent step is to replicate it. When the user initiates the simulation, the trasgu component will invoke the jMatlink layer to generate an MDL file. This layer initiates the Matlab server and utilizes the Matlab engine to generate the mdl file. Ultimately, the Matlab engine invokes Simulink, resulting in the creation of the mdl file. After the creation of the mdl file, the Matlab engine initiates the Simulink software to begin the simulation. Once the simulation is completed, the simulation values are sequentially transmitted to trasgu, who then proceeds to visualize them.

Functional simulation necessitates a parallel model derived from the functional layer in Simulink. This model is identical to the XML files that define resources in the information system utilized by the environment. Every resource specified in the environment is matched by a corresponding resource described in Simulink. The Simulink model comprises a block set composed of input blocks, controller blocks, output blocks, and input-output blocks. Additionally, there are two sorts of blocks: discrete and continuous. The discrete blocks are specifically designed for digital devices, whereas the continuous blocks are intended for analog equipment.

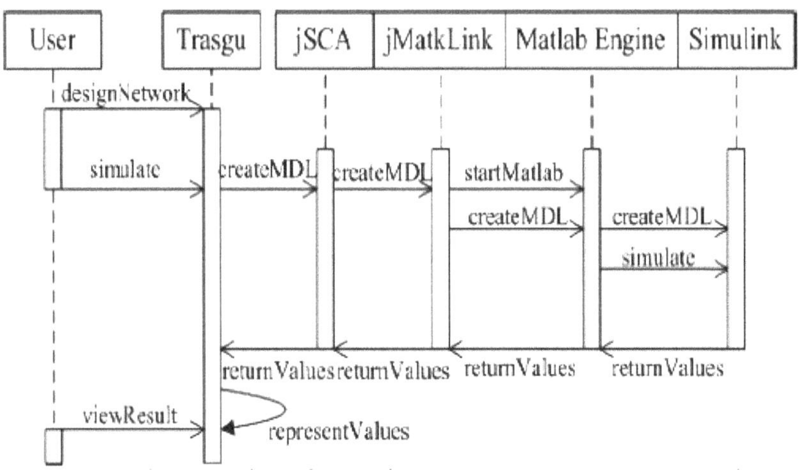


**Fig. 10. Sequence architecture diagram that shows the interaction among user and simulation**

The simulation initiates the dynamic creation of a comparable control network in Simulink. The Simulink control network is stored in a mdl file. In order to generate the mdl file, we incorporate the appropriate network blocks. The parameters of each block are configured and the connections between the blocks are established. Subsequently, it is necessary to incorporate probability or variation events for every individual sensor. These events activate the input resources and allow us to observe the network's response. The final phase involves providing feedback by transmitting the signal from the actuators back to the sensors. This is done by combining the outputs of the actuators with the original event, resulting in the acquisition of the actual value.

An illustration of this Simulink network is presented in Figure 11. Section I displays events generators and an adding block that represents feedback. Every input resource is associated with an event generator, which can be either discrete or continuous. Section II contains input resources. Section III illustrates the services that combine input and output resources. Service definitions are recursively created by other MDL files. The mdl files correspond to the services specified in the environment. In section IV, we possess output resources. Section V presents feedback created by actuators, which alter observed magnitudes. The feedback is calibrated based on the configuration of the actuator in the environment. Every part displays a white box linked to each individual resource. These boxes include a list of values received from a simulation. This list consists of distinct values that are recorded at specific intervals.

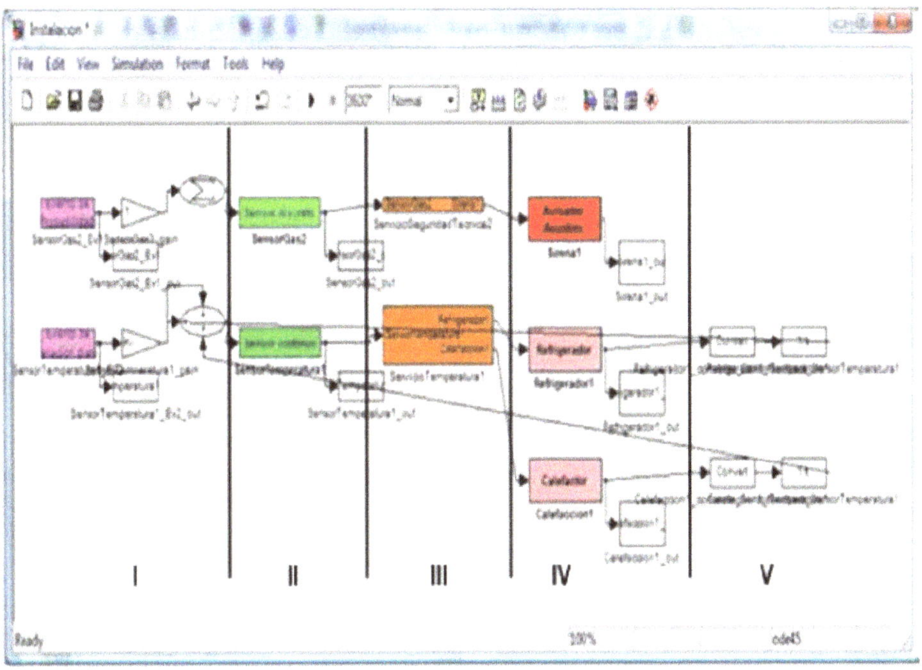


**Fig. 11. Simulink control network generated dynamically from a previous design made at the environment**

Once the mdl file is generated, Simulink initiates the simulation process. The environment then retrieves all variable lists from Simulink and displays them in the previously configured view at the specified frequency. In this manner, the simulation lacks interactivity, as the user must patiently await the simulation results from Simulink after the simulation process is initiated.

Figure 12 shows the mdl file that defines a HVAC service designed previously at environment on figure 6. At left side, input resources connection can be viewed as white circles, in this case temperature sensor. In the middle, we there are conversion data type blocks for avoiding Simulink data type errors and comparison operators that defines when the actuators will start to work. Actuators are drawn as white circle on the right.

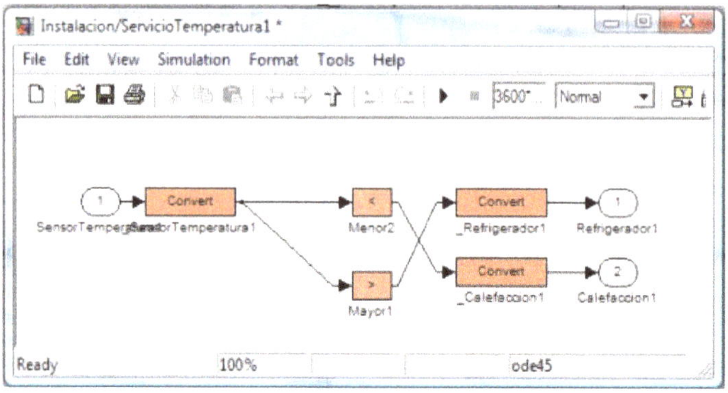


**Fig. 12. HVAC service created in Simulink from the design made at environment**

## 5. CONCISE SUMMARY

This chapter introduces a prototyping environment that allows for the creation and modeling of control networks in specific domains, such as digital house and intelligent building. The goal is to streamline the process of creating and validating control networks. We advocate for a top-down approach, in which the selection of technology implementation is deferred until the last stage of the design process. The environment kernel is a control network paradigm that encompasses several levels of abstraction, including functional, structural, and technological layers.

The model serves as the foundation for various applications in the field, such as network design, simulation, architecture specification, report development, budgeting, and more. Simulation is a crucial step in the design control network process. It lowers expenses and enables us to create well-functioning and streamlined control networks.

The future objective is to offer competitive solutions in the design of control networks. The challenge necessitates addressing the progressively scientific and technological parts of the issue. Thus, in the immediate term, we will focus on further exploring the generalization of the results. This includes developing new models for the structural and technological layers, implementing an interactive simulation module, and simulating the technological layer. In the long term, we plan to extend our approach to other contexts such as industrial buildings and office buildings.

## REFERENCES

1. Balci, O., 1998. Verification, validation, and testing. In The Handbook of Simulation. John Wiley & Sons.
2. Bravo, C., Redondo, M.A., Bravo, J., and Ortega, M., 2000. DOMOSIMCOL: A Simulation Collaborative Environment for the Learning of Domotic Design. ACM SIGCSE Bulletin, Vol. 32, No. 2, 65-67.

3. Bravo, C., Redondo, M., Ortega, M.., and Verdejo, M.F., 2006. Collaborative environments for the learning of design: a model and a case study in Domotics. Computers & Education, Vol. 46, No. 2, 152-173.
4. Breslau, L. Estrin, D. Fall, K. Floyd, S. Heidemann, J. Helmy, A. Huang, P. McCanne, S. Varadhan, K. Ya Xu Haobo Yu. 2000. Advances in network simulation. Computer, Vol. 33, Issue: 5, pp: 59-67. ISSN: 0018-9162.
5. Conte, G. Scaradozzi, D. Perdon, A. Cesaretti, M. Morganti, G. 2007. A simulation environment for the analysis of home automation systems. Control & Automation, 2007. MED '07. Mediterranean Conference on, pp: 1-8, ISBN: 978-1-4244-1282-2
6. Denning, P. J., Comer, D. E., Gries, D., Michael Mulder, C., Tucker, A., Turner, A. J., Young, P. R., 1989. Computing as a discipline. Communications of the ACM, Vol.32 No.1, 9-23
7. Fuster, A. and Azorín, J., 2005. Arquitectura de Sistemas para el hogar digital. Hogar digital. El camino de la domótica a los ambientes inteligentesI Encuentro Interdisciplinar de Domótica 2005, pp 35-44
8. Fuster, A., de Miguel, G. and Azorín, J., 2005. Tecnología Middleware para pasarelas residenciales. Hogar digital. El camino de la domótica a los ambientes inteligentes.I Encuentro Interdisciplinar de Domótica 2005, 87-102.
9. González, V. M., Mateos, F., López, A.M., Enguita, J.M., García M.and Olaiz, R., 2001. Visir, a simulation software for domotics installations to improve laboratory training, Frontiers in Education Conference, 31st Annual Vol. 3, F4C-6-11.
10. Haenselmann, T., King, T., Busse, B., Effelsberg, W. and Markus Fuchs, 2007. Scriptable Sensor Network Based Home-Automation. Emerging Directions in Embedded and Ubiquitous Computing, 579-591.
11. Jung, J., Park, K. and Cha J., 2005. Implementation of a Network-Based Distributed System Using the CAN Protocol. Knowledge-Based Intelligent Information and Engineering Systems. Vol. 3681.
12. Kawamura, R. and Maeomichi, 2004. Standardization Activity of OSGi (Open Services Gateway Initiative), NTT Technical Review, Vol. 2, No. 1, 94-97.
13. Mellor, S., Scott K., Uhl, A. and Weise, D., 2004. MDA Distilled, Principles of Model Driven Architecture, Addison-Wesley Professional.
14. Min, W., Hong, Z., Guo-ping, L. and Jin-hua, S., 2007. Networked control and supervision system based on LonWorks fieldbus and Intranet/Internet. Journal of Central South University of Technology, Vol. 14, No.2, 260-265.
15. Moya, F. and López, J.C., 2002 .SENDA: An Alternative to OSGi for Large Scale Domotics. Networks, pp. 165-176.

16. Müller, S. and Waller, H., 1999. Efficient Integration of Real-Time Hardware and Web Based Services Into MATLAB. 11th European Simulation Symposium and Exhibition, 26- 28.

17. Muñoz, J., Fons, J., Pelechano, V. and Pastor, O., 2003. Hacia el Modelado Conceptual de Sistemas Domóticos, Actas de las VIII Jornadas de Ingeniería del Software y Bases de Datos. Universidad de Alicante, 369-378

18. Newcomer, E. and Lomow, G., 2005. Understanding SOA with Web Services. Addison Wesley.

19. Norton S. and Suppe F., 2001. Why atmospheric modeling is good science. MIT Press. p. 88- 133.

20. Pan, M. and Tseng, Y. 2007. ZigBee and Their Applications. Sensor Networks and Configuration. Ed. Springer Berlin Heidelberg, pp:349-368 ISBN:978-3-540-37364-3

21. Rhee, S., Yang, S., Park, S., Chun, J.and Park, J., 2004. UPnP Home Networking-Based IEEE1394 Digital Home Appliances Control. Advanced Web Technologies and Applications, Vol. 3007, 457-466.

22. Sommerville, I. 2004. Software Engineering, 7th ed. Boston: Addison-Wesley.

23. Sun Microsystems Inc, 1999. JINI Architectural Overview, Technical White Paper, 1999

24. Valdivieso, R.J., Sánchez, J.G., Azorín, J. and Fuster, A., 2007. Entorno para el desarrollo y

25. simulación de arquitecturas de redes de control en el hogar. II Internacional Simposium Ubiquitous Computing and Ambient Intelligence.

# 06 COMPARATIVE ANALYSIS OF DEFUZZIFICATION TECHNIQUES: AUTOMATING TEMPERATURE AND FLOW CONTROL IN A HEAT EXCHANGER

## 1. INTRODUCTION: AN OVERVIEW

The objective of this study is to examine the performance of traditional and fuzzy control techniques in regulating the flow and temperature of a heat exchanger, as well as to analyze various defuzzification methods used in fuzzy control. Modifications are made to the fuzzy controller structure in order to achieve an optimal response. The study focuses on the conventional controller and compares the approaches used in these two types of controllers. Within the modifications made to the fuzzy controller, there is a change in the way of defuzzification, specifically the methods used for defuzzification are exchanged. This allows for comparisons to be made on the performance of each approach.

The existence of heat exchangers is crucial in various sectors of the business, particularly those involving thermal processes (Shah & Sekulic, 2003; Kuppan, 2000). The aforementioned activities are integral parts of an engineer's daily life, particularly those working in the field of control systems. Consequently, it is intriguing to explore the implementation of control mechanisms for such systems. This study examines two prominent facets: This text discusses the comparison between traditional control and fuzzy control, as well as the analysis of various defuzzification methods used in fuzzy logic. The analysis considers the contributions made by other authors and emphasizes that the results obtained will only be applicable when controlling a heat exchanger. The sources cited for further reading are Kovacic & Bogdan (2006), Harris (2006), Xia et al. (1991), Fischer et al. (1998), and Alotaibi et al. (2004). The test system consists of two heat exchangers: one with concentric pipes and the other with a hull and pipes. Both heat exchangers are equipped with a temperature and flow automatic control system that adjusts the heating load current (see Fig. 1).

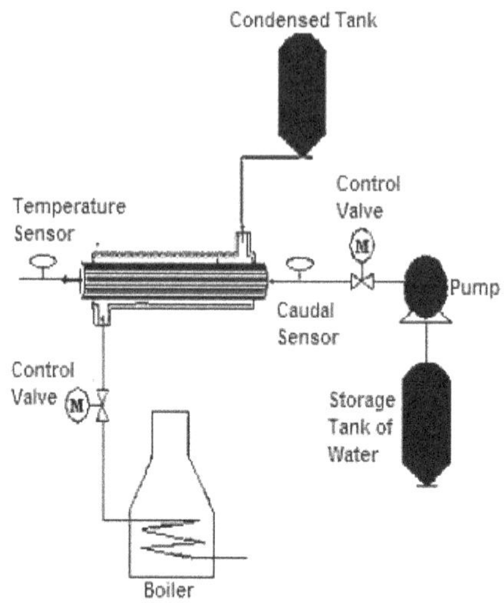


**Fig. 1. Thermal pilot plant scheme**

The control is achieved using two proportional valves. One valve regulates the input of water to maintain the desired water level, while the second valve controls the input of vapor (heat source) to maintain the required amount of vapor for achieving the goal temperature. Measurements of any physical property condition in physical experimentation are often associated with the concept of uncertainty (Youden, 1998). The flow is measured using a rotating palette sensor, while the temperature is measured using thermocouples (Eckert & Goldstein, 1970, McMillan & Considine, 1999, Morris, 2001). The National Instruments® FieldPoint module acquires the signals provided by sensors (Kehtarnavaz & Kim, 2005, Archila et al., 2006, Blume, 2007). After the data is analyzed by the controller, the module takes charge and sends the signal to the control valves.

## 2. FUZZY AND CLASSIC CONTROL SOFTWARE

The control software utilizes two distinct sections: the fuzzy control program and the PID (Proportional/Integral/Derivative) control program. The controllers are developed using the LabWindows/CVI environment by National Instruments Company®, enabling the execution of relevant operations on data captured using FieldPoint modules, which are used in control systems. The fuzzy control interface is responsible for receiving sensor data, such as temperature for temperature control or flow sensor data for flow control. It processes this data and, based on a predefined set of rules, determines a response to be transmitted to the actuators. This program is designed to represent fuzzy sets based on user-defined criteria, perform defuzzification of input data, apply inference rules to these inputs, aggregate the output sets, and execute the defuzzification process to determine the system's desired response. The PID control classic interface is similar to the fuzzy control interface, but the distinction is in its responsibility for executing the three control actions: proportional, derivative, and integral. These actions are used to generate the system's reactions, which help it achieve its target state.

The general representation of the PID control system can be seen in figure 2. In this system, R(s) represents the signal target or set point, U(s) represents the output of the PID controller that is directed towards the plant G(s), and Y(s) represents the value of the variable being controlled. Y(s) is reduced to the reference value and the error is determined, which serves as the input for the controller.

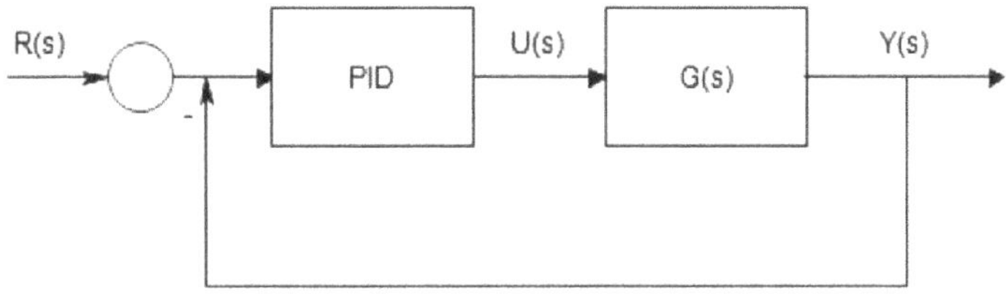


**Fig. 2. General PID Control System**

The goal of temperature control is to ensure that the heat exchanger surpasses the desired temperature value for water and maintains it consistently, even in the presence of external disturbances. The control valve operates to supply a specific amount of vapor that is used to heat the water. The system control receives the temperature error, which is measured by a thermocouple placed at the exit of the exchanger. The control also determines the amount of current required to open or close the proportional valve in order to regulate the system. This control is achieved by utilizing a PID controller.

The goal of flow control is to attain and maintain the desired water mass flow entering the heat exchanger, even in the presence of disturbances. This implies that one should manipulate the control valve, which is responsible for regulating the amount of water that enters the system. The system will receive the error signal from the flow sensor located at the input of the system. The PID controller will then regulate the amount of current required to adjust the proportional valve. Both operations commence by computing the disparity between the actual temperature and the intended temperature, or the measured flow and the desired flow. Identify the error in this form. The control parameters are extracted and used to determine the output, which is then applied to the plant. This output measures values ranging from 0 to 20 mA, which correspond to the aperture angles of a proportional valve.

## 3. FUZZY CONTROL SYSTEM

The control system receives inputs of temperature error and gradient, which are collected from a sensor located at the outlet of the exchanger. The control system then adjusts the current quantity required to open the proportional valve at the outlet. The rules and membership function system can be found in table 1 and figure 3, respectively.

**Table 1. Temperature control rules assembly**

| Error Δ Error | Negative | Zero | Positive |
|---|---|---|---|
| Negative | Open | Open | Not operation |
| Zero | Open | Not operation | Close |
| Positive | Not operation | Close | Close |

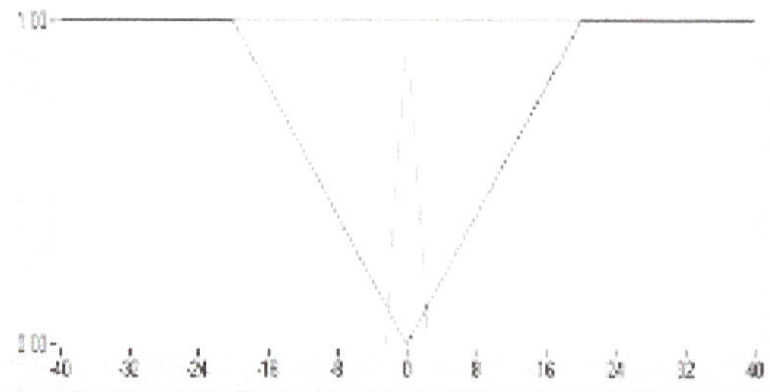

**Membership functions - current at theoutset**

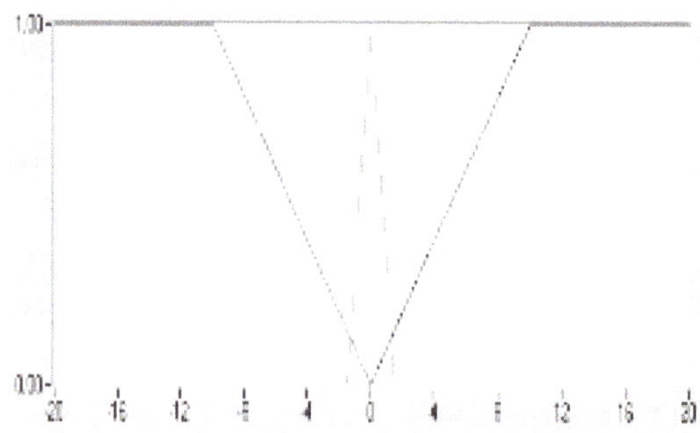

**Membership functions – derived variable from Error**

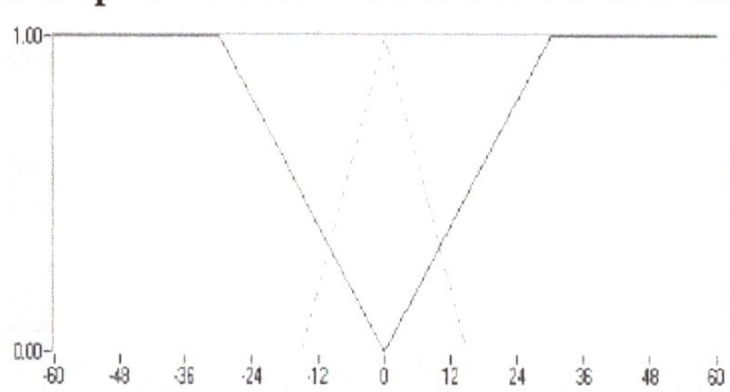


**Membership functions - variable Error**

**Fig. 3. Membership Functions - temperature control**

The goal of flow control is to regulate the mass flow of water entering the heat exchanger, ensuring that it reaches and maintains the desired value throughout its operation, even in the event of disruptions. This means that the control valve is responsible for regulating the amount of water that enters the system. The system input refers to the errors detected by the flow sensor located at the system entry, as well as the rate at which these errors change over time. The output of the system is the amount of current required to regulate the proportional valve. Both operations commence by computing the disparity between the actual temperature and the intended temperature, or the measured flow and the desired flow. Identify the error in this form. Compute the gradient while minimizing the error compared to the previous iteration. After identifying these variables, which serve as the inputs for the fuzzy logic controller, the process of fuzzification, inference, and defuzzification is carried out to determine the controller's output. This output measures values ranging from 0 to 20 mA, which correspond to the aperture angles of a proportional valve. The system rules and membership function are derived from table 2 and depicted in figure 4, respectively.

**Table 2. Flow control rules assembly**

| Error Δ Error | Negative | Zero | Positive |
|---|---|---|---|
| Negative | Close | Close | Not operation |
| Zero | Close | Not operation | Open |
| Positive | Not operation | Open | Open |

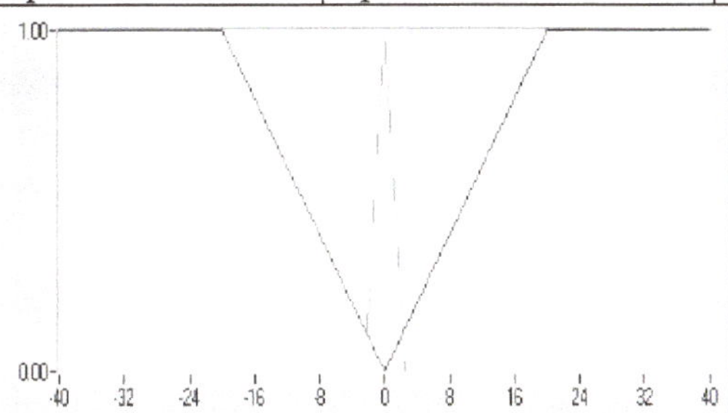


**Functions of membership - variable error**

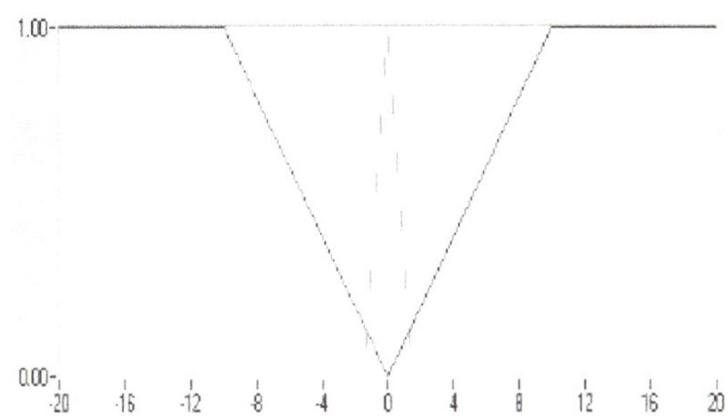


**Functions of membership – derived variable from error**

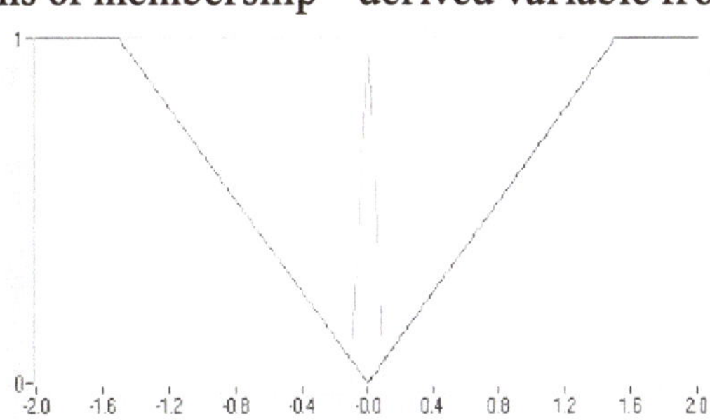


**Functions of membership - current at the outset**

**Fig. 4. Membership Functions - flow control**

## 3.1 Comparison of implemented defuzzification methods and their respective results

The defuzzification methods selects were five; identify in the control area by center of gravity weighted by height, center gravity weighted by area, average of centers, points of maximum criterion weighted by height and points of maximum criterion weighted by area (Zhang & Edmunds, 1991, Guo et al., 1996, Saade & Diab, 2000). In systems control, the main term is the stability that can offer the system, for this is necessary the delayed time that the system in being stabilized, error margin between value desired (Vc), and system stabilization values (Ve) and inertial influence of system. For temperature and flow control tests, is defined a set point 25 [Lts/min] and 40 [ºC]. The parameters and equations used for different responses in each one of the methods are shows in table 3, table 4 and table 5, according the parameters established in table 3.

**Table 3. Methods and models defuzzification**

| METHODS | EQUATIONS | |
|---|---|---|
| 1. Center of gravity weighted by height | $\bar{x} = \frac{\sum_{i=1}^{n} h_i * w_i}{\sum_{i=1}^{n} h_i}$ | Where, w is gravity center of resultant assembly after fuzzy operation select, and h is the height of the same assembly. |
| 2. Center of gravity weighted by area. | $\bar{x} = \frac{\sum_{i=1}^{n} s_i * w_i}{\sum_{i=1}^{n} s_i}$ | Where, w is gravity center ofresultant assembly after fuzzy operation select, and *s* is the area of the same assembly. |
| 3. Points of maximum criterion weighted by area. | $\bar{x} = \frac{\sum_{i=1}^{n} s_i * G_i}{\sum_{i=1}^{n} s_i}$ | Where, *G* is the point of maximum criterion of resultant set after to realize fuzzy operation select and *s* is the area of the same set. |
| 4. Points of maximum criterion weighted by height. | $\bar{x} = \frac{\sum_{i=1}^{n} h_i * G_i}{\sum_{i=1}^{n} h_i}$ | Where, *G* is the point of maximum criterion of resultant set after to realize fuzzy operation select and *h* is height of the same set. |
| 5. Average of centers | $y = \frac{\sum_{l=1}^{M} y^{-l}(\mu_{B'}(y^{-l}))}{\sum_{l=1}^{M}(\mu_{B'}(y^{-l}))}$ | Where *y-l* represents the center of fuzzy set Gl (defined as the point *V* in which *μGl (y)* reaches its value maximum), and *μB (y)* defined for the degrees of membership resultant by fuzzy inference. |

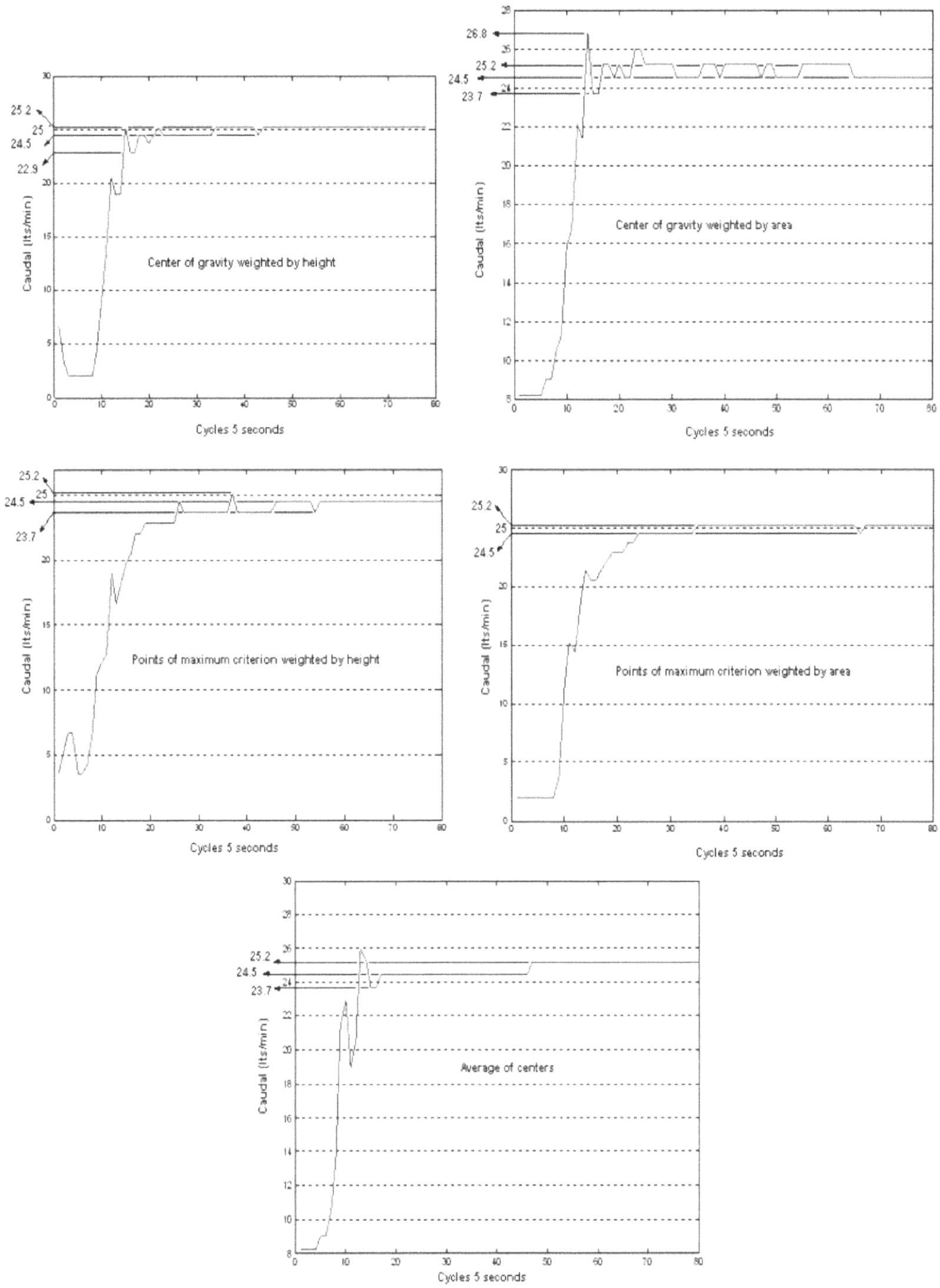


**Table 5. Flow control response**

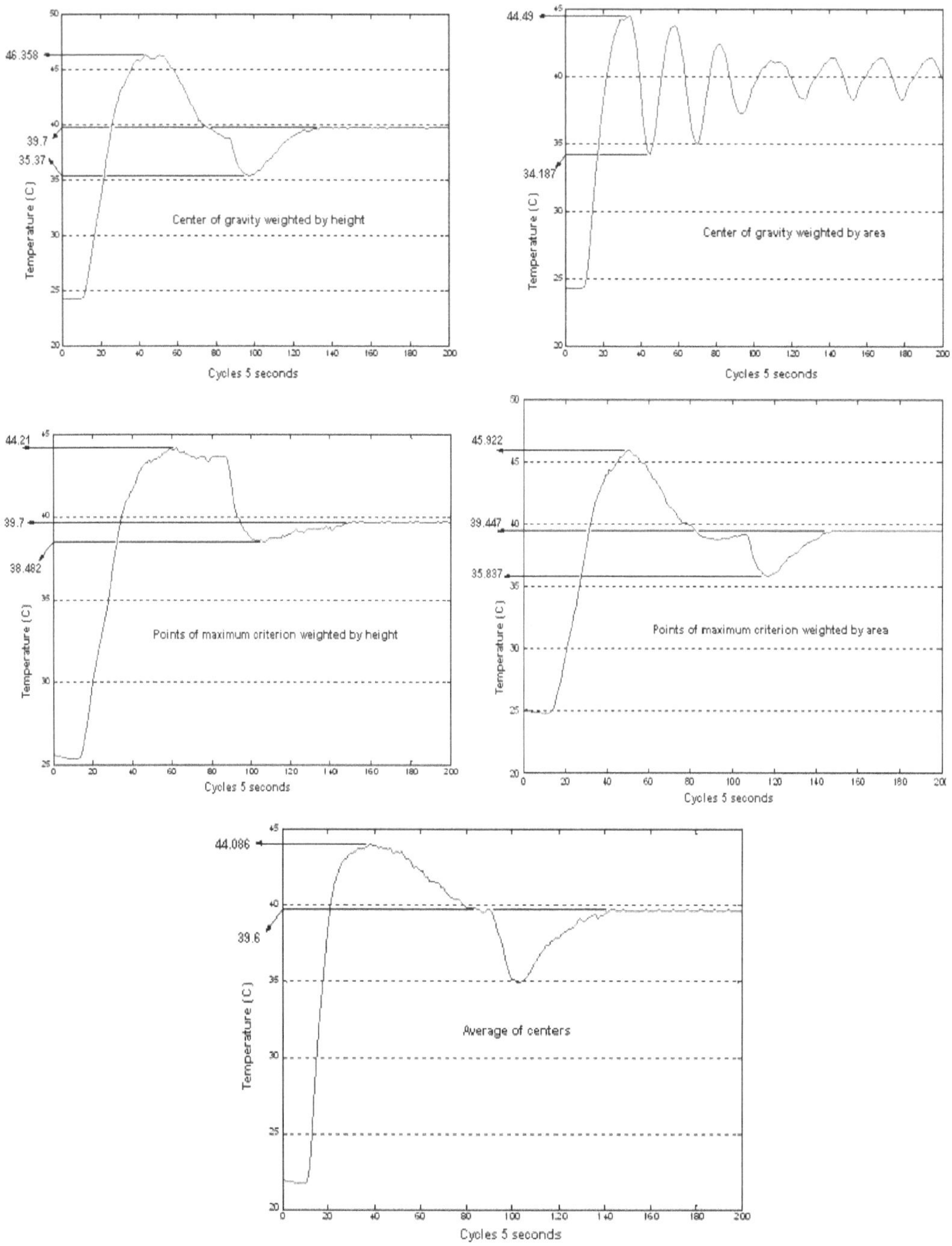


**Table 6. Temperature control response**

A summary of results obtained on different methods is shown in table 6 for flow control and table 7 for temperature control.

**Table 6. Response of defuzzification methods - flow control**

| Defuzzification method | Stability time [sec] | Error margin (Vc - Ve) | Inertial influence of system |
|---|---|---|---|
| Center gravity weighted by height | 105 | 0.8% above of the set point 2% below of the set point | 0.8% above of the set point 8.4% below of the set point |
| Center gravity weighted by area | 125 | 0.8% above of the set point 2% below of the set point | 7.2% above of the set point 5.2% below of the set point |
| Average of centers | 85 | 0.8% above of the set point 2% below of the set point | 4% above of the set point 5.2% below of the set point |
| Points of maximum criterion weighted by height | 230 | 2% below of the set point | 0.8% above of the set point 5.2% below of the set point |
| Points of maximum criterion weighted by area | 120 | 0.8% above of the set point 2% below of the set point | 0.8% above of the set point 2% below of the set point |

**Table 7. Response of defuzzification methods - temperature control**

| Defuzzification method | Stability time [sec] | Error margin (Vc - Ve) | Inertial influence of system |
|---|---|---|---|
| Center gravity weighted by height | 670 | 0.75% below of the set point | 40.89% above of the set point 11.57% below of the set point |
| Center gravity weighted by area | Not stabilized | Not stabilized | 11.25% above of the set point 14.53% below of the set point |
| Average of centers | 710 | 1% below of the set point | 10.21% above of the set point 12.5% below of the set point |
| Points of maximum criterion weighted by height | 745 | 0.75% below of the set point | 10.52% above of the set point 3.79% below of the set point |
| Points of maximum criterion weighted by area | 735 | 1.38% below of the set point | 14.80% above of the set point 10.40% below of the set point |

## 3.2 Comparative analysis of the classic and fuzzy controller

To be able to realize this analysis should make use of fundamentals concepts at the moment of to evaluate the controller efficiency. The concepts in this case are: systems delayed time in being stabilized, error margin between order value (Vc) and stabilization values (Ve) and inertial Influence of system. For the comparative analysis between fuzzy controller and PID controller, in the flow control use of tests realized to each one of these controllers with set point 25 [Lts/min] and 40 [ºC]. The results obtained are shown in the table 8.

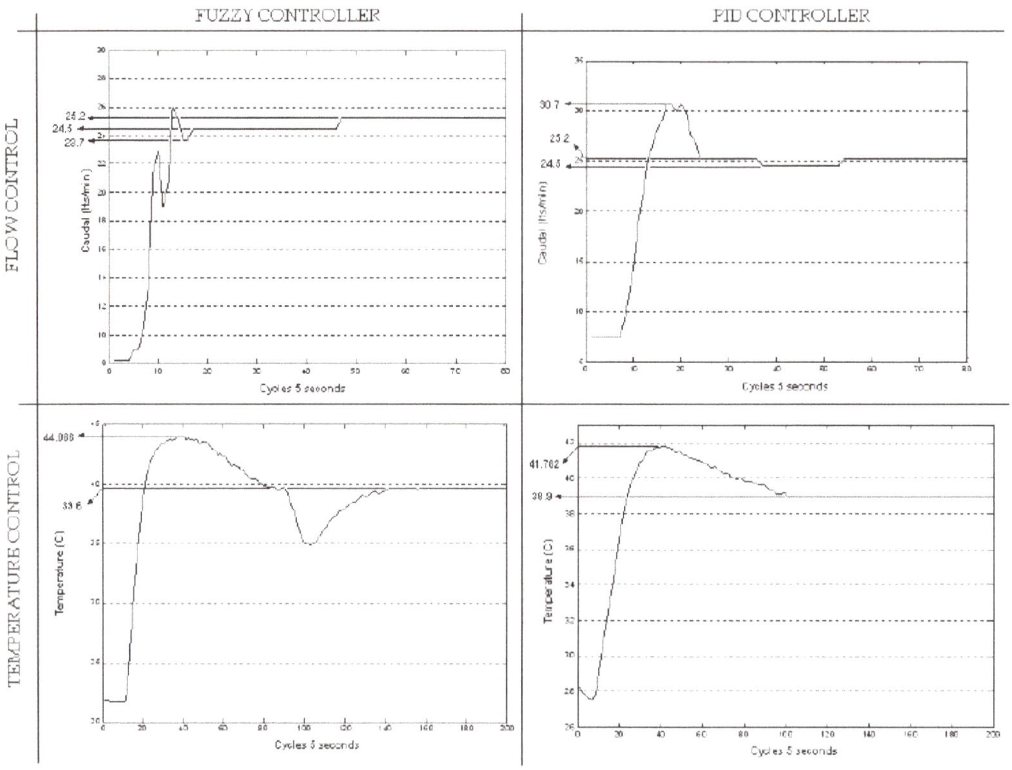


**Table 7. Controllers response**

A summary of results obtained on different methods is shown for flow control (table 9) and temperature control (table 10).

**Table 9. Response of controllers - flow control**

| Controller | Stability time [sec] | Error margin (Vc - Ve) | Inertial influence of system |
|---|---|---|---|
| FUZZY CONTROL | 85 | 0.8% below of the set point 2% above of the set point | 4% below of the set point 5.2% above of the set point |
| PID CONTROL | 115 | 0.8% below of the set point 2% above of the set point | 22.8% below of the set point 2% above of the set point |

**Table 10. Response of controllers - temperature control**

| Controller | Stability time [sec] | Error margin (Vc - Ve) | Inertial influence of system |
|---|---|---|---|
| FUZZY CONTROL | 710 | 1% below of the set point | 10.21% above of the set point 12.5% below of the set point |
| PID CONTROL | 505 | 2.75% below of set point | 4.45% above of set point 2.75% below of set point |

## 4. CONCISE SUMMARY

The findings of this study demonstrate the practical feasibility of employing fuzzy logic for regulating the flow and temperature of the heat exchanger's warming-up current input. The implementation of flow and temperature control using fuzzy logic offers the advantage of not requiring a precise mathematical model for the control system. One downside is that the design must typically be achieved through trial and error. It is feasible to utilize fuzzy techniques to effectively control industrial processes, minimizing errors and accurately identifying their overall behavior. This involves constructing a set of fuzzy sets and their corresponding rules. The tuning of the fuzzy controller is determined by the rules matrix as well as the size of the variable sets, whether they are input or output. This is contingent upon the identical behavioral system. In order to create fuzzy control, it is essential to establish the methods and alternatives used in each of its constituent blocks. Optimal results can be achieved by using this form while tuning the system. The response of the fuzzy controller is not influenced by the defuzzification method used. Instead, it depends on the proper use of the membership functions and the amount of linguistic variables used for each input and output variable of the system. Additionally, the outcome is contingent upon the specific type and dimensions of the sets that are employed.

## REFERENCES

1. Alotaibi, S.; Sen, M.; Goodwine, B. & Yang, K.T. (2004). Controllability of Cross-flow Heat Exchanger. *International Journal of Heat and Mass Transfer*, Vol. 4, No. 5, February 2004, pp. 913–924.
2. Archila, J.F.; Dutra, M.S.; Lengerke, O. & Vega, T.J. (2006). Design and implementation of a control system in closed-loop using the communication system FIELD POINT for a hydrostatic transmission. *RESET: Journal Specialized in Informatics and Electronics Systems of Telecommunications Systems.* Vol. 1, pp. 48–53.
3. Blume, P.A. (2007). *The LabVIEW Style Book (National Instruments Virtual Instrumentation Series).* Prentice Hall PTR, ISBN-10/ASIN: 0-1314-5835-3, Upper Saddle River, New Jersey, USA.

4. Eckert, E.R.G. & Goldstein, R.J. (1970). *Measurements in Heat Transfer*. Technivision Services, ISBN-10/ASIN: 0-8510-2026-7, Slough, England.
5. Fischer, M.; Nelles, O. & Isermann, R. (1998). Adaptive Predictive Control of a Heat Exchanger Based on a Fuzzy Model. *Control Engineering Practice*, Vol. 6, No. 2, pp. 259-269.
6. Guo, S.; Peters, L. & Surmann, H. (1996). Design and application of an analog fuzzy logic controller. *IEEE Transactions on Fuzzy Systems*, Vol. 4, No. 4, November 1996, pp. 429-438 ISSN: 1063-6706.
7. Harris, J. (2006). *Fuzzy Logic Applications in Engineering Science*, Springer, ISBN-10 1-4020- 4078-4, Netherland.
8. Kehtarnavaz, N. & Kim, N. (2005). *Digital Signal Processing System-Level Design Using LabVIEW*. Newnes Elsevier, ISBN-10/ASIN: 0-7506-7914-X, Burlington, USA.
9. Kovacic, Z. & Bogdan, S. (2006). *Fuzzy Controller Design: Theory and Applications*, Control Engineering Series, CRC Press, Taylor & Francis Group, ISBN 0-8493-3747-X, Boca Raton.
10. Kuppan, T., (2000). *Heat Exchanger Design Handbook*, Marcel Dekker, Inc., ISBN-10: 0-8247- 9787-6, New York.
11. McMillan, G.K. & Considine, D.M. (1999). *Process/Industrial Instruments and Controls Handbook*. 5th Edition, McGraw-Hill, ISBN-10/ASIN: 0-0701-2582-1, New York, USA.
12. Morris, A.S. (2001). *Measurement and Instrumentation Principles*. Third Edition, Butterworth- Heinemann, ISBN-10: 0-7506-5081-8, Great Britain.
13. Saade, J.J. & Diab, H.B. (2000). Defuzzification Techniques for Fuzzy Controllers. *IEEE Transactions on Systems, Man, and Cybernetics*, Part B, Vol. 30, No. 1, February 2000, pp. 223-229,
14. Shah, R.K. & Sekulic, D.P. (2003). *Fundamentals of Heat Exchanger Design*, John Wiley & Sons, ISBN-10/ASIN: 0-4713-2171-0, New Jersey.
15. Xia, L.; De Abreu-Garcia, J.A. & Hartley, T.T. (1991). Modeling and Simulation of a Heat Exchanger. *IEEE International Conference on Systems Engineering*, August 1991, pp. 453–456, ISBN: 0-7803-0173-0.
16. Youden, W. J. (1998). *Experimentation and Measurement*. Dover Publications, Mineola, New York.
17. Zhang, B.S. & Edmunds, J.M. (1991). On Fuzzy Logic Controllers. *Proceedings of International Conference Control'91*, Vol. 2, pp. 961-965, ISBN: 0-85296-509-5.

# 07 ANALYSIS AND DESIGN OF PHASE-LOCKED LOOPS USING NONLINEAR METHODS

## 1. INTRODUCTION: AN OVERVIEW

Phase-locked loops (PLLs) have extensive applications in telecommunication and computer architectures. PLLs were first invented in the 1930s-1940s by De Bellescize (1932) and Wendt & Fredentall (1943). Subsequently, there were extensive studies conducted on the theory and practice of PLLs by Viterbi (1966), Lindsey (1972), Gardner (1979), Lindsey and Chie (1981), Leonov et al. (1992), Encinas (1993), Nash (1994), Brennan (1996), and Stensby (1997).

One of the initial uses of phase-locked loop (PLL) is associated with the challenges of transmitting data via radio signals. In the field of radio engineering, PLL (Phase-Locked Loop) is utilized for many purposes such as synchronizing the carrier signal, recovering the carrier signal, demodulating the signal, and synthesizing frequencies. This application is discussed in works by Stephens (2002) and Ugrumov (2000).

Phase-locked loops are utilized to generate internal frequencies of chips and synchronize the operation of various devices and data buses in an architecture that incorporates chips operating on different frequencies (Young et al., 1992; Egan, 2000; Kroupa, 2003; Razavi, 2003; Shu & Sanchez-Sinencio, 2005; Manassewitsch, 2005). Modern computer motherboards consist of several devices and data busses that operate at different frequencies. These components often require synchronization (Wainner & Richmond, 2003; Buchanan & Wilson, 2001).

Another practical application of PLL is in the field of energy conservation. An effective way for processors to address this problem is to reduce the kernel frequency in proportion to the processor load. The use of independent phase-locked loops allows for a more even distribution of the kernel load, resulting in energy savings and reduced heat generation. This is achieved by allowing each kernel to operate at its own frequency. Phase-locked loops are now extensively utilized to address clock skew and synchronization issues in computer architectures and chip microarchitecture. One essential aspect of CPUs is clock skew, as demonstrated by Xanthopoulos (2001) and Bindal (2003).

Engineers have developed various methods for analyzing phase-locked loops, which are discussed in many publications (Banerjee, 2006; Best, 2003; Kroupa, 2003; Bianchi, 2005; Egan, 2007). However, the challenges of constructing appropriate nonlinear models and conducting nonlinear analysis on these models are still unresolved. To address these challenges, special methods from the qualitative theory of differential, difference, integral, and integro-differential equations are required (Gelig et al., 1978; Leonov et al., 1996a; Leonov et al., 1996b; Leonov & Smirnova, 2000; Abramovitch, 2002; Suarez & Quere, 2003; Margaris, 2004; Kudrewicz & Wasowicz, 2007; Kuznetsov, 2008; Leonov, 2006). Due to space limitations, we are unable to provide an exhaustive list of references in the field of design and the study of the PLL requires readers to consult to the mentioned papers, books, and the references cited within them.

## 2. MATHEMATICAL MODEL OF PLL

In this work three levels of PLL description are suggested:

- the level of electronic realizations,
- the level of phase and frequency relations between inputs and outputs in block diagrams,
- the level of difference, differential and integro-differential equations.

The second level, involving the asymptotical analysis of high-frequency oscillations, is nec- essary for the well-formed derivation of equations and for the passage to the third level of description.

Consider a PLL on the first level (Fig. 1)

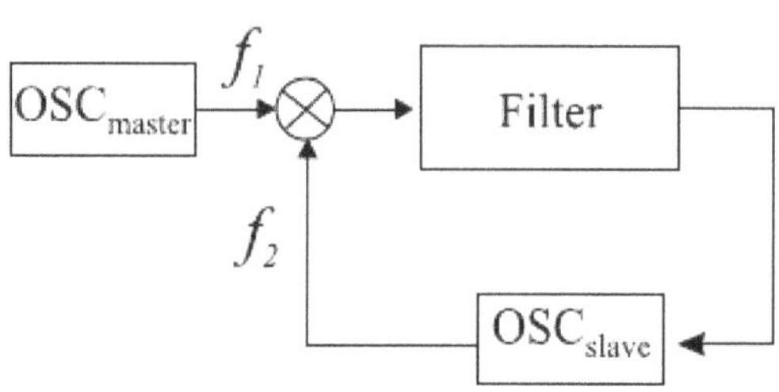


**Fig. 1. Block diagram of PLL on the level of electronic realizations.**

Here OSCmaster is a master oscillator, OSCslave is a slave (tunable) oscillator, which generate high-frequency "almost harmonic oscillations"

$$f_j(t) = A_j \sin(\omega_j(t)t + \psi_j) \quad j = 1, 2, \tag{1}$$

where Aj and ψj are some numbers, ωj (t) are differentiable functions. Block is ⊗ a multiplier of oscillations of $f_1$(t) and f2(t) and the signal $f_1$(t) $f_2$(t) is its output. The relations between the input ξ(t) and the output σ(t) of linear filter have the form

$$\sigma(t) = \alpha_0(t) + \int_0^t \gamma(t-\tau)\xi(\tau)\, d\tau. \tag{2}$$

Here γ(t) is an impulse transient function of filter, α0(t) is an exponentially damped function, depending on the initial data of filter at the moment t = 0. The electronic realizations of generators, multipliers, and filters can be found in (Wolaver, 1991; Best, 2003; Chen, 2003; Giannini & Leuzzi, 2004; Goldman, 2007; Razavi, 2001; Aleksenko, 2004). In the simplest case it is assumed that the filter removes from the input the upper sideband with frequency

ω1(t)+ ω2(t) but leaves the lower sideband ω1(t) - ω2(t) without change.

Now we reformulate the high-frequency property of oscillations *fj (t)* and essential assump- tion that γ*(t)* and ωj *(t)* are functions of «finite growth». For this purpose we consider the great fixed time interval *[0, T]*, which can be partitioned into small intervals of the form *[τ, τ + δ]*, *(τ ∈ [0, T])* such that the following relations

$$\begin{aligned} &|\gamma(t) - \gamma(\tau)| \le C\delta, \quad |\omega_j(t) - \omega_j(\tau)| \le C\delta, \\ &\forall t \in [\tau, \tau+\delta], \quad \forall \tau \in [0, T], \end{aligned} \tag{3}$$

$$|\omega_1(\tau) - \omega_2(\tau)| \le C_1, \quad \forall \tau \in [0, T], \tag{4}$$

$$\omega_j(t) \ge R, \quad \forall t \in [0, T] \tag{5}$$

are satisfied. Here we assume that the quantity δ is sufficiently small with respect to the fixed

numbers *T, C, C1,* the number R is sufficiently great with respect to the number δ. The latter means that on the small intervals [τ, τ + δ] the functions γ(t) and ωj *(t)* are "almost constants" and the functions *fj* (t) rapidly oscillate as harmonic functions.

Consider two block diagrams shown in Fig. 2 and Fig. 3.

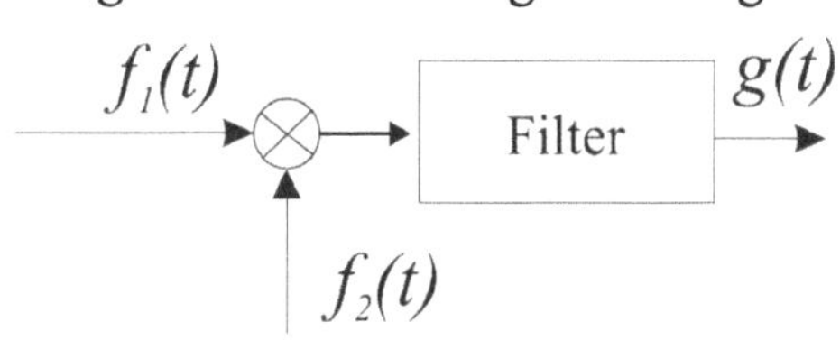


**Fig. 2. Multiplier and filter.**

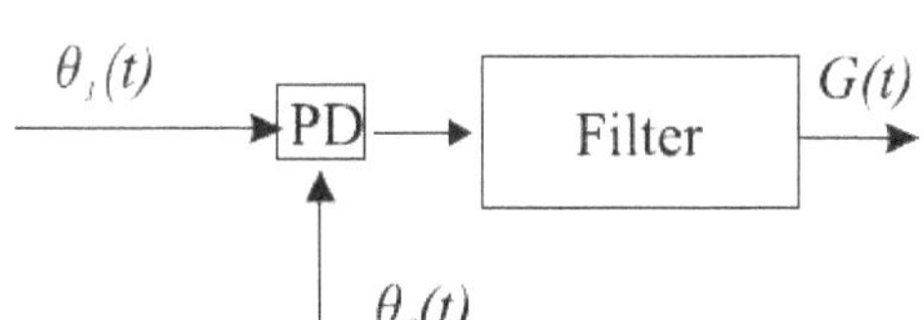


**Fig. 3. Phase detector and filter.**

Here θj (t) = ωj (t)t + ψj are phases of the oscillations fj (t), PD is a nonlinear block with the characteristic ϕ(θ) (being called a phase detector or discriminator). The phases θj (t) are the inputs of PD block and the output is the function ϕ(θ1(t) - θ2(t)). The shape of the phase detector characteristic is based on the shape of input signals.

The signals f1(t) f2(t) and ϕ(θ1(t) - θ2(t)) are inputs of the same filters with the same impulse transient function γ(t). The filter outputs are the functions g(t) and G(t), respectively. A classical PLL synthesis is based on the following result:

**Theorem 1. (Viterbi, 1966) If conditions (3)–(5) are satisfied and we have**

$$\varphi(\theta) = \frac{1}{2} A_1 A_2 \cos\theta,$$

then for the same initial data of filter, the following relation

$$|G(t) - g(t)| \leq C_2\delta, \ \forall t \in [0, T]$$

is satisfied. Here C2 is a certain number being independent of δ.

**Proof of Theorem 1** (Leonov, 2006)

For t ∈ [0, T] we obviously have

$$g(t) - G(t) =$$

$$= \int_0^t \gamma(t-s)\Big[A_1 A_2\Big(\sin(\omega_1(s)s+\psi_1)\sin(\omega_2(s)s+\psi_2)\Big) -$$

$$- \varphi\Big(\omega_1(s)s - \omega_2(s)s + \psi_1 - \psi_2\Big)\Big] ds =$$

$$= -\frac{A_1 A_2}{2}\int_0^t \gamma(t-s)\Big[\cos\Big((\omega_1(s)+\omega_2(s))s + \psi_1 + \psi_2\Big)\Big] ds.$$

Consider the intervals [kδ, (k + 1)δ], where k = 0, . . . , m and the number m is such that t ∈

[mδ, (m + 1)δ]. From conditions (3)–(5) it follows that for any s ∈ [kδ, (k + 1)δ] the relations

$$\gamma(t-s) = \gamma(t-k\delta) + O(\delta) \tag{6}$$

$$\omega_1(s) + \omega_2(s) = \omega_1(k\delta) + \omega_2(k\delta) + O(\delta) \tag{7}$$

are valid on each interval [kδ, (k + 1)δ]. Then by (7) for any s ∈ [kδ, (k + 1)δ] the estimate

$$\cos\Big((\omega_1(s)+\omega_2(s))s+\psi_1+\psi_2\Big)=\cos\Big((\omega_1(k\delta)+\omega_2(k\delta))s+\psi_1+\psi_2\Big)+O(\delta) \qquad (8)$$

is valid. Relations (6) and (8) imply that

$$\begin{aligned}&\int_0^t \gamma(t-s)\left[\cos\Big((\omega_1(s)+\omega_2(s))s+\psi_1+\psi_2\Big)\right]ds=\\&=\sum_{k=0}^{m}\gamma(t-k\delta)\int_{k\delta}^{(k+1)\delta}\left[\cos\Big((\omega_1(k\delta)+\omega_2(k\delta))s+\psi_1+\psi_2\Big)\right]ds+O(\delta).\end{aligned} \qquad (9)$$

From (5) we have the estimate

$$\int_{k\delta}^{(k+1)\delta}\left[\cos\Big((\omega_1(k\delta)+\omega_2(k\delta))s+\psi_1+\psi_2\Big)\right]ds=O(\delta^2)$$

and the fact that R is sufficiently great as compared with δ. Then

$$\int_0^t \gamma(t-s)\left[\cos\Big((\omega_1(s)+\omega_2(s))s+\psi_1+\psi_2\Big)\right]ds=O(\delta).$$

Theorem 1 is completely proved. ■

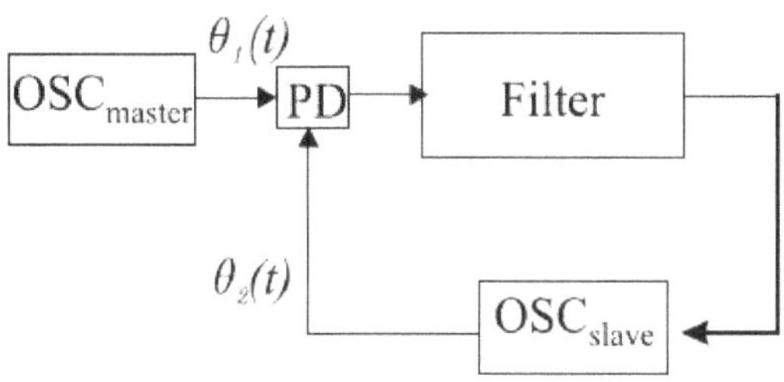


**Fig. 4. Block diagram of PLL on the level of phase relations**

Thus, the outputs g(t) and G(t) of two block diagrams in Fig. 2 and Fig. 3, respectively, differ little from each other and we can pass (from a standpoint of the asymptotic with respect to δ) to the following description level, namely to the second level of phase relations. In this case a block diagram in Fig. 1 becomes the following block diagram (Fig. 4).

Consider now the high-frequency impulse oscillators, connected as in diagram in Fig. 1. Here

$$f_j(t)=A_j\text{sign}\left(\sin(\omega_j(t)t+\psi_j)\right). \qquad (10)$$

We assume, as before, that conditions (3)– (5) are satisfied.

Consider $2\pi$-periodic function $\phi(\theta)$ of the form

$$\varphi(\theta)=\begin{cases}A_1A_2(1+2\theta/\pi) \text{ for } \theta\in[-\pi,0],\\ A_1A_2(1-2\theta/\pi) \text{ for } \theta\in[0,\pi].\end{cases} \qquad (11)$$

and block diagrams in Fig. 2 and Fig. 3.

**Theorem 2.** *(Leonov, 2006)*

If conditions (3)–(5) are satisfied and the characteristic of phase detector $\phi(\theta)$ has the form (11), then for the same initial data of filter the following relation

$$|G(t) - g(t)| \leq C_3\delta, \ \forall t \in [0, T]$$

Here $C_3$ is a certain number being independent of $\delta$.

**Proof of Theorem 2**

In this case we have

$$g(t) - G(t) = \\ = \int_0^t \gamma(t-s)\Big[A_1A_2\text{sign}\Big(\sin(\omega_1(s)s + \psi_1)\sin(\omega_2(s)s + \psi_2)\Big) - \\ - \varphi(\omega_1(s)s - \omega_2(s)s + \psi_1 - \psi_2)\Big]\,ds.$$

Partitioning the interval [0, t] into the intervals [kδ, (k + 1)δ] and making use of assumptions

(5) and (10), we replace the above integral with the following sum

$$\sum_{k=0}^{m} \gamma(t-k\delta)\Big[\int_{k\delta}^{(k+1)\delta} A_1A_2\text{sign}\Big[\cos\Big((\omega_1(k\delta)\omega_2(k\delta))k\delta + \psi_1 - \psi_2\Big) - \\ - \cos\Big((\omega_1(k\delta) + \omega_2(k\delta))s + \psi_1 + \psi_2\Big)\Big]\,ds - \\ - \varphi\Big((\omega_1(k\delta) - \omega_2(k\delta))k\delta + \psi_1 - \psi_2\Big)\delta\Big].$$

The number m is chosen in such a way that $t \in [m\delta, (m + 1)\delta]$. Since $(\omega1(k\delta)+ \omega2(k\delta))\delta \gg 1$, the relation

$$\int_{k\delta}^{(k+1)\delta} A_1A_2\text{sign}\Big[\cos\Big((\omega_1(k\delta) - \omega_2(k\delta))k\delta + \psi_1 - \psi_2\Big) - \\ - \cos\Big((\omega_1(k\delta) + \omega_2(k\delta))s + \psi_1 + \psi_2\Big)\Big]\,ds \approx \\ \approx \varphi\Big((\omega_1(k\delta) - \omega_2(k\delta))k\delta + \psi_1 - \psi_2\Big)\delta, \tag{12}$$

is satisfied. Here we use the relation

$$A_1A_2\int_{k\delta}^{(k+1)\delta} \text{sign}\big[\cos\alpha - \cos(\omega s + \psi_0)\big]\,ds \approx \varphi(\alpha)\delta$$

for ωδ » 1, α ∈ [-π, π], ψ0 ∈ R1.

Thus, Theorem 2 is completely proved. ■

Theorem 2 is a base for the synthesis of PLL with impulse oscillators. For the impulse clock oscillators it permits one to consider two block diagrams simultaneously: on the level of elec- tronic realization (Fig. 1) and on the level of phase relations (Fig. 4), where general principles of the theory of phase synchronization can be used (Leonov & Seledzhi, 2005b; Kuznetsov et al., 2006; Kuznetsov et al., 2007; Kuznetsov et al., 2008; Leonov, 2008).

## 3. DIFFERENTIAL EQUATIONS OF PLL

Let us make a remark necessary for derivation of differential equations of PLL. Consider a quantity

$$\dot{\theta}_j(t) = \omega_j(t) + \dot{\omega}_j(t)t.$$

For the well-synthesized PLL such that it possesses the property of global stability, we have exponential damping of the quantity ω˙ j (t):

$$|\dot{\omega}_j(t)| \leq Ce^{-\alpha t}.$$

Here C and α are certain positive numbers being independent of t. Therefore, the quantity ω˙ j (t)t is, as a rule, sufficiently small with respect to the number R (see conditions (3)– (5)). From the above we can conclude that the following approximate relation θ˙j (t) ≈ ωj (t) is

valid. In deriving the differential equations of this PLL, we make use of a block diagram in Fig. 4 and exact equality

$$\dot{\theta}_j(t) = \omega_j(t). \tag{13}$$

Note that, by assumption, the control law of tunable oscillators is linear:

$$\omega_2(t) = \omega_2(0) + LG(t). \tag{14}$$

Here ω2(0) is the initial frequency of tunable oscillator, L is a certain number, and G(t) is a control signal, which is a filter output (Fig. 4). Thus, the equation of PLL is as follows

$$\dot{\theta}_2(t) = \omega_2(0) + L\left(\alpha_0(t) + \int_0^t \gamma(t-\tau)\varphi(\theta_1(\tau) - \theta_2(\tau))d\tau\right).$$

Assuming that the master oscillator is such that ω1(t) ≡ ω1(0), we obtain the following rela- tions for PLL

$$(\theta_1(t) - \theta_2(t))' + L\left(\alpha_0(t) + \int_0^t \gamma(t-\tau)\varphi(\theta_1(\tau) - \theta_2(\tau))d\tau\right) = \omega_1(0) - \omega_2(0). \tag{15}$$

This is an equation of standard PLL. Note, that if the filter (2) is integrated with the transfer function W(p)= (p + α)-1

$$\dot{\sigma} + \alpha\sigma = \varphi(\theta)$$

then for $\phi(\theta)= \cos(\theta)$ instead of equation (15) from (13) and (14) we have

$$\ddot{\tilde{\theta}} + \alpha\dot{\tilde{\theta}} + L\sin\tilde{\theta} = \alpha(\omega_1(0) - \omega_2(0)) \tag{16}$$

with $\tilde{\theta} = \theta_1 - \theta_2 + \frac{\pi}{2}$. So, if here phases of the input and output signals mutually shifted by

$\pi/2$ then the control signal G(t) equals zero.

Arguing as above, we can conclude that in PLL it can be used the filters with transfer functions

of more general form

$$K(p) = a + W(p),$$

where a is a certain number, W(p) is a proper fractional rational function. In this case in place of equation (15) we have

$$(\theta_1(t) - \theta_2(t))' + L\left[a\varphi(\theta_1(t) - \theta_2(t)) + \alpha_0(t) + \int_0^t \gamma(t-\tau)\varphi(\theta_1(\tau) - \theta_2(\tau))d\tau\right] = \tag{17}$$

$$= \omega_1(0) - \omega_2(0).$$

In the case when the transfer function of the filter a + W(p) is non-degenerate, i.e. its numer- ator and denominator do not have common roots, equation (17) is equivalent to the following system of differential equations

$$\begin{aligned} \dot{z} &= Az + b\psi(\sigma) \\ \dot{\sigma} &= c^* z + \rho\psi(\sigma). \end{aligned} \tag{18}$$

Here $\sigma = \theta 1 - \theta 2$, A is a constant $(n \times n)$-matrix, b and c are constant (n)-vectors, $\rho$ is a number, and $\psi(\sigma)$ is $2\pi$-periodic function, satisfying the relations:

$$\begin{aligned} \rho &= -aL, \\ W(p) &= L^{-1}c^*(A - pI)^{-1}b, \\ \psi(\sigma) &= \varphi(\sigma) - \frac{\omega_1(0) - \omega_2(0)}{L(a + W(0))}. \end{aligned}$$

The discrete phase-locked loops obey similar equations

$$\begin{aligned} z(t+1) &= Az(t) + b\psi(\sigma(t)) \\ \sigma(t+1) &= \sigma(t) + c^* z(t) + \rho\psi(\sigma(t)), \end{aligned} \tag{19}$$

where $t \in Z$, Z is a set of integers. Equations (18) and (19) describe the so-called standard PLLs (Shakhgil'dyan & Lyakhovkin, 1972; Leonov, 2001). Note that there exist many other modifications of PLLs and some of them are considered below.

## 4. MATHEMATICAL ANALYSIS METHODS OF PLL

The theory of phase synchronization was formulated in the latter half of the previous century, drawing upon three practical theories: the theory of synchronous and induction electrical motors, the theory of auto-synchronization of unbalanced rotors, and the theory of phase-locked loops. The main principle of this approach focuses on addressing the issue of phase synchronization at three different levels: (i) in mechanical, electromechanical, or electronic models, (ii) in terms of phase relationships, and (iii) in the context of differential, difference, integral, and integro-differential equations. In this scenario, the discrepancy in oscillation phases is converted into the control action, resulting in synchronization. These fundamental concepts provided momentum for the development of universal techniques for investigating phase synchronization systems. The direct Lyapunov approach was enhanced by incorporating the production of periodic Lyapunov-like functions, the method of positively invariant cone grids, and the method of nonlocal reduction, resulting in a highly effective modification. The final approach, which integrates the components of the direct Lyapunov method and the bifurcation theory, enables the expansion of the traditional findings of F. Tricomi and his successors to encompass multidimensional dynamical systems.

## 4.1 Method of periodic Lyapunov functions

Here we formulate the extension of the Barbashin–Krasovskii theorem to dynamical systems with a cylindrical phase space (Barbashin & Krasovskii, 1952). Consider a differential inclusion

$$\dot{x} \in f(x), \quad x \in R^n, \quad t \in R^1, \tag{20}$$

where f (x) is a semicontinuous vector function whose values are the bounded closed convex set f (x) ⊂ Rn. Here Rn is an n-dimensional Euclidean space. Recall the basic definitions of the theory of differential inclusions.

**Definition 1.** We say that Uε (Ω) is an ε-neighbourhood of the set Ω if

$$U_\varepsilon(\Omega) = \{x \mid \inf_{y \in \Omega} |x - y| < \varepsilon\},$$

where |·| is an Euclidean norm in Rn.

**Definition 2.** A function f (x) is called semicontinuous at a point x if for any ε > 0 there exists a number δ(x, ε) > 0 such that the following containment holds:

$$f(y) \in U_\varepsilon(f(x)), \quad \forall y \in U_\delta(x).$$

**Definition 3.** A vector function x(t) is called a solution of differential inclusion if it is absolutely continuous and for the values of t, at which the derivative x˙ (t) exists, the inclusion

x˙ (t) ∈ f (x(t))

holds.

Under the above assumptions on the function f (x), the theorem on the existence and con- tinuability of solution of differential inclusion (20) is valid (Yakubovich et al., 2004). Now we assume that the linearly independent vectors d1,. . ., dm satisfy the following relations:

$$\dot{x}(t) \in f(x(t))$$

Usually, $d_j^* x$ is called the phase or angular coordinate of system (20). Since property (21) allows us to introduce a cylindrical phase space (Yakubovich et al., 2004), system (20) with property (21) is often called a system with cylindrical phase space.

The following theorem is an extension of the well-known Barbashin-Krasovskii theorem to differential inclusions with a cylindrical phase space.

**Theorem 3.** Suppose that there exists a continuous function V(x) : Rn → R1 such that the following conditions hold:

$$1)\ V(x+d_j) = V(x), \quad \forall x \in R^n, \quad \forall j = 1, \dots, m;$$

$$2)\ V(x) + \sum_{j=1}^{m} (d_j^* x)^2 \to \infty \text{ as } |x| \to \infty;$$

3) for any solution x(t) of inclusion (20) the function V(x(t)) is nonincreasing;

4) if V(x(t)) ≡ V(x(0)), then x(t) is an equilibrium state.

Then any solution of inclusion (20) tends to stationary set as t → +∞.

Recall that the tendency of solution to the stationary set Λ as t means that

$$\lim_{t \to +\infty} \inf_{z \in \Lambda} |z - x(t)| = 0.$$

A proof of Theorem 3 can be found in (Yakubovich et al., 2004).

## 4.2 Method of positively invariant cone grids. An analog of circular criterion

This method was proposed independently in the works (Leonov, 1974; Noldus, 1977). It is suf- ficiently universal and "fine" in the sense that here only two properties of system are used such as the availability of positively invariant one-dimensional quadratic cone and the invariance of field of system (20) under shifts by the vector dj (see (21)).

Here we consider this method for more general nonautonomous case

$$\dot{x} = F(t, x), \quad x \in \mathbb{R}^n, \, t \in \mathbb{R}^1,$$

where the identities F(t, x + dj ) = F(t, x) are valid ∀x ∈ $\mathbb{R}$ n, ∀t ∈ $\mathbb{R}$ 1 for the linearly inde- pendent vectors dj ∈ $\mathbb{R}$ n (j = 1, ..., m). Let x(t)= x(t, $t_0$, $x_0$) is a solution of the system such that x($t_0$, $t_0$, $x_0$)= $x_0$.

We assume that such a cone of the form Ω = {x* Hx ≤ 0}, where H is a symmetrical matrix such that one eigen value is negative and all the rest are positive, is positively invariant. The latter means that on the boundary of cone ∂Ω = {xHx = 0} the relation

$$\dot{V}(x(t)) < 0$$

is satisfied for all x(t) such that {x(t) ⊃= 0, x(t) ∈ ∂Ω} (Fig. 5).

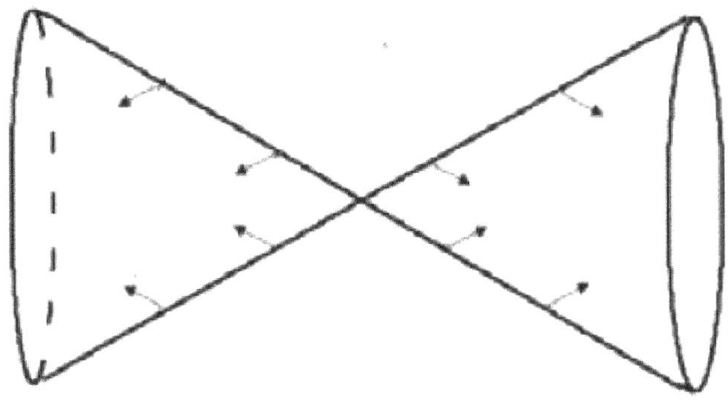

**Fig. 5. Positively invariant cone.**

By the second property, namely the invariance of vector field under shift by the vectors kdj,

k ∈ Z, we multiply the cone in the following way

$$\Omega_k = \{(x - kd_j)H(x - kd_j) \leq 0\}.$$

Since it is evident that for the cones Ωk the property of positive invariance holds true, we obtain a positively invariant cone grid shown in Fig. 6. As can be seen from this figure, all the

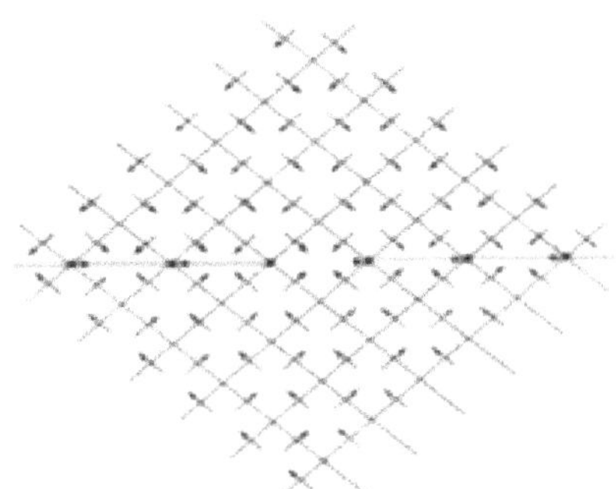

**Fig. 6. Positively invariant cone grid.**

solutions x(t, t0, x0) of system, having these two properties, are bounded on [t0, +∞).

If the cone Ω has only one point of intersection with the hyperplane {d∗x = 0} and all solutions x(t), for which at the time t the inequality

$$x(t)^* Hx(t) \geq 0$$

is satisfied, have property V˙ (x(t)) ≤ -ε|x(t)|2 (here ε is a positive number), then from Fig. 6 it is clear that the system is Lagrange stable (all solutions are bounded on the interval [0, +∞)).

Thus, the proposed method is simple and universal. By the Yakubovich–Kalman frequency theorem it becomes practically efficient (Gelig et al., 1978; Yakubovich et al., 2004).

Consider, for example, the system

$$\dot{x} = Px + q\varphi(t,\sigma), \quad \sigma = r^* x, \tag{22}$$

where P is a constant singular n × n-matrix, q and r are constant n-dimensional vectors, and

ϕ(t, σ) is a continuous 2π-periodic in σ function R1 × R1 → R1, satisfying the relations ϕ(t, σ) 1

$$\mu_1 \le \frac{\varphi(t,\sigma)}{\sigma} \le \mu_2, \ \forall t \in R^1, \ \forall \sigma \ne 0, \quad \varphi(t,0) = 0.$$

Here μ1 and μ2 are some numbers, which by virtue of periodicity of ϕ(t, σ) in σ, without loss of generality, can be assumed to be negative, μ1 < 0, and positive, μ2 > 0, respectively. We introduce the transfer function of system (22)

$$\chi(p) = r^*(P - pI)^{-1}q,$$

which is assumed to be nondegenerate. Consider now quadratic forms V(x)= x* Hx and

$$G(x,\xi) = 2x^* H[(P+\lambda I)x + q\xi] + (\mu_2^{-1}\xi - r^* x)(\mu_1^{-1}\xi - r^* x),$$

where λ is a positive number.

By the Yakubovich–Kalman theorem, for the existence of the symmetrical matrix H with one negative and n - 1 positive eigenvalues and such that the inequality G(x, ξ) < 0, ∀x ∈ Rn, ξ ∈ R1, x ≠ 0 is satisfied, it is sufficient that (C1) the matrix (P + λI) has (n - 1) eigenvalues with negative real part and (C2) the frequency inequality

$$\mu_1^{-1}\mu_2^{-1} + (\mu_1^{-1} + \mu_2^{-1})\mathrm{Re}\chi(i\omega - \lambda) + |\chi(i\omega - \lambda)|^2 < 0, \ \forall \omega \in R^1$$

is satisfied.

It is easy to see that the condition G(x, ξ) < 0, ∀x ≈= 0, ∀ξ implies the relation

$$\dot{V}(x(t)) + 2\lambda V(x(t)) < 0, \ \forall x(t) \ne 0.$$

This inequality assures the positive invariance of the considered cone Ω. Thus, we obtain the following analog of the well-known circular criterion.

**Theorem 4. ( Leonov, 1974; Gelig et al., 1978; Yakubovich et al., 2004)** *If there exists a positive number λ such that the above conditions (C1) and (C2) are satisfied, then any solution x(t, t0, x0) of system (22) is bounded on the interval* (t0, +∞).

A more detailed proof of this fact can be found in (Leonov & Smirnova 2000; Gelig et al., 1978; Yakubovich et al., 2004). We note that this theorem is also true under the condition of nonstrict inequality in (C2) and in the cases when μ1 = -∞ or μ2 = +∞ (Leonov & Smirnova 2000; Gelig et al., 1978; Yakubovich et al., 2004).

We apply now an analog of the circular criterion, formulated with provision for the above remark, to the simplest case of the second-order equation

$$\ddot{\theta} + \alpha\dot{\theta} + \varphi(t, \theta) = 0, \tag{23}$$

where $\alpha$ is a positive parameter (equation (16) can be transformed into (23) by $\tilde{\theta} = \theta + \arcsin\ [\alpha\ (\omega 1(0) - \omega 2(0))L]$). This . This equation can be represented as system (22) with n = 2

and the transfer function

$$\chi(p) = \frac{1}{p(p+\alpha)}.$$

Obviously, condition (C1) of theorem takes the form $\lambda \in (0, \alpha)$ and for $\mu_1 = -\infty$ and $\mu_2 = \alpha_2/4$ condition ($C_2$) is equivalent to the inequality

$$-\omega^2 + \lambda^2 - \alpha\lambda + \alpha^2/4 \leq 0, \ \forall\omega \in R^1.$$

This inequality is satisfied for $\lambda = \alpha/2$. Thus, if in equation (23) the function $\phi(t, \theta)$ is periodic with respect to $\theta$ and satisfies the inequality

$$\frac{\varphi(t, \theta)}{\theta} \leq \frac{\alpha^2}{4}, \tag{24}$$

then any its solution $\theta(t)$ is bounded on $(t0, +\infty)$.

It is easily seen that for $\phi(t, \theta) \equiv \phi(\theta)$ (i.e. $\phi(t, \theta)$ is independent of t) equation (23) is di-chotomic. It follows that in the autonomous case if relation (24) is satisfied, then any solution of (23) tends to certain equilibrium state as $t \to +\infty$.

Here we have interesting analog of notion of absolute stability for phase synchronization sys- tems. If system (22) is absolutely stable under the condition that for any nonlinearity $\phi$ from the sector $[\mu 1, \mu 2]$ any its solution tends to certain equilibrium state, then for equation (23) with $\phi(t, \theta) \equiv \phi(\theta)$ this sector is $(-\infty, \alpha 2/4]$.

At the same time, in the classical theory of absolute stability (without the assumption that $\phi$ is periodic), for $\phi(t, \theta) \equiv \phi(\theta)$ we have two sectors: the sector of absolute stability $(0, +\infty)$ and the sector of absolute instability $(-\infty, 0)$.

Thus, the periodicity alone of $\phi$ allows one to cover a part of sector of absolute stability and a complete sector of absolute instability: $(-\infty, \alpha 2/4] \supset (-\infty, 0) \cup (0, \alpha 2/4]$ (see Fig. 7).

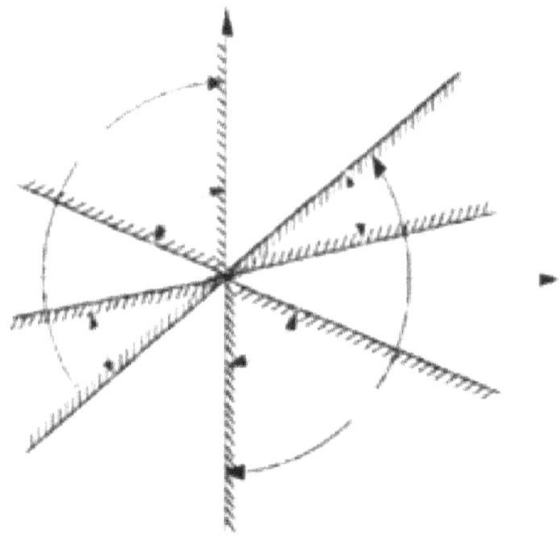

**Fig. 7. Sectors of stability and instability.**

More complex examples of using the analog of circular criterion can be found in (Leonov & Smirnova 2000; Gelig et al., 1978; Yakubovich et al., 2004).

## 4.3 Method of nonlocal reduction

We describe the main stages of extending the theorems of Tricomi and his progenies, obtained for the equation

$$\ddot{\theta} + \alpha\dot{\theta} + \psi(\theta) = 0, \tag{25}$$

to systems of higher dimensions.

Consider first the system

$$\begin{aligned} \dot{z} &= Az + b\psi(\sigma) \\ \dot{\sigma} &= c^{*}z + \rho\psi(\sigma), \end{aligned} \tag{26}$$

describing a standard PLL. We assume, as usual, that ψ(σ) is 2π-periodic, A is a stable n × n-matrix, b and c are constant n-vectors, and ρ is a number. Consider the case when any solution of equation (25) or its equivalent system

$$\begin{aligned} \dot{\eta} &= -\alpha\eta - \psi(\theta) \\ \dot{\theta} &= \eta \end{aligned} \tag{27}$$

tends to the equilibrium state as t → +∞. In this case it is possible to demonstrate (Barbashin & Tabueva, 1969) that for the equation

$$\frac{d\eta}{d\theta} = \frac{-\alpha\eta - \psi(\theta)}{\eta} \tag{28}$$

equivalent to (27) there exists a solution η(θ) such that η(θ0)= 0, η(θ) ⊃= 0, ∀θ ⊃= θ0,

$$\lim_{\theta\to+\infty} \eta(\theta) = -\infty, \quad \lim_{\theta\to-\infty} \eta(\theta) = +\infty. \tag{29}$$

Here θ0 is a number such that ψ(θ0)= 0, ψ≈ (θ0) < 0. We consider now the function

$$V(z,\sigma) = z^* Hz - \frac{1}{2}\eta(\sigma)^2,$$

which induces the cone Ω = {V(z, σ) ≤ 0} in the phase space {z, σ}. This is a generaliza- tion of quadratic cone shown in Fig. 5. We prove that under certain conditions this cone is positively invariant. Consider the expression

$$\frac{dV}{dt} + 2\lambda V = 2z^* H\left[(A+\lambda I)z + b\psi(\sigma)\right] - \lambda\eta(\sigma)^2 - \eta(\sigma)\frac{d\eta(\sigma)}{d\sigma}(c^* z + \rho\psi(\sigma)) =$$
$$= 2z^* H\left[(A+\lambda I)z + b\psi(\sigma)\right] - \lambda\eta(\sigma)^2 + \psi(\sigma)(c^* z + \rho\psi(\sigma)) + \alpha\eta(\sigma)(c^* z + \rho\psi(\sigma)).$$

Here we make use of the fact that η(σ) satisfies equation (28). We note that if the frequency inequalities

$$\begin{aligned} &\operatorname{Re} W(i\omega - \lambda) - \varepsilon|K(i\omega - \lambda)|^2 > 0, \\ &\lim_{\omega\to\infty} \omega^2(\operatorname{Re} K(i\omega - \lambda) - \varepsilon|K(i\omega - \lambda)|^2) > 0, \end{aligned} \qquad (30)$$

where K(p) = c∗ (A - pI)-1b - ρ, are satisfied, then by the Yakubovich–Kalman frequency theorem there exists H such that for ξ and all z ⊃= 0 the following relation

$$2z^* H[(A+\lambda I)z + b\xi] + \xi(c^* z + \rho\xi) + \varepsilon|(c^* z + \rho\xi)|^2 < 0$$

is valid. Here ε is a positive number. If A + λI is a stable matrix, then H > 0.

Thus, if (A + λI) is stable, (30) and α2 ≤ 4λε are satisfied, then we have

$$\frac{dV}{dt} + 2\lambda V < 0, \quad \forall z(t) \neq 0$$

and, therefore, Ω is a positively invariant cone.

We can make a breeding of the cones $\Omega_k = \{z^* Hz - \frac{1}{2}\eta_k(\sigma)^2 \leq 0\}$ in the same way as in the last section and construct a cone grid (Fig. 6), using these cones. Here ηk (σ) is the solution η(σ), shifted along the axis σ by the quantity 2kπ. The cone grid is a proof of boundedness of solutions of system (26) on the interval (0, +∞). Under these assumptions there occurs a dichotomy. This is easily proved by using the Lyapunov function

$$z^* Hz + \int_0^\sigma \psi(\sigma)d\sigma.$$

Thus we prove the following

Theorem 5. If for certain λ > 0 and ε > 0 the matrix A + λI is stable, conditions (30) are satisfied, and system (27) with α = $2\sqrt{\lambda\varepsilon}$ is a globally stable system (all solutions tend to stationary set as t → +∞), then system (26) is also a a globally stable system.

Various generalizations of this theorem and numerous examples of applying the method of nonlocal reduction, including the applying to synchronous machines, can be found in the works (Leonov, 1975; Leonov, 1976; Gelig et al., 1978; Leonov et al., 1992; Leonov et al., 1996a). Various criteria for the existence of circular solutions and second-kind cycles were also ob- tained within the framework of this method (Gelig et al., 1978; Leonov et al., 1992; Leonov et al., 1996a; Yakubovich et al., 2004).

## 5. FLOATING PLL

A crucial criterion for PLLs in digital signal processors is that they must exhibit phase floating. This implies that the system must totally erase the discrepancy in time between clocks. Allow us to elucidate the issue of mitigating clock skew in multiprocessor systems during the implementation of parallel algorithms. Imagine a clock, labeled as C, that sends clock pulses across a bus to multiple processors, labeled as Pk, which are functioning simultaneously in parallel. This is illustrated in Figure 8.

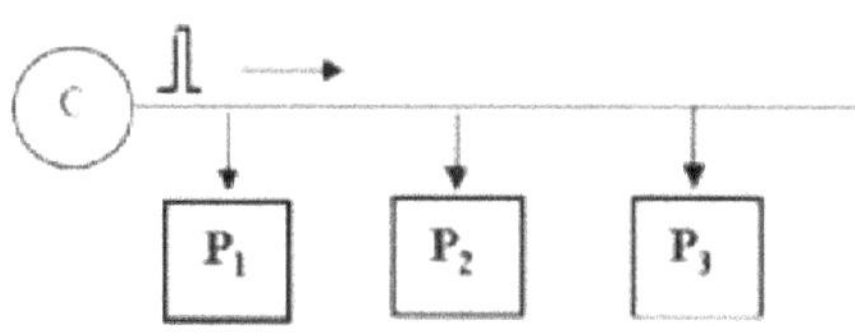


**Fig. 8. Clock C that transmits clock pulses through a bus to processors Pk working in parallel.**

When implementing parallel algorithms, the processors need to execute a specific sequence of actions concurrently. These processes are to be initiated upon the arrival of clock pulses to processors. Due to the varying lengths of the routes over which the pulses travel from the clock to each processor, a timing discrepancy occurs between the processors. The term used to describe this occurrence is known as clock skew.

Eliminating clock skew is a crucial challenge in parallel computing and information processing, as well as in the design of array processors (Kung, 1988). Various methods have been developed over the past three decades to address the issue of reducing clock skew.

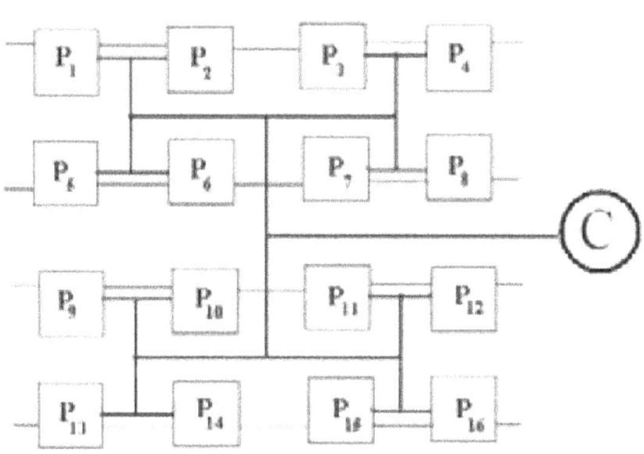


**Fig. 9. Join of processors in the form of an H-tree.**

When designing multiprocessor systems, Kung (1988) and Saint-Laurent et al. (2001) proposed using an H-tree structure to connect the processors. This structure ensures that the pathways from the clock to each processor are of equal length, as shown in Figure 9.

However, in this case the clock skew is not eliminated completely because of heterogeneity of the wires (Kung, 1988). Moreover, for a great number of processors, the configuration of communication wires is very complicated. This leads to difficult technological problems.

The solution of the clock skew problem at a hard- and software levels has resulted in the invention of asynchronous communication protocols, which can correct the asynchronism of operations by waiting modes (Kung, 1988). In other words, the creation of these protocols per- mits one not to distort the final results by delaying information at some stages of the execution of a parallel algorithm. As an advantage of this approach, we may mention the fact that we need not develop a special complicated hardware support system. Among the disadvantages we note the deceleration of performance of parallel algorithms.

In addition to the problem of eliminating the clock skew, one more important problem arose. An increase in the number of processors in multiprocessor systems required an increase in the power of the clock. But powerful clocks lead to produce significant electromagnetic noise.

About ten years ago, a new method for eliminating the clock skew and reducing the gener- ator's power was suggested. It consists of introducing a special distributed system of clocks controlled by PLLs. An advantage of this method, in comparison with asynchronous commu- nication protocols, is the lack of special delays in the performance of parallel algorithms. This approach allows one to reduce significantly the power of clocks.

Consider the general scheme of a distributed system of oscillators (Fig. 10). By Theorem 2, we can make the design of a block diagram of floating PLL, which plays a role of the function of frequency synthesizer and the function of correction of the clock-skew (see parameter τ in Fig. 11). Its block diagram differs from that shown in Fig. 4 with the phase

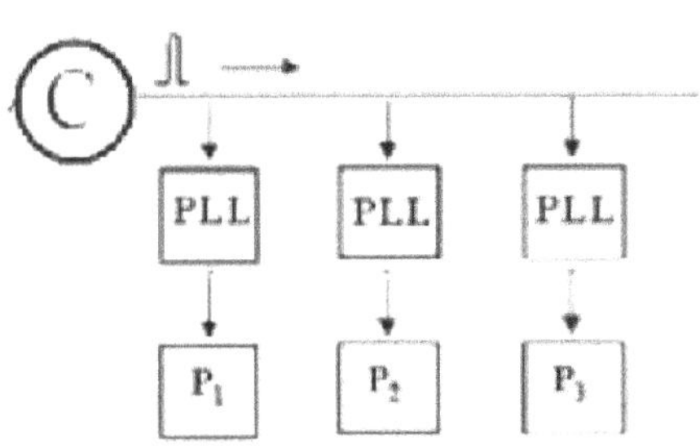


**Fig. 10. General scheme of a distributed system of oscillators with PLL.**

detector characteristic (11) only in that a relay element with characteristic sign G is inserted after the filter.

Such a block diagram is shown in Fig. 11.

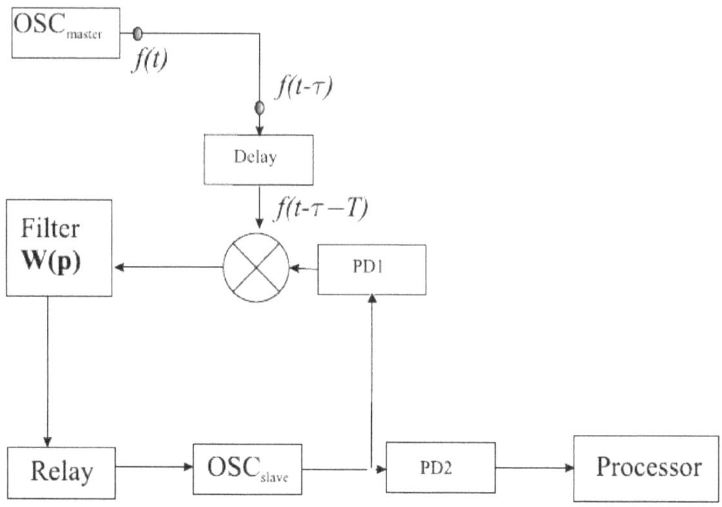


**Fig. 11. Block diagram of floating PLL**

Here OSCmaster is a master oscillator, Delay is a time-delay element, Filter is a filter with transfer function

$$W(p) = \frac{\beta}{p+\alpha},$$

$OSC_{slave}$ is a slave oscillator, PD1 and PD2 are programmable dividers of frequencies, and Processor is a processor.

The Relay element plays a role of a floating correcting block. The inclusion of it allows us to null a residual clock skew, which arises for the nonzero initial difference of frequencies of master and slave oscillators.

We recall that it is assumed here that the master oscillator $\theta'1(t) \equiv \omega1(t) \equiv \omega1(0) = \omega1$ is highly stable. The parameter of delay line T is chosen in such a way that $\omega1(0)(T + \tau) = 2\pi k + 3\pi/2$. Here k is a certain natural number, $\omega1(0)\tau$ is a clock skew.

By Theorem 2 and the choice of T the block diagram, shown in Fig. 11, can be changed by the close block diagram, shown in Fig. 12. Here $\phi(\theta)$ is a $2\pi$-periodic characteristic of phase detector. It has the form

$$\varphi(\theta) = \begin{cases} 2A_1A_2\theta/\pi & \text{for } \theta \in [-\frac{\pi}{2}, \frac{\pi}{2}] \\ 2A_1A_2(1-\theta/\pi) & \text{for } \theta \in [\frac{\pi}{2}, \frac{3\pi}{2}], \end{cases} \tag{31}$$

$\theta_2(t) = \frac{\theta_3(t)}{M}$, $\theta_4(t) = \frac{\theta_3(t)}{N}$, where the natural numbers M and N are parameters of pro grammable divisions PD1 and PD2, respectively.

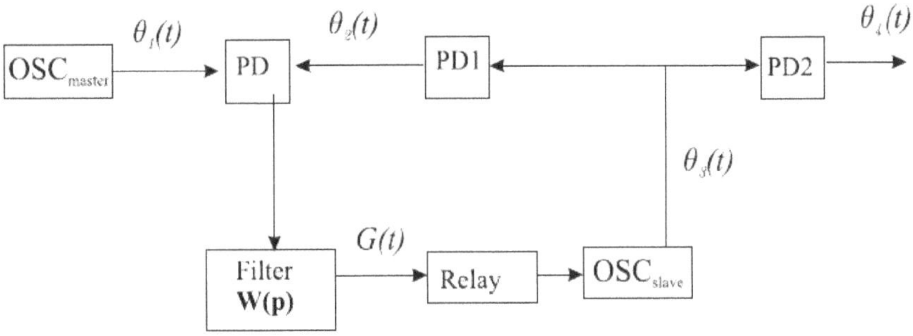


**Fig. 12. Equivalent block diagram of PLL**

For a transient process (a capture mode) the following conditions

$$\lim_{t\to+\infty}\left(\theta_4(t)-\frac{M}{N}\theta_1(t)\right)=\frac{2\pi kM}{N} \tag{32}$$

(a phase capture) and

$$\lim_{t\to+\infty}\left(\dot\theta_4(t)-\frac{M}{N}\dot\theta_1(t)\right)=0 \tag{33}$$

(a frequency capture), must be satisfied.

Relations (32) and (33) are the main requirements to PLL for array processors. The time of transient processors depends on the initial data and is sufficiently large for multiprocessor system ( Kung, 1988; Leonov & Seledzhi, 2002).

Assuming that the characteristic of relay is of the form $\Psi(G)$ = signG and the actuating element of slave oscillator is linear, we have

$$\dot\theta_3(t)=R\mathrm{sign}G(t)+\omega_3(0), \tag{34}$$

where R is a certain number, ω3(0) is the initial frequency, and θ3(t) is a phase of slave oscillator.

Taking into account relations (34), (1), (31) and the block diagram in Fig. 12, we have the following differential equations of PLL

$$\begin{aligned}&\dot G+\alpha G=\beta\varphi(\theta),\\&\dot\theta=-\frac{R}{M}\mathrm{sign}G+\left(\omega_1-\frac{\omega_3(0)}{M}\right).\end{aligned} \tag{35}$$

Here θ(t)= θ1(t) - θ2(t). In general case, we get the following PLL equations:

$$\begin{aligned}&\dot z=Az+b\varphi(\sigma)\\&\dot\sigma=g(c^*z),\end{aligned} \tag{36}$$

where σ = θ1 - θ2, the matrix A and the vectors b and c are such that

$$W(p)=c^*(A-pI)^{-1}b,$$

$$g(G)=-L(\mathrm{sign}\,G)+(\omega_1(0)-\omega_2(0)).$$

Rewrite system (35) as follows

$$\dot{G} = -\alpha G + \beta\varphi(\theta), \quad \dot{\theta} = -F(G), \tag{37}$$

Where

$$F(G) = \frac{R}{M}\text{sign}G - \left(\omega_1 - \frac{\omega_3(0)}{M}\right).$$

**Theorem 6. If the inequality**

$$|R| > |M\omega_1 - \omega_3(0)| \tag{38}$$

is valid, then any solution of system (37) tends to a certain equilibrium state as t → +∞. If the inequality

$$|R| < |M\omega_1 - \omega_3(0)| \tag{39}$$

is valid, then all the solutions of system (37) tend to infinity as t → +∞.

Consider equilibrium states for system (37). For any equilibrium state we have

$$\dot{\theta}(t) \equiv 0, \quad G(t) \equiv 0, \quad \theta(t) \equiv \pi k.$$

**Theorem 7.** We assume that relation (38) is valid. In this case if R > 0, then the following equilibria

$$G(t) \equiv 0, \quad \theta(t) \equiv 2k\pi \tag{40}$$

are locally asymptotically stable and the following equilibria

$$G(t) \equiv 0, \quad \theta(t) \equiv (2k+1)\pi \tag{41}$$

are locally unstable. If R < 0, then equilibria (41) are locally asymptotically stable and equilibria (40) are locally unstable.

Thus, for relations (32) and (33) to be satisfied it is necessary to choose the parameters of system in such a way that the inequality holds

$$R > |M\omega_1 - \omega_3(0)|. \tag{42}$$

Proofs of Theorems 6 and 7. Let R > |Mω1 - ω3(0)|. Consider the Lyapunov function

$$V(G,\theta) = \int_0^G \Phi(u)du + \beta\int_0^\theta \varphi(u)du,$$

where Φ(G) is a single-valued function coinciding with F(G) for G /= 0. For G = 0, the function Φ(G) can be defined arbitrary. At points t such that G(t) /= 0, we have

$$\dot{V}(G(t),\theta(t)) = -\alpha G(t)F(G(t)). \tag{43}$$

Note that, for G(t)= 0, the first equation of system (35) yields

$$\dot{G}(t) \neq 0 \text{ for } \theta(t) \neq k\pi.$$

It follows that there are no sliding solutions of system (35). Then, relation (43) and the inequal- ity F(G)G > 0, ∀G /= 0 imply that conditions 3) and 4) of Theorem 3 are satisfied. Moreover, $V(G, \theta + 2\pi) \equiv V(G, \theta)$ and $V(G, \theta) \to +\infty$ as $G \to +\infty$. Therefore, conditions (1) and (2)

of Theorem 3 are satisfied. Hence, any solution of system (35) tends to the stationary set as $t \to +\infty$. Since the stationary set of system (35) consists of isolated points, any solution to system (35) tends to equilibrium state as $t \to +\infty$.

If the inequality

$$-R > |M\omega_1 - \omega_3(0)|, \qquad (44)$$

is valid, then, in place of the function $V(G, \theta)$, one should consider the Lyapunov function

$W(G, \theta) = -V(G, \theta)$ and repeat the above considerations.

Under inequality (39), we have the relation $F(G) \neq 0$, $\forall G \in R1$. Together with the second

equation of system (37), this implies that

$$\lim_{t \to +\infty} \theta(t) = \infty.$$

Thus, Theorem 6 is completely proved.

$\theta(t) = \infty$.

To prove Theorem 7, we note that if condition (42) holds in a neighbourhood of points $G = 0$,

$\theta = 2\pi k$, then the function $V(G, \theta)$ has the property

$$V(G, \theta) > 0 \quad \text{for} \quad |G| + |\theta - 2k\pi| \neq 0.$$

Together with equality (43), this implies the asymptotic stability of these equilibrium states. In a neighbourhood of points $G = 0$, $\theta = (2k + 1)\pi k$, the function $V(G, \theta)$ has the property $V(0, \theta) < 0$ for $\theta \neq (2k + 1)\pi$. Together with equality (43), this implies the instability of these equilibrium states.

If inequality (44) holds, then, in place of the function $V(G, \theta)$, we can consider the function $W(G, \theta) = -V(G, \theta)$ and repeat the considerations. ■

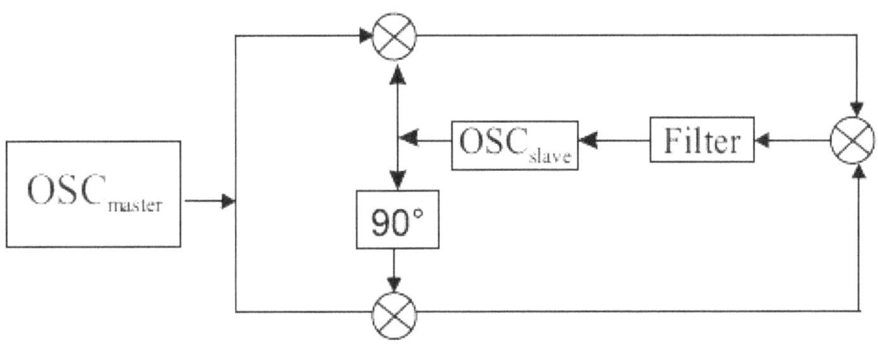


**Fig. 13. Costas loop**

## 6. COSTAS LOOP

Consider now a block diagram of the Costas loop (Fig. 13). Here all denotations are the same as in Fig. 1, 90o is a quadrature component. As before, we consider here the case of the high- frequency harmonic and impulse signals fj (t).

However together with the assumption that conditions (3) and (5) are valid we assume also

that (4) holds for the signal of the type (1) and the relation

$$|\omega_1(\tau) - 2\omega_2(\tau)| \leq C_1, \quad \forall \tau \in [0, T], \tag{45}$$

is valid for the signal of the type (10). Applying the similar approach we can obtain differen- tial equation for the Costas loop, where

$$\begin{aligned} \dot{z} &= Az + b\Psi(\sigma), \\ \dot{\sigma} &= c^*z + \rho\Psi(\sigma). \end{aligned} \tag{46}$$

Here A is a constant n × n-matrix, b and c are constant n-vectors, ρ is a number, and Ψ(σ) is a 2π- periodic function, satisfying the following relations

$$\psi(\sigma) = \frac{1}{8}A_1^2 A_2^2 \sin\sigma - \frac{\omega_1(0) - \omega_2(0)}{L(a + W(0))},$$

$$\sigma = 2\theta_1 - 2\theta_2$$

in the case of harmonic oscillations (1) and

$$\psi(\sigma) = P(\sigma) - \frac{\omega_1(0) - 2\omega_2(0)}{2L(a + W(0))},$$

$$P(\sigma) = \begin{cases} -2A_1^2 A_2^2 \left(1 + \frac{2\sigma}{\pi}\right), & \sigma \in [0, \pi] \\ -2A_1^2 A_2^2 \left(1 - \frac{2\sigma}{\pi}\right), & \sigma \in [-\pi, 0] \end{cases}$$

$$\sigma = \theta_1 - 2\theta_2$$

in the case of impulse oscillations (10), where ρ = -2aL, W(p) = $(2L)^{-1}$c∗ (A - pI)-1b. This implies that for deterministic (when the noise is lacking) description of the Costas loops the conventional introduction of additional filters turns out unnecessary. Here a central filter plays their role.

## 7. BIFURCATIONS IN DIGITAL PLL

In the analysis of the stability of discrete systems, similar to continuous systems, the direct Lyapunov approach can be employed. The application of the direct Lyapunov method can provide frequency inequalities, comparable to the frequency theorem used in Yakubovich-Kalman's approach for discrete systems (Leonov & Seledzhi, 2002). Similar to the continuous instance, phase systems exhibit distinct traits that need the utilization of the direct Lyapunov approach in conjunction with other study techniques. Discrete systems have counterparts of the cone grids approach and the Bakaev-Guzh reduction procedure (Bakaev, 1959; Leonov & Smirnova, 2000).

Here we consider bifurcation effects, arising in discrete models of PLL (Osborne, 1980; Leonov & Seledzhi, 2002).

Discrete phase-locked loops (Simpson, 1994; Lapsley et al., 1997; Smith, 1999; Solonina et al., 2000; Solonina et al., 2001; Aleksenko, 2002; Aleksenko, 2004) with sinusoidal characteristic of phase discriminator are described in details in (Banerjee & Sarkar, 2008). Here a description of bifurcations of a filter-free PLL with a sine-shaped characteristic of phase detector (see (Belykh & Lebedeva, 1982; Leonov & Seledzhi, 2002; Lindsey & Chie, 1981; Osborne, 1980)) is considered. If the initial frequencies of the master and slave oscillators coincide, then the equation of the PLL is of the form

$$\sigma(x+1) = \sigma(x) - r\sin\left(\sigma(x)\right), \qquad (47)$$

where r is a positive number. It is easy to see (Abramovich et al., 2005; Leonov & Seledzhi, 2002; Leonov & Seledzhi, 2005a) that this system is globally asymptotically stable for r ∈ (0, 2). Now we study equation (47) for r > 2. Let r ∈ (2, r1), where r1 is a root of the equation $\sqrt{r^2-1} = \pi + arccos\frac{1}{r}$. Then (47) maps [-π, π] into itself, that is σ(t) ∈ [-π, π] for σ(0) ∈ [-π, π] t = 1, 2, . . . .

In the system (47) there is transition to chaos via the sequence of period doubling bifurcations (Fig. 14). Equation (47) is not unimodal, so we can not directly apply the usual Renorm-Group method for its analytical investigation.

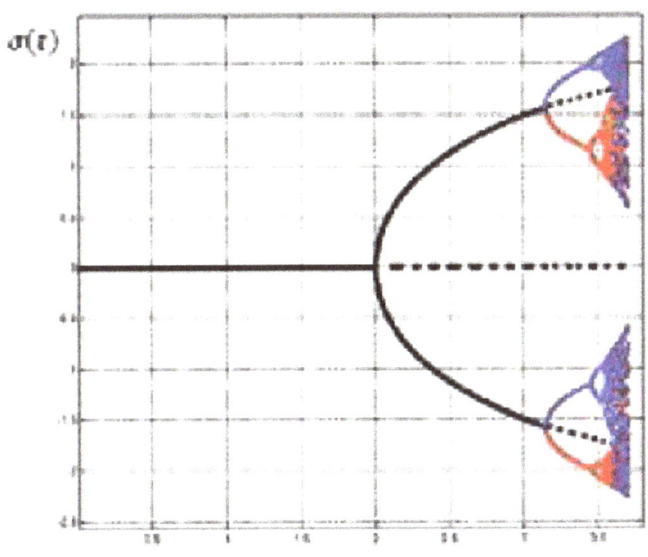


**Fig. 14. Bifurcation tree.**

Some first bifurcation parameters can be calculated analytically (Osborne, 1980), the others can be found only by means of numerical calculations (Abramovich et al., 2005; Leonov & Seledzhi, 2005a).

The first 13 calculated bifurcation parameters of period doubling bifurcations of (47) are the following

$$
\begin{array}{ll}
r_1 = 2 & r_2 = \pi \\
r_3 = 3.445229223301312 & r_4 = 3.512892457411257 \\
r_5 = 3.527525366711579 & r_6 = 3.530665376391086 \\
r_7 = 3.531338162105000 & r_8 = 3.531482265584890 \\
r_9 = 3.531513128976555 & r_{10} = 3.531519739097210 \\
r_{11} = 3.531521154835959 & r_{12} = 3.531521458080261 \\
r_{13} = 3.531521523045159 &
\end{array}
$$

Here r2 is bifurcation of splitting global stable cycle of period 2 into two local stable cycles of period 2. The other rj correspond to period doubling bifurcations.

We have here following Feigenbaum's effect of convergence for $\delta_{n+1} = \dfrac{r_{n+1} - r_n}{r_{n+2} - r_{n+1}}$:

$$
\begin{array}{ll}
\delta_2 = 3.759733732581654 & \delta_3 = 4.487467584214882 \\
\delta_4 = 4.624045206680584 & \delta_5 = 4.660147831971297 \\
\delta_6 = 4.667176508904449 & \delta_7 = 4.668767988303247 \\
\delta_8 = 4.669074658227896 & \delta_9 = 4.669111696537520 \\
\delta_{10} = 4.669025736544542 & \delta_{11} = 4.668640891299296 \\
\delta_{12} = 4.667817727564633. &
\end{array}
$$

## 8. CONCISE SUMMARY

The theory of phase synchronization was formulated in the latter half of the previous century, drawing upon three practical theories: the theory of synchronous and induction electrical motors, the theory of auto-synchronization of unbalanced rotors, and the theory of phase-locked loops. The primary tenet of this approach is to address the issues related to phase synchronization across three distinct levels:

- At the level of mechanical, electromechanical, or electronic model,
- At the level of phase relations,
- At the level of difference, differential, and integro-differential equations.

In this scenario, the discrepancy in the oscillation phases is converted into the control action, resulting in synchronization. These fundamental concepts provided momentum for the development of universal techniques for investigating phase synchronization systems. The direct Lyapunov approach was enhanced by incorporating the production of periodic Lyapunov-like functions, the method of positive invariant cone grids, and the method of nonlocal reduction, resulting in a highly effective modification. The final approach, which integrates the components of the direct Lyapunov method and the bifurcation theory, enables the expansion of the classical findings of F. Tricomi and his successors to the multi-dimensional dynamical systems.

## REFERENCES

1. Abramovitch, D. (2002). Phase-Locked Loops: A control Centric Tutorial, Proceedings of the American Control Conference 2002. vol. 1, pp. 1–15
2. Abramovich, S., Kudryashova, E., Leonov, G.A. & Sugden, S. (2005). Discrete Phase-Locked Loop Systems and Spreadsheets, Spreadsheets in Education (eJSiE),vol. 2, Issue 1
3. Aleksenko, A.G. (2004). Digital engineering, Unimedstyle, Moscow (in Russian)
4. Alekseenko, A.G. (2002). Fundamentals of Microcurcuit, Yunimediastail, Moscow (in Russian) Bakaev, Yu.N. (1959). Some Questions of the Nonlinear Theory of Phase Systems, Proc. Zhukovskii VVIA, vol. 800
5. Banerjee, D. (2006). Pll Performance, Simulation and Design, Ed. 4. Dog Ear Publishing Banerjee, T. & Sarkar, B.C. (2008). Chaos and bifurcation in a third-order digital phase-locked loop, Int. J. Electron. Commun., 62, pp. 86–91
6. Barbashin, E.A. & Krasovskii, N.N. (1952). On Motion Stability in the Large, Dokl. Akad. Nauk SSSR, vol. 86, no. 3, pp. 453–459
7. Barbashin, E.A. & Tabueva, V.A. (1969). Dynamic Systems with Cylindrical Phase Space, Nauka, Moscow (in Russian) De Bellescize, H. (1932). La Reseption synchrone, Onde Electrique
8. Belykh, V.N. & Lebedeva, L.N. (1982). Studying Map of Circle, Prikl. Mat. Mekh., no. 5, pp.611–615
9. Best Ronald, E. (2003). Phase-Lock Loops: Design, Simulation and Application 5ed, McGraw Hill Bianchi, G. (2005). Phase-locked Loop Synthesizer Simulation, McGraw-Hill
10. Bindal, N., Kelly, T., Velastegui, N. & Wong, K.L. (2003). Scalable sub-10ps Skew Global Clock Distribution for a 90nm Multi-GHz IA Microprocessor, Proceedings of IEEE Interna- tional Solid-State Circuits Conference, Digest of Technical Papers. ISSCC. vol. 1, pp. 346– 498
11. Brennan, P.V. (1996). Phase-locked loops: principles and practice, Macmillan Press Buchanan, W. & Wilson, A. (2001). Advanced PC architecture, Addison-Wesley Chen, Wai-Kai. (2003). The circuits and filters handbook, CRC Press
12. Gardner, F. (1979). Phase-lock techniques, 2ed, John Wiley & Sons Giannini, F. & Leuzzi, G. (2004). Nonlinear microwave circuit design, John Wiley & Sons

13. Gelig, A.Kh., Leonov, G.A. & and Yakubovich, V.A. (1978). Stability of Nonlinear Systems with Nonunique Equilibrium State, Nauka, Moscow (in Russian)

14. Goldman, S.J. (2007). Phase-locked Loop Engineering Handbook for Integrated Circuits, Artech House Egan, W. (2007). Phase-Lock Basics, Wiley-IEEE

15. Egan, W. (2000). Frequency Synthesis by Phase Lock, 2nd ed., John Wiley & Sons Encinas, J. (1993). Phase Locked Loops, Springer

16. Kroupa, V. (2003). Phase Lock Loops and Frequency Synthesis, John Wiley & Sons

17. Kudrewicz, J. & Wasowicz S. (2007). Equations of Phase-Locked Loops: Dynamics on the Circle, Torus and Cylinder, World Scientific

18. Kung, S. (1988). VLSI Array Processors, Prentice Hall

19. Kuznetsov, N.V. (2008). Stability and Oscillations of Dynamical Systems: Theory and Applications, Jyväskylä University Printing House, Jyväskylä

20. Kuznetsov, N.V., Leonov, G.A. & Seledzhi, S.M. (2008). Phase Locked Loops Design And Anal- ysis, Int. Conference on Informatics in Control, Automation and Robotics, Proceedings, pp. 114–118

21. Kuznetsov, N.V., Leonov, G.A. & Seledzhi S.M. (2007). Global stability of phase-locked loops. 3rd Int. IEEE conference on Physics and Control (http://lib.physcon.ru/?item= 1192)

22. Kuznetsov, N.V., Leonov, G.A. & Seledzhi, S.M. (2006). Analysis of phase-locked systems with discontinuous characteristics of the phase detectors. Preprints of 1st IFAC conference on Analysis and control of chaotic systems, pp. 127–132

23. Lapsley, P., Bier, J., Shoham, A. & Lee, E. (1997). DSP Processor Fundamentals Architecture and Features, IEE Press

24. Leonov, G.A. (1974). On Boundedness of the Trajectories of the Phase Systems, Sib. Mat. Zh., vol. 15, no. 3, pp. 687–692

25. Leonov, G.A. (1975). Stability and Oscillaitons of the Phase Systems, Sib. Mat. Zh., vol. 16, no. 5, pp. 1031–1052

26. Leonov, G.A. (1976). Second Lyapunov Method in the Theory of Phase Synchronization, Prikl. Mat. Mekh., 1976, vol. 40, no. 2, pp. 238–244

27. Leonov, G., Reitmann, V. & Smirnova, V. (1992). Nonlocal Methods for Pendulum-Like Feedback Systems, Teubner Verlagsgesselschaft, Stuttgart–Leipzig

28. Leonov, G., Ponomarenko, D. & Smirnova, V. (1996a). Frequency-Domain Methods for Non-linear Analysis. Theory and Applications, World Scientific

29. Leonov, G.A., Burkin, I.M. & Shepeljavy, A.I. (1996b). Frequency Methods in Oscillation Theory, Kluver, Dordrecht
30. Leonov, G.A. & Smirnova, V.B. (2000). Mathematical Problems of the Phase Synchronization The- ory, Nauka, St. Petersburg
31. Leonov, G.A. (2001). Mathematical Problems of Control Theory, World Scientific
32. Leonov, G. A. & Seledzhi, S.M. (2002). Phase-locked loops in array processors, Nevsky dialekt, St.Petersburg (in Russian)
33. Leonov, G.A. & Seledghi, S.M. (2005a). Stability and bifurcations of phase-locked loops for digital signal processors, International journal of bifurcation and chaos, 15(4), pp. 1347– 1360
34. Leonov, G.A. & Seledghi, S.M. (2005). Design of phase-locked loops for digital signal proces- sors, International Journal of Innovative Computing, Information Control, 1(4), pp. 779– 789
35. Leonov, G.A. (2006). Phase-Locked Loops. Theory and Application, Automation and remote control, 10, pp. 47–55
36. Leonov, G.A. (2008). Computation of phase detector characteristics in phase-locked loops for clock synchronization, Doklady Mathematics, vol. 78, no. 1, pp. 643–64
37. Lindsey, W. (1972). Sinchronization systems in communication and control, Prentice-Hall Lindsey, W. & Chie, C. (1981). A Survey of Digital Phase Locked Loops. Proceedings of the IEEE, vol. 69, pp. 410–431
38. Manassewitsch, V. (2005). Frequency synthesizers: theory and design, Wiley
39. Margaris, N.I. (2004). Theory of the Non-Linear Analog Phase Locked Loop, Springer Verlag Nash, G. (1994). Phase Locked Loop, Design Fundamentals, Motorola, Phoenix
40. Noldus, E., (1977). New Direct Lyapunov-type Method for Studying Synchronization Prob- lems, Automatika, 1977, vol. 13, no. 2, pp. 139–151
41. Osborne, H.C. (1980). Stability Analysis of an N-th Power Digital Phase Locked Loop. Part. I: First-order DPLL, IEEE Trans. Commun., vol. 28, pp. 1343–1354
42. Razavi, B. (2003). Phase-Locking in High-Performance Systems: From Devices to Architectures, John Wiley & Sons
43. Razavi, B. (2001). Design of Analog CMOS Integrated Circuits, McGraw Hill
44. Saint-Laurent, M., Swaminathan, M. & Meindl, J.D. (2001). On the Micro-Architectural Impact of Clock Distribution Using Multiple PLLs, In Proceedings of International Conference on Computer Design, pp. 214–220
45. Shakhgildyan, V.V. & Lyakhovkin, A.A. (1972). Sistemy fazovoi avtopodstroiki chastoty (Phase Locked Loops), Svyaz, Moscow (in Russian)

46. Shu, K. & Sanchez-Sinencio, E. (2005). CMOS PLL Synthesizers: Analysis and Design, Springer Simpson, R. (1994). Digital Signal Processing Using The Motorola DSP Family, Prentice Hall Smith, S.W. (1999). The Scientist and Engineers Guide to Digital Signal Processing, California Technical Publishing, San Diego
47. Solonina, A., Ulakhovich, D. & Yakovlev, L. (2001). Algorithms and Processors for Digital Signal Processing, BHV, St. Petersburg (in Russian)
48. Solonina, A., Ulahovich, D. & Jakovlev, L. (2000). The Motorola Digital Signal Processors, BHV, St. Petersburg (in Russian)
49. Stensby, J.L. (1997). Phase-locked Loops: Theory and Applications. CRC Press
50. Stephens, D.R. (2002). Phase-Locked Loops For Wireless Communications (Digital, Analog and Op- tical Implementations), Second Edition, Kluwer Academic Publishers
51. Suarez, A. & Quere, R. (2003). Stability Analysis of Nonlinear Microwave Circuits, Artech House Ugrumov, E. (2000). Digital engineering, BHV, St.Petersburg (in Russian)
52. Viterbi, A. (1966). Principles of coherent communications, McGraw-Hill
53. Wainner, S. & Richmond, R. (2003). The book of overclocking: tweak your PC to unleash its power. No Starch Press
54. Wendt, K. & Fredentall, G. (1943). Automatic frequency and phase control of synchronization in TV receivers, Proc. IRE, vol. 31, no. 1, pp. 1–15
55. Wolaver, D.H. (1991). Phase-locked Loop Circuit Design, Prentice Hall
56. Xanthopoulos, T., et al. (2001). The Design and Analysis of the Clock. Distribution. Network for a 1.2GHz Alpha Microprocessor, IEEE ISSCC Tech. Dig., pp. 232–233
57. Yakubovich, V., Leonov, G. & Gelig, A. (2004). Stability of Systems with Discontinuous Nonlin- earities, World Scientisic
58. Young, I.A., Greason, J. & Wong, K. (1992). A PLL clock generator with 5 to 110MHz of lock range for microprocessors. IEEE J. Solid-State Circuits, vol. 27, no. 11, pp. 1599–1607

# 08 TECHNIQUES FOR ESTIMATING PARAMETERS AND CONTROLLING THE FREQUENCY OF PIEZOELECTRIC TRANSDUCERS

## 1. INTRODUCTION: AN OVERVIEW

This chapter discusses the considerations regarding parameter identification and frequency control in piezoelectric transducers utilized in high power ultrasonic applications such as metal or plastic welding, ultrasonic cleaning, and others. Piezoelectric transducers function as actuators, transforming electrical energy into acoustic energy at ultrasonic frequencies. Having knowledge of the corresponding electrical circuit characteristics and resonance frequency is crucial in practical design. The primary objective of study is to enhance the efficiency of energy conversion and improve the frequency stability of control systems for piezoelectric transducers. World literature contains numerous works that discuss the field of transducer measuring and frequency regulation. The following references, pertaining to the theme of the chapter, are presented as follows.

Piezoelectric transducers have demonstrated their significant effectiveness in many high power ultrasonic applications such as cleaning, welding, chemical or biological activations, and others for a considerable period of time. Furthermore, these applications remain highly essential. The power ultrasonic transducers are supplied via power inverters, which utilize high-frequency commutation of transistors (Bose, 1992). Practical applications rely on a wide range of electronic equipment, utilizing either analogue or digital technology, for control purposes (Marchesoni, 1992). Ensuring a high level of energy conversion efficiency in power ultrasound equipment is crucial. Various control techniques are employed in practical applications to regulate the frequency of the signal in power inverters, as documented by Senchenkov (1991), Khemel et al. (2001), and numerous other sources. An analysis of high-power ultrasonic piezoelectric transducers is conducted utilizing complicated structures and equivalent circuits. The analysis begins with Mason's model and is executed on circuit analysis programs, such as Morris (1986). Impedance measurement is conducted using various methodologies (Chen et al., 2008). The wave guides are examined during the construction of ultrasonic systems, specifically in the arrangement of transducer-wave guide-process (Mori, 1989). Books on ultrasonics discuss the techniques and principles of modeling and designing (Prokic, 2004), (Radmanovici & all, 2004). Several technical solutions have been documented in practice, such as those proposed by Fuichi and Nose in 1981, Khmelev et al. in 2001, Kirsh and Berens in 2006, and Sulivan in 1983.

This chapter starts by discussing the overall structure of the electrical circuit of piezoelectric transducers and their practical features. The second paragraph introduces a parameter estimate approach that utilizes the translation from the Nyquist hodograph in the frequency domain to a parameter model. This parameter model is represented by the transfer function of the transducer's impedance. The third paragraph describes a second approach for estimating parameters. This method involves automatically measuring the impedance of a piezoelectric transducer using a deterministic convergence system with a gradient method and continuous adjustment. The chapter concludes by presenting a technique for regulating the frequency of ultrasonic high power piezoelectric transducers. This technique involves utilizing a feedback control system that relies on the first derivative of the movement current.

## 2. ULTRASONIC PIEZOELECTRIC TRANSDUCERS

### 2.1 Constructive and functional characteristics

The ultrasonic piezoelectric transducers are made in a large domain of power from ten to thousand watts, in a frequency range of 20 kHz – 2 MHz. Example of characteristics of some commercial transducers are given in Tab. 1.

**Table 1. Characteristics of some piezoelectric transducers made at I.F.T.M. Bucharest**

| Transducer type | *P* [W] | *fs* [KHz] | *fp* [KHz] | *m* [Kg] | *I* [mA] | Constr. type | C0 [nF] |
|---|---|---|---|---|---|---|---|
| TGUS 100-020-2 | 100 | 20±1 | 22±2 | 0,65 | 300 | 2 | 4,2±0,6 |
| TGUS 100-025-2 | 100 | 25±1 | 27±2 | 0,6 | 300 | 2 | 4,2±0,6 |
| TGUS 150-040-1 | 150 | 40±2 | 43±2 | 0,26 | 300 | 1, 2 | 4,1±0,6 |
| TGUS 500-020-1 | 500 | 20±1 | 22±2 | 1,1 | 500 | 1 | 5,8±0,6 |

The 1st type is for general applications and the 2nd type is for ultrasonic cleaning to be mounted on membranes. Two examples of piezoelectric transducers TGUS 150-040-1 and TGUS 500-25-1 are presented in Fig. 1.

**Fig. 1. Piezoelectric transducer of 150 W at 40 kHz (left) and 500 W at 20 kHz (right)**

They have small losses, a good coupling coefficient *k*ef, a good quality mechanical coefficient

$Q_{m0}$ and a high efficiency η0: in normal operating conditions of temperature, humidity and atmospheric pressure.

$$k_{ef} = \sqrt{1-\left(\frac{f_s}{f_p}\right)^2}, \quad Q_{m0} = \frac{f_p}{\Delta f_p}, \quad \eta_0 = 1-\frac{2}{k_{ef}}\sqrt{\frac{tg\delta}{Q_{m0}}} \tag{1}$$

## 3. ELECTRICAL PROPERTIES

The high power ultrasonic installations have as components ultrasonic generator piezoelectric transducers, which are accomplish some technical conditions. They have the electrical equivalent circuit from Fig. 2.

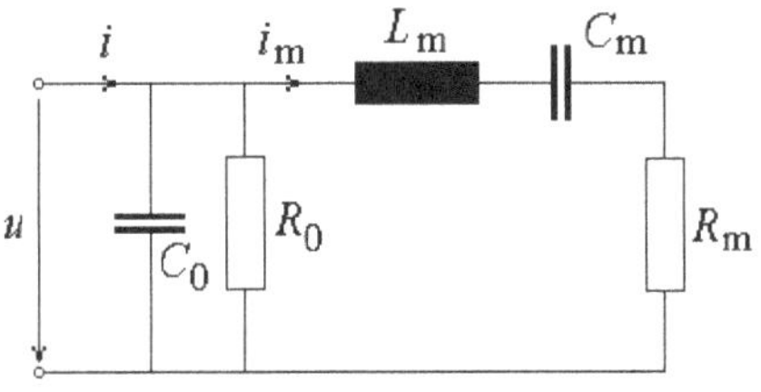


**Fig. 2. The simplified linear equivalent electrical circuit**

Their magnitude-frequency characteristic is presented in Fig. 3.

We may notice on this characteristic a series resonant frequency *fs* and a parallel resonant frequency *fp*, placed at the right. The magnitude has the minimum value Zm at the series frequency and the maximum value ZM at the parallel resonant frequency, on bounded domain of frequencies. The piezoelectric transducer is used in the practical applications working at the series resonant frequency.

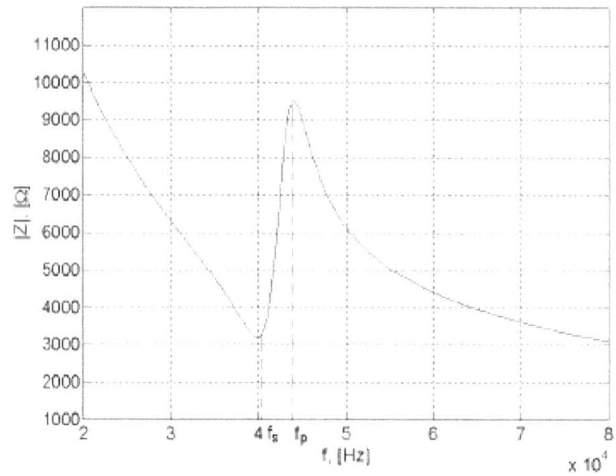


**Fig. 3. The impedance magnitude-frequency characteristic**

The most important aspect of this magnitude characteristic is the fact that the frequency characteristic is modifying permanently in the transient regimes, being affected by the load applied to the transducer, in the following manner: the minimum impedance Zm is increasing, the maximum impedance ZM is decreasing and also the frequency bandwidth [*fs*, *fp*] is modifying in specific ways according to the load types. So, when at the transducer a concentrator is coupled, as in Fig. 4, the frequency bandwidth [*fs*, *fp*] became very narrow, as

*fp*- *fs* ≈ 1-2 Hz.

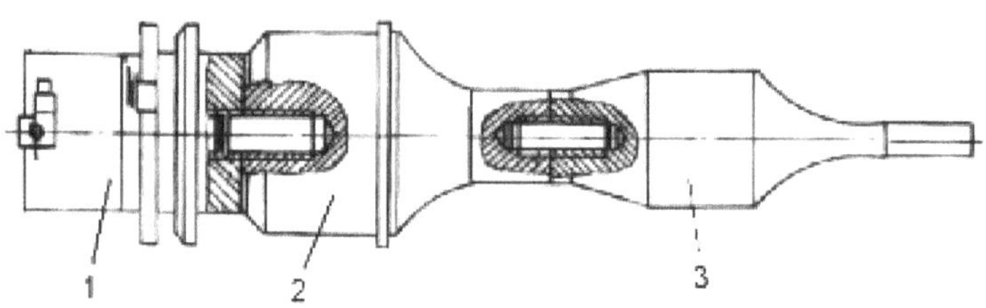


**Fig. 4. A transducer 1 with a concentrator 2 and a welding tool 3**

This is a great impediment because in this case a very précised and stable frequency control circuit is necessary at the electronic ultrasonic power generator for the feeding voltage of the transducer.

When at the transducer a horn or a membrane is mounted, as in Fig. 5, the frequency bandwidth [*fs*, *fp*] increases for 10 times, *fp*- *fs* ≈ *n* kHz.

The resonance frequencies are also modifying by the coupling of a concentrator on the transducer. In this case, to obtain the initial resonance frequency of the transducer the user must adjust mechanically the concentrator at the transducer own resonance frequency. At the ultrasonic blocks with three components (Fig. 4) a transducer 1, a mechanical concentrator 2 and a processing tool 3, the resonance frequency is given by the entire assembled block (1, 2, 3) and in the ultimate instance by the processing tool 3. At cleaning equipments the series resonance frequency is decreasing with 3÷4 KHz.

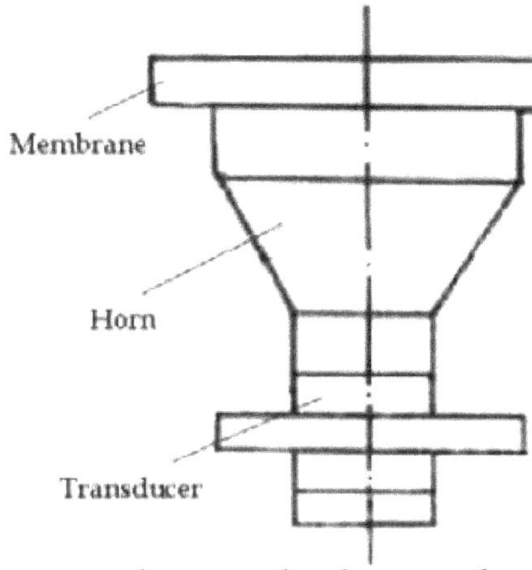


**Fig. 5. A transducer with a horn and a membrane**

The transducers are characterised by a Nyquist hodograph of the impedance present in Fig. 6, which has the theoretical form of a circle. In reality, due to the non-linear character of the transducer, especially at high power, this circle is deformed.

The movement current *im* of piezoelectric transducer is important information related to the maximum power conversion efficiency at resonance frequency. It is the current passing through the equivalent RLC series circuit, which represents the mechanical branch of the equivalent circuit. It is obtained as the difference:

$$i_m = i - i_{C0} \tag{2}$$

An example of the measured movement current is presented in Fig. 7.

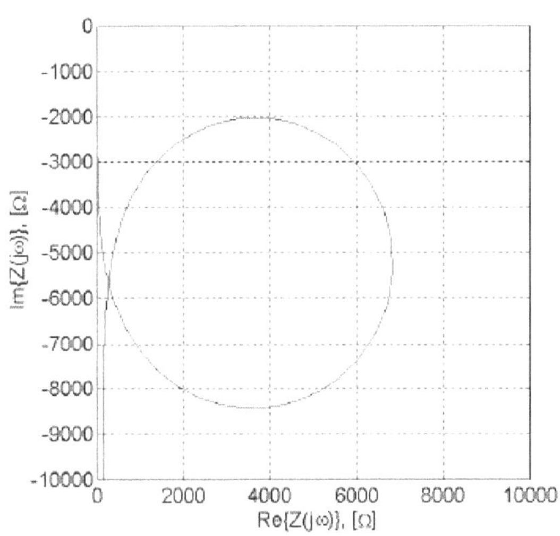


**Fig. 6. Impedance hodograph around the resonant frequency**

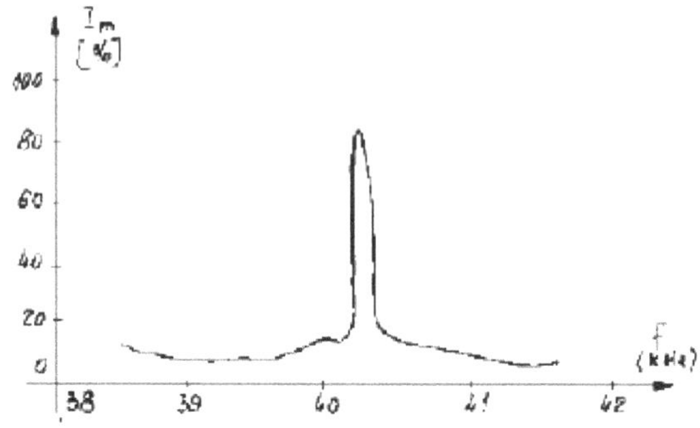


**Fig. 7. Movement current frequency characteristic**

# 4. FREQUENCY-BASED IDENTIFICATION

## 4.1 Generalities

To achieve a well-designed ultrasonic equipment, it is necessary to have a thorough understanding of the equivalent models of ultrasonic components. The transducer, which serves as an electromechanical power generator for mechanical oscillations of ultrasonic frequency, is the key component in this regard. The model is theoretically demonstrated and practically estimated with a high level of accuracy. In practice, estimating involves selecting a model that closely simulates the actual measurements. The identification process considers several factors, including the model type, the type of test signal, and the assessment criterion for the error between the model and the transducer being analyzed. We begin by selecting a preferred model and then fine-tuning the parameters until the discrepancy between the transducer's behavior and the model is minimized. The transducer's construction is assumed to be known, and it corresponds to the equivalent circuit shown in Figure 2. The objective of the identification process is to determine the corresponding parameters of this electrical circuit. The model is derived from empirical data. A parametric model that can be used is the complex impedance of the transducer, which is expressed using a Laplace transformation. Another paradigm, but in the frequency domain, is shown by the Nyquist hodograph of impedance shown in Figure 6. The frequency model is determined by a finite number of observed and unrelated values. The preferred approach for the piezoelectric transducer is to convert the frequency model into a parameter model known as the complex impedance (Tertisco & Stoica, 1980). One significant drawback of this method is the requirement for sophisticated estimating equipment and the need to identify the transducer type and the complex impedance of the comparable electrical circuit. The frequency characteristic can be easily determined by testing the transducer with a sinusoidal test signal of varying frequency. The transition from a frequency model to a parameter model involves determining the parameters of the transfer impedance. The technique for identifying the transducer involves the following steps: organizing and collecting experimental data, interpreting the measured data, deducing a model and defining its structure, and validating the model.

## 4.2 Identification method

Frequency representation of a transducer was presented before. The frequency characteristics may be obtained applying a sinusoidal test voltage signal to the transducer and obtaining a current with the same frequency, but with other magnitude and phase, variables with the applied frequency. The theoretic complex impedance is:

$$|Z(j\omega) = |Z(j\omega)|e^{\varphi(\omega)} = \mathrm{Re}\{Z(j\omega)\} + j\,\mathrm{Im}\{Z(j\omega)\} \quad (3)$$

Its parameter representation is:

$$Z(s) = \frac{B(s)}{A(s)} = \frac{\sum_{i=0}^{m} b_i s^i}{1 + \sum_{i=1}^{n} a_i s^i} = \frac{b_0 + b_1 s + b_2 s^2 + \ldots + b_m s^m}{1 + a_1 s + a_2 s^2 + \ldots + a_n s^n} \quad (4)$$

A general dimensional structure for identification with the orders {*n*, *m*} is considered, where *n* and *m* follow to be estimated.

The model that must be obtained by identification is given by:

$$Z_M(j\omega_k) = \frac{b_0 + b_1(j\omega_k) + \ldots + b_m(j\omega_k)^m}{1 + a_1(j\omega_k) + \ldots + a_n(j\omega_k)^n} = \frac{\alpha + j\omega_k\beta}{\delta + j\omega_k\gamma} = \frac{B(j\omega_k)}{A(j\omega_k)} \quad (5)$$

We presume the existence of the experimental frequency characteristic, as samples:

$$\begin{aligned} Z_e(j\omega_k) &= \mathrm{Re}\{Z_e(j\omega_k)\} + j\,\mathrm{Im}\{Z_e(j\omega_k)\} \\ R_k &= \mathrm{Re}\{Z_e(j\omega_k)\} \\ I_k &= \mathrm{Im}\{Z_e(j\omega_k)\},\ k = 1,2,3,\ldots,n_p \end{aligned} \quad (6)$$

For any particular value ωk the error ε(ωk) is defined as:

$$\varepsilon(\omega_k) = |Z_e(j\omega_k) - Z_M(j\omega_k)| = \left|Z_e(j\omega_k) - \frac{B(j\omega_k)}{A(j\omega_k)}\right| \quad (7)$$

The error criterion is defined as:

$$E = \sum_{k=0}^{n_p} |\varepsilon(\omega_k)|^2 \quad (8)$$

The estimation of orders {*n*, *m*} and parameters is formulated as a parametric optimisation:

$$p = [a_0 \quad a_1 \quad \ldots \quad a_n \quad b_1 \quad b_2 \quad \ldots \quad b_m] = \arg\min_p \sum_{k=0}^{n_p} |\varepsilon(\omega_k)|^2 \quad (9)$$

The error criterion is non-linear in parameters and the direct has practical difficulties: a huge computational effort, local minima, instability and so on. To simplify the algorithm, the error ε(ωk) is weighted with A(jωk). A new error function is obtain:

$$B(j\omega_k)\varepsilon(j\omega_k) = A(j\omega_k).Z(j\omega_k) - B(j\omega_k) = X(\omega_k) + jY(\omega_k) \tag{10}$$

The weighted error function $e(\omega\ k)$ is given by

$$e(\omega_k) = A(j\omega_k).\varepsilon(j\omega_k) \tag{11}$$

The new approximation error, corresponding to the weighted error is:

$$E = \sum_{k=1}^{n_p} |e(\omega_k)|^2 = \sum_{k=1}^{n_p} |A(j\omega_k).\varepsilon(j\omega_k)|^2 = \sum_{k=1}^{n_p} \left[X^2(\omega_k) + Y^2(\omega_k)\right] \tag{12}$$

The minimization of $E$ is done based on the weighted least squares criterion, in which the weighting function $[A(j\omega k)]2$ was chosen so E to be square in model parameters:

$$p = \arg\min_p \sum_{k=1}^{n_p} |e(\omega_k)|^2 \tag{13}$$

$$E^i = \sum_{k=1}^{n_p} \left|\frac{A_i(j\omega_k)}{A_{i-1}(j\omega_k)}\right| \varepsilon_i^2(\omega_k) \tag{14}$$

But, also this method is not good in practice. The frequency characteristic must be approximated on the all frequency domain. The low frequencies are not good weighted, so the circuit gain will be wrong approximated. To eliminate this disadvantage the criterion is modified in the following way:

$$p^i = \arg\min_p \sum_{k=1}^{n_p} \left|\frac{A_i(j\omega_k)}{A_{i-1}(j\omega_k)}\right|^2 \varepsilon_i^2(\omega_k) \tag{15}$$

where $i$ represents the iteration number, $pi$ is the vector of the parameters at the iteration $i$. The error $\varepsilon i(\omega k)$ is given by:

$$\varepsilon_i(\omega_k) = \left|Z_e(j\omega_k) - \frac{B_i(j\omega_k)}{A_i(j\omega_k)}\right| \tag{16}$$

At the algorithm initiation:

$$B_0(j\omega_k) = 1 \tag{17}$$

The criterion is quadratic in $pi$, so the parameter vector at the iteration $i$ may be analytically determinate.

In the same time the method converges, because there is the condition:

$$\lim_{i\to\infty} \frac{A_i(j\omega_k)}{A_{i-1}(j\omega_k)} = 1 \tag{18}$$

The estimation accuracy will have the same value on the entire frequency spectre. The procedure is an iterative variant of the least weighted squares method. At each iteration the criterion is minimized and the linear equation system is obtained:

$$\left(\frac{\partial E_i}{\partial a_k}\right)_i = 0$$
$$\left(\frac{\partial E_i}{\partial b_k}\right)_i = 0 \quad (19)$$

To obtain an explicit relation for *pi* we notice that:

$$\mathrm{Re}\{A_i(j\omega_k)\} = \sum_{i=0}^{r_1}(-1)^i a_{2i}\omega_k^{2i}$$
$$\mathrm{Im}\{A_i(j\omega_k)\} = \sum_{i=0}^{r_2}(-1)^i a_{2i+1}\omega_k^{2i+1} \quad (20)$$

where *r*1 = *n*/2 and *r*2=*n*/2-1, if *n* is odd and *r*1 = (*n*-1)/2 şi *r*2=(*n*-1)/2, if *n* is even. By analogy Re{*B*(*j*ωk)} and Im{*B*(*j*ωk)} may be represented in the same way, for *r*3 and *r*4, function of *m*.

From the linear relations the following linear system is obtained:

$$E \times p_i = F \quad (21)$$

where the matrix *E*, *pi*, *F* are given by the relations (24), in which *k* takes the values from 1, 0, 0 and 0 until *r*1,2,3,4 for rows, from up to down, and *j* takes values from 1, 0, 0 and 0 until *r*1,2,3,4 for columns from the left to the right.

$$E = \begin{bmatrix} (-1)^{j+1}\Gamma_{2(j+k)} & 0 & (-1)^j\Delta_{2(j+k)} & (-1)^j\Phi_{2(j+k)+1} \\ 0 & (-1)^j U_{2(j+k+1)} & (-1)^j\Phi_{2(j+k)+1} & (-1)^{j+1}\Delta_{2(j+k+1)} \\ (-1)^{j+1}\Delta_{2(j+k)} & (-1)^j\Phi_{2(j+k)+1} & (-1)^j\Pi_{2(j+k)+1} & 0 \\ (-1)^{j+1}\Phi_{2(j+k)+1} & (-1)^{j+1}\Delta_{2(j+k+1)} & 0 & (-1)^j\Gamma_{2(j+k+1)} \end{bmatrix} \quad (22)$$

$$p_i = \begin{bmatrix} \hat{a}_2 & \dots & \hat{a}_{2r1} & \hat{a}_1 & \dots & \hat{a}_{2r2+1} & \hat{b}_0 & \dots & \hat{b}_{2r3} & \hat{b}_1 & \dots & \hat{b}_{2r4+1} \end{bmatrix}^T$$
$$F = [\Gamma_2 \;\dots\; \Gamma_{2r1} \;\; 0 \;\; \Delta_0 \;\dots\; \Delta_{2r3} \;\; \Phi_1 \;\dots\; \Phi_{2r4+1}]^T \quad (23)$$

$$\Gamma_i = \sum_{k=1}^{n_p}\frac{(R_k^2 + I_k^2)\omega_k^i}{|A(j\omega_k)|_{i-1}^2},\; \Delta_i = \sum_{k=1}^{n_p}\frac{R_k \cdot \omega_k^i}{|A(j\omega_k)|_{i-1}^2}, \quad (24)$$

$$\Phi_i = \sum_{k=1}^{n_p}\frac{I_k \cdot \omega_k^i}{|A(j\omega_k)|_{i-1}^2},\; \Pi_i = \sum_{k=1}^{n_p}\frac{\omega_k^i}{|A(j\omega_k)|_{i-1}^2} \quad (25)$$

The values of *n* and *m* are determinate after iterative modifications and iterative estimations.

The block diagram of the estimation procedure is given in Fig. 8.

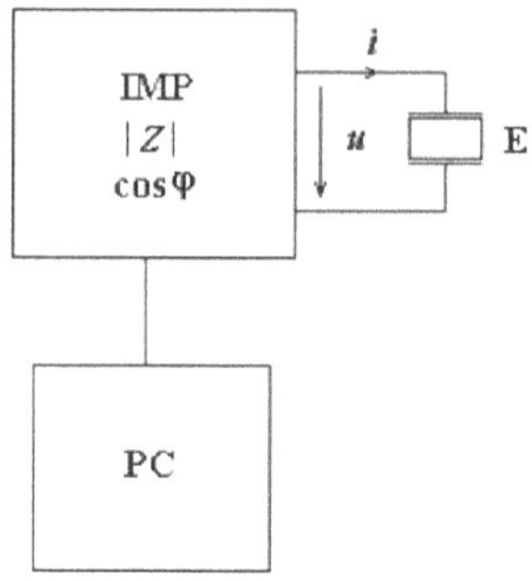


**Fig. 8. Estimation equipment**

The digital impedance meter IMP is used to measure the frequency characteristic of the piezoelectric transducer E. A personal computer PC use an estimating program to analyze and manipulate measured data. In practical applications, estimated parameters are obtained with a relative tolerance of 10%.

## 5. AUTOMATIC PARAMETER ESTIMATION

The method determines the values of the parameters in the equivalent circuit shown in Figure 2. These parameters include the mechanical inductance (Lm), the mechanical capacitor (Cm), the resistance associated with acoustic dissipation (Rm), the input capacitor (C0), and other characteristics such as the mechanical resonance frequency (fm), the movement current (im), and the efficiency (η). The estimation is conducted comprehensively and systematically, considering both the loaded and empty states of the transducer, while it is installed on various equipment. By minimizing the ultrasonic procedure at the transducer, we may analyze the overall features of the ultrasonic assembling block transducer-process using the aforementioned factors and variables.

The identification process relies on an automated measurement technique that analyzes complex impedances using the system identification theory (Eyikoff, 1974). It involves implementing a generalized model of a piezoelectric transducer and minimizing an error criterion using the deepest descent method, a gradient-based approach.

Piezoelectric transducers are utilized in the design of industrial ultrasonic equipment, positioned between the electronic generators and the adapter mechanical components. A multitude of electrical and mechanical forces are exerting and straining over the transducer. A thorough understanding of electrical characteristics is crucial for ensuring effective process control and maximizing the efficiency of ultrasonic processes.

Based on the equivalent circuit, considered as a physical model for the transducer, we may determine a mathematic model, the integral-differential equation:

$$L_m \frac{di}{dt} + R_m i + \frac{1}{C_m}\int i dt = L_m C_0 \frac{d^2 u}{dt^2} + L_m C_0 \frac{R_m}{R_p}\frac{du}{dt} + \frac{R_m}{R_0} u + \frac{1}{C_m R_0}\int u dt \quad (26)$$

This model represents a relation between the voltage *u* applied at the input, as an acting force and the current *i* through transducer. The model is in continuous time. We do not know the parameters and the state variables of the model. This model assures a good representation. A complex one will make a heavier identification. The classical theory of identification is using different methods as: frequency methods, stochastic methods and other. This method has the disadvantage that it determines only the global transfer function.

Starting from equation (28) we obtain the linear equation in parameters:

$$\sum_0^2 \alpha_i i_i + \sum_0^3 \beta_i u_i = 0 \quad (27)$$

$$i_0 = i,\ i_1 = di/dt,\ i_2 = \int i dt,$$
$$u_0 = -u,\ u_1 = -du/dt,\ u_2 = -d^2u/dt^2,\ u_3 = -\int u dt \quad (28)$$

$$\alpha_0 = R_m,\ \alpha_1 = L_m,\ \alpha_2 = 1/C_m,$$
$$\beta_0 = R_m/R_0,\ \beta_1 = L_m/R_d + C_0 R_m,\ \beta_2 = L_m C_0,\ \beta_3 = 1/(C_m R_0) \quad (29)$$

The relation gives the transducer generalized model, with the generalized error:

$$e = \sum_0^2 \alpha_i i_i + \sum_0^3 \beta_i u_i \quad (32)$$

The estimation is doing using a signal continuous in time, sampled, sinusoidal, with variable frequency. For an accurate determination of parameters there are necessary the following knowledge: the magnitude order of the parameters and some known values of them.

The error criterion is imposed as a quadratic one:

$$E = e^2 \quad (30)$$

which influences in a positive sense at negative and positive variations of error. To minimize this error criterion we may adopt, for example a gradient method in a scheme of continuous adjustment of parameters, with the deepest descent method. In this case the model is driven to a tangential trajectory, what for a certain adjusted speed it gives the fastest error decreasing. The trajectory is normally to the curves with E=ct. The parameters are adjusted with the relation:

$$\begin{bmatrix} \dot{\alpha}_i \\ \dot{\beta}_i \end{bmatrix} = -\Gamma \begin{bmatrix} -\dfrac{\partial E}{\partial \alpha_i} \\ -\dfrac{\partial E}{\partial \beta_i} \end{bmatrix} = -\Gamma \begin{bmatrix} 2e\dfrac{\partial e}{\partial \alpha_i} \\ 2e\dfrac{\partial e}{\partial \beta_i} \end{bmatrix} = -2\Gamma \begin{bmatrix} i_i \\ u_i \end{bmatrix} \quad (31)$$

where G is a constant matrix, which together with the partial derivatives determines parameter variation speed. Derivative measuring is not instantaneously, so a variation speed limitation must be maintained. To determine the constant G we may apply Lyapunov stability method. Based on the generalized model and of equation (34) the estimation algorithm may be implemented digitally. The block diagram of the estimator is presented in Fig. 9.

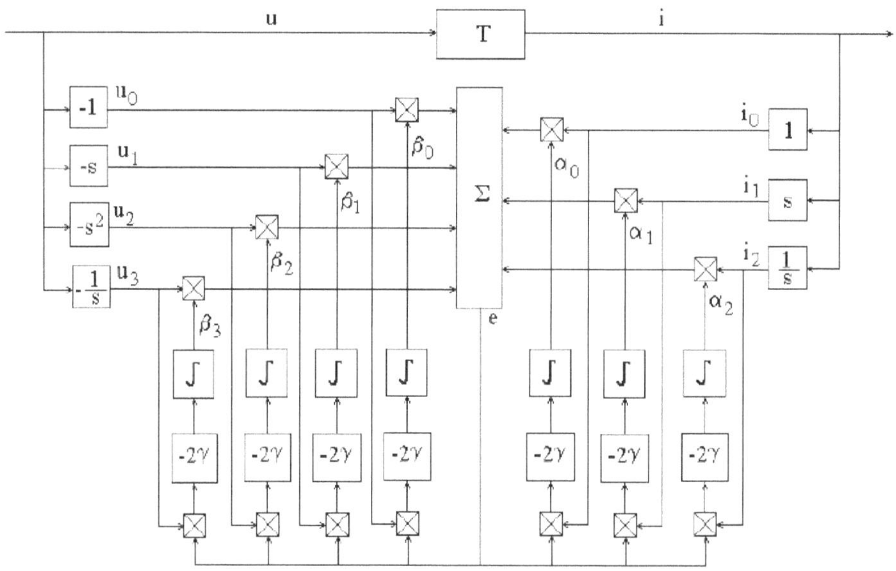


**Fig. 9. The block diagram of parameter estimator**

Shannon condition must be accomplished in sampling. We may notice some identical blocks from the diagram are repeating themselves, so they may be implemented using the same procedures. Based on differential equation:

$$u = L_m \frac{di_m}{dt} + R_m i_m + \frac{1}{C_m}\int i_m dt = \alpha_1 i_1 + \alpha_2 i_2 + \alpha_0 i_0 \quad (32)$$

which is characterising the mechanical branch of transducer with the parameters obtained with the above scheme, we may determine the movement current with the principle block diagram from Fig. 10.

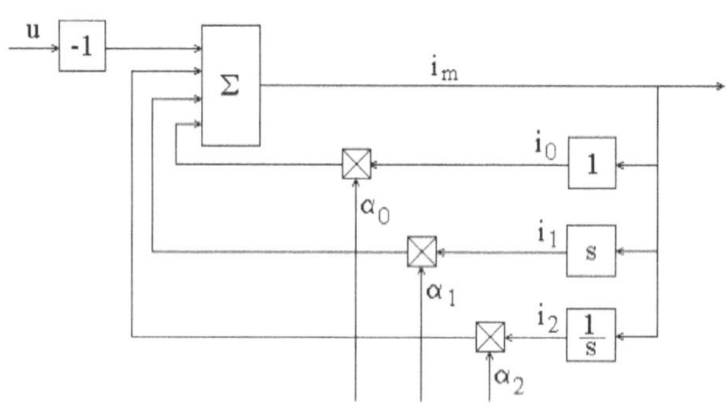


**Fig. 10. The block diagram of movement current estimation**

The variation of the error criterion *E* in practical tests is presented in Fig. 11, for 1000 samples.

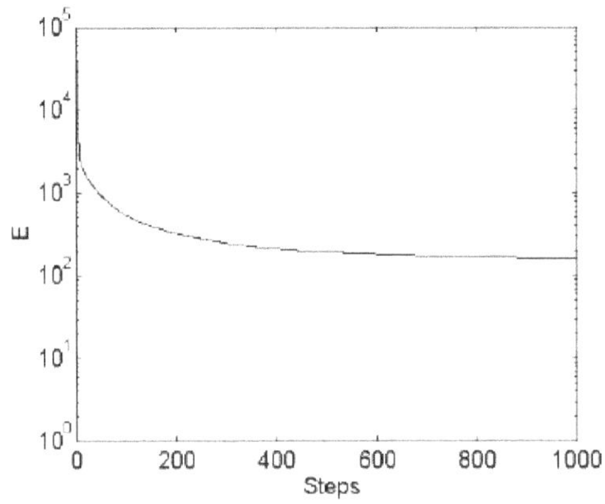


**Fig. 11. Error criterion variation**

Using the model parameters we may compute the mechanical resonance frequency with the relation:

$$f_m = \frac{1}{2\pi\sqrt{L_m C_m}} \tag{33}$$

The efficiency of conversion as a rapport from the acoustic power *Pm* and total power *Pt* is

$$\eta = \frac{P_m}{P_t} \tag{34}$$

where *Pm* is the power on the resistance *Rm*:

$$P_m = I_m{}^2 R_m \tag{35}$$

and the total power is the power consumed from the source:

$$P_t = P_m + P_0 \tag{36}$$

where *P0* is the power consumed by the unloaded transducer:

$$P_0 = I_{mo}{}^2 R_p \tag{37}$$

where *Im0* is the movement current through the unloaded transducer and *Rp* is the resistance corresponding to mechanical circuit unloaded.

Using the estimator from Fig. 9 and 10 we may do an identification of mechanical adapters. A mechanical adaptor coupled to the transducer influences the equivalent electrical circuit, modifying the equivalent parameters, the resonance frequency and the movement current. We may do the same measuring several times over the unloaded and then over the loaded transducer. Knowing the characteristics of the unloaded transducer we may find the way how the adapter influences the equivalent circuit. So, we may determine the parameters of the assemble transducer - adapter, reduced to the transducer: resonance frequency, movement current and efficiency. To determine efficiency we must take in consideration the power of the unloaded transducer and the power of the unloaded adapter.

Also, the process may be identified using the same estimator. Considering the transducer coupled with an adapter and introduced into a ultrasonic process, as welding, cleaning and other, we may determine by an identification for the loaded functioning the way that the process influences the equivalent parameters. We may determine the resonant frequency of the ultrasonic process and the global acoustic efficiency of ultrasonic system transducer-adapter- process. We may determine the mechanical resonant frequency of the entire assemble, which is the frequency at what the electronic power generator must functioning to obtain maximum efficiency, the movement current of the loaded transducer and total efficiency, including the power given to the ultrasonic process.

This estimation method has the following advantages: easy to be treated mathematically; easy to implement; generally applicable to all the transducers which have the same equivalent circuit; it assures an optimal estimation with a know error; it offers a good convergence speed,

The method may be implemented digitally, on DSPs, or on PCs, for example using Simulink and dSpace, or using LabView. We present an example of a simple virtual instrument in Fig. capable to be developed to implement the block diagram from Fig. 12.

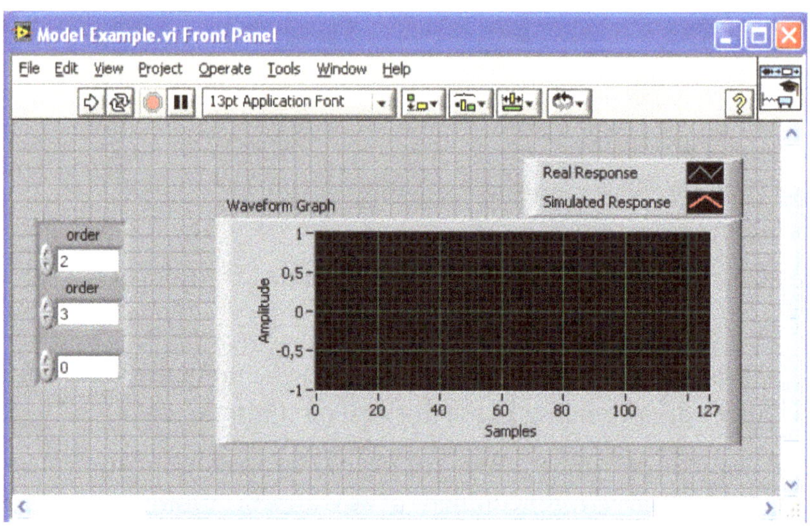


**Fig. 12. Example of a front panel for a virtual instrument**

The instantaneous variation of parameters and variables of the equivalent circuit may be presented on waveform graphs, data values may be introduced using input controllers. Behind the panel a LabView block diagram similar may be developed using existent virtual instruments from the LabView toolboxes.

## 6. FREQUENCY CONTROL

### 6.1 Control principle

To perform an effective function of an ultrasonic device for intensification of different technological processes a generator should have a system for an automatic frequency searching and tuning in terms of changes of the oscillation system resonance frequency. The present method is based on a feedback made using the estimated movement current from the transducer. The following presentation has at its basic the paper (Volosencu, 2008).

In the general case the ultrasonic piezoelectric transducers have a non-linear equivalent electric circuit from Fig. 13.

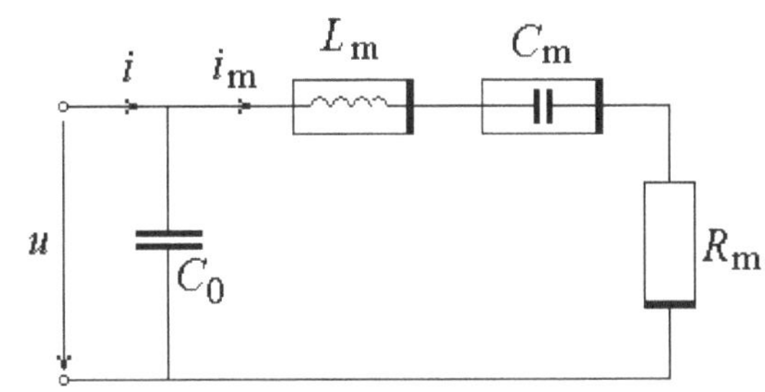


**Fig. 13. The non-linear equivalent circuit**

In this circuit there is emphasized the mechanical part, seen as a series RLC circuit, with the equivalent parameters *Rm*, *Lm* and *Cm*, which are non-linear, depending on transducer load. The current through mechanical part *im* is the movement current. The input capacitor *C0* of the transducer is consider as a constant parameter. The equations (14) are describing the time variation of the signals and the mechanical parameters, where j is the magnetic flux through the mechanical inductance *Lm* and *q* is the electric load over the mechanical capacitor *Cm*:

$$
\begin{aligned}
u_{Lm} &= \frac{d\varphi_{Lm}}{dt} = \frac{dL_m}{dt} i_m + L_m \frac{di_m}{dt} \\
i_{Cm} &= \frac{dq_{Cm}}{dt} == \frac{dC_m}{dt} u_{Cm} + C_m \frac{du_{Cm}}{dt} \\
R_m &= \frac{du_{Rm}}{di_m} \\
i &= i_0 + i_m \\
u &= u_{Lm} + u_{Cm} + u_{Rm} \\
C_0 \frac{du}{dt} &= i_{C0}
\end{aligned}
\qquad (38)
$$

The piezoelectric traducer has a frequency characteristic of its impedance Z with a series and a parallel resonance, as it is presented in Fig. 14. The movement current *im* has the frequency characteristic from Fig. 15.

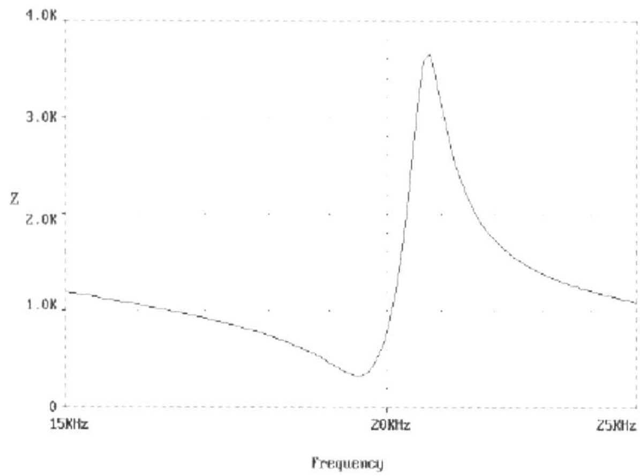


**Fig. 14. The magnitude-frequency characteristic of transducer impedance**

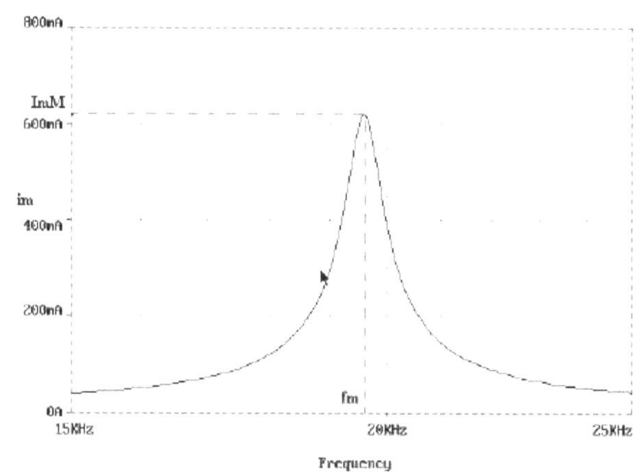


**Fig. 15. The frequency characteristic of the transducer movement current**

The maximum mechanical power developed by the transducer is obtained when it is fed at the frequency *fm*, were the maximum movement current *im*=*Im*M is obtained. Of course, the maximum of the movement current *im* is obtained when the movement current derivative *dim* is zero:

$$\mathrm{dim}(t) = \frac{di_m}{dt} = 0 \tag{39}$$

So, a frequency control system, functioning after the error of the derivative of movement current may be developed, is using a PI frequency controller, to assure a zero value for this error in the permanent regime.

## 6.2 Control system

The block diagram of the frequency control system based on the above assumption is presented in Fig. 16.

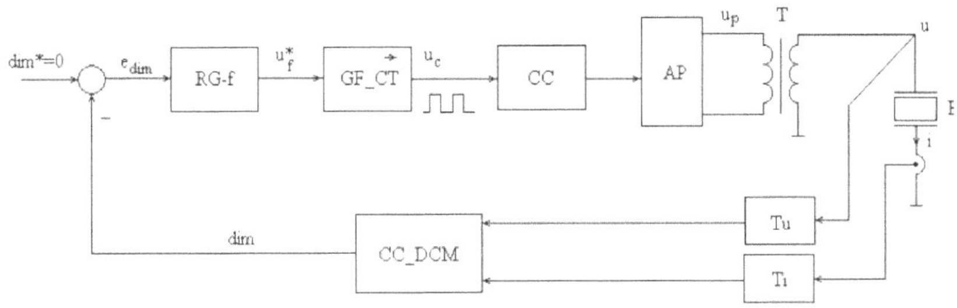


**Fig. 16. The block diagram of the frequency control system**

A power amplifier AP, working in commutation, at high frequency, feeds a piezoelectric transducer E, with a rectangular high voltage *u*, with the frequency *f*. An output transformer T assures the high voltage *u* for the ultrasonic transducer E. A command circuit CC assures the command signals for the power amplifier AP. The command signal *uc* is a rectangular signal, generated by a voltage controlled frequency generator GF_CT. The rectangular command signal *uc* has the frequency *f* and equal durations of the pulses. The frequency of the signal *uc* is controlled with the voltage *uf**. The frequency control system from Fig. 16 is based on the derivative movement current error *e*dim as the difference between the reference value *dim**=0 and the computed value of the derivative *dim*:

$$e_{\dim} = \dim^{*} - \dim \qquad (40)$$

A PI controller RG-f is used to control the frequency, with the following transfer function:

$$u_f^{*}(s) = K_R\left(1 + \frac{1}{T_R s}\right)e_{\dim}(s) \qquad (41)$$

The frequency controller is working after the error of the derivative of the movement current *e*dim. The derivative of the movement current *dim* is computed using a circuit CC_DCM, based on the following relation with Laplace transformations:

$$\dim(s) = s[i(s) - sC_0 u(s)] \qquad (42)$$

where C0 is the known constant value of the capacitor from transducer input and *u* and *i* are the measured values of the transducer voltage and current. The voltage upon the transducer *u* and the current *i* through the transducer are measured using a voltage sensor Tu and respectively a current sensor Ti.

## 6.3 Modelling and simulation

Two models for the transducer and for the block diagram from Fig. 16 were developed to test the control principle by simulation. In the first model the parameters of the mechanical part are considered with a fix static value and a dynamical variation. In the second model the electromechanical transducer is considered coupled with a mechanical concentrator and the equivalent circuits are coupled in series. Approximating the relations (41), the following relations are used to model the behaviour of the piezoelectric transducer:

$$
\begin{aligned}
u_{Lm}(s) &= sL_m(s)i_m(s) + sL_m(s)i_m(s)\\
i_{Cm}(s) &= sC_m(s)u_{Cm}(s) + sC_m(s)su_{Cm}(s)\\
u_{Rm}(s) &= \frac{1}{s}R_m[si_m(s)]\\
u(s) &= u_{Lm}(s) + u_{Cm}(s) + u_{Rm}(s)
\end{aligned}
\tag{43}
$$

The mechanic parameters from the above relations have the following variations, in the vicinity of the stationary points *Rm0*, *Lm0* and *Cm0*:

$$R_m = R_{m0} + \Delta R_m.\ L_m = L_{m0} + \Delta L_m.\ C_m = C_{m0} + \Delta C_m \tag{44}$$

The movement current *im* (*s*) is modelled, based on the above relations, with the following relation:

$$i_m(s) = \frac{1}{s}\left[\frac{1}{L_m} \times \left(u(s) - \frac{1}{s}.\frac{1}{C_m} \times i_m(s) - R_m \times i_m(s)\right)\right] \tag{45}$$

The block diagram of the movement current model is presented in Fig. 17.

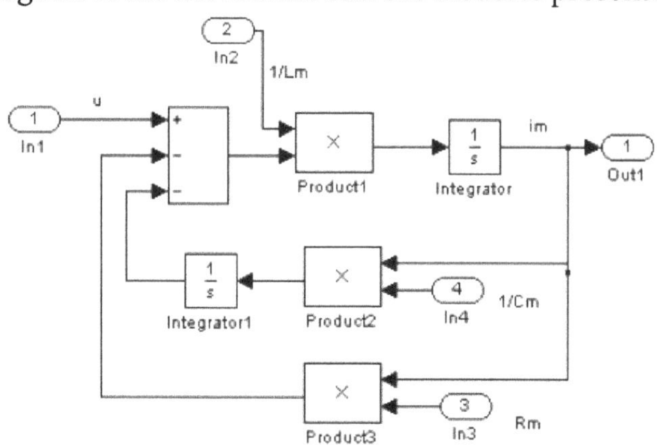


**Fig. 17. Simulation model for mechanical part**

A second model is taken in consideration. The transducer is considered coupled with the concentrator and the equivalent circuit is presented in Fig. 18.

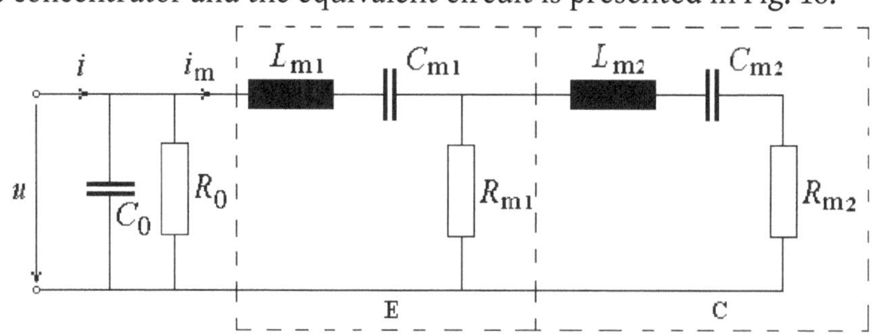


**Fig. 18. Equivalent circuit of the transducer with concentrator**

In this model there is a series RLC circuit with the parameters *Lm*1, *Cm*1 and *Rm*1 for the transducer T and a series RLC circuit with the parameters *Lm*2, *Cm*2 and *Rm*2 for the concentrator C, coupled in cascade. The parts of the control block diagram are modelled using Simulink blocks. A transient characteristic of the frequency control system is presented in Fig. 19.

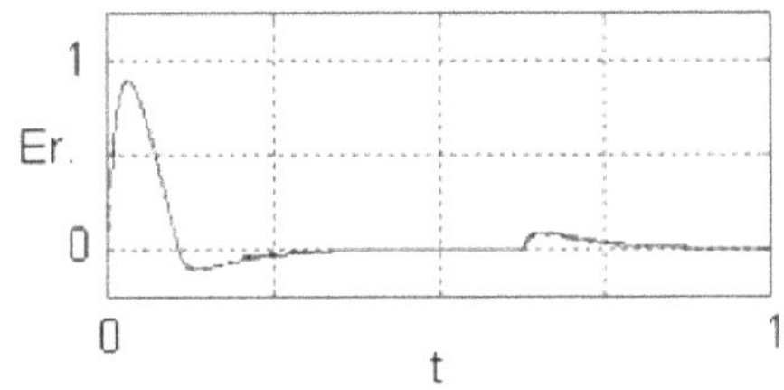


**Fig. 19. Transient characteristic for the error, obtained by simulation**

The simulation is made considering for the first model the variation with 10 % at the transducer parameters. The deviation in frequency is eliminated fast. The frequency response has a small overshoot.

## 6.4 Implementation and test results

The frequency control system is designed for universal use and utilizes analog circuits with high and low power capabilities. The power amplifier AP is constructed utilizing four power IGBT transistors arranged in a full bridge configuration, operating in high frequency commutation. Figure 20 displays the schematic diagram of the power inverter's electrical circuit.

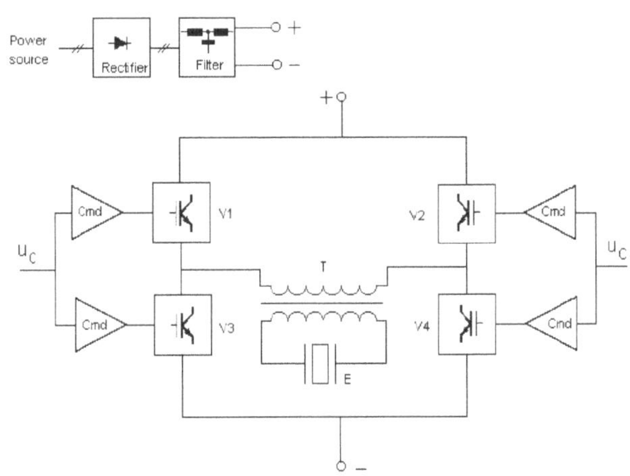


**Fig. 20. The power inverter**

The four Insulated Gate Bipolar Transistor (IGBT) transistors, labeled V1 to V4, are equipped with voltage and current protection circuits. The power transistors are controlled by four command circuits. The power inverter is supplied by the electrical system through a rectifier and a filter. The command circuits are being supplied with the command voltage uc from the VC_FG circuit. The GF_CT voltage-controlled frequency generator is constructed by utilizing a phase lock loop (PLL) circuit and a comparator. The computing circuit CC_DCM, which employs analog operational amplifiers, is utilized to implement the relations and the frequency controller RG-f. The transformer T is constructed utilizing ferrite cores and operates at a high frequency. Figure 21 displays the electronic generator.

**Fig. 21. The electronic generator**

The following are examples of transitory signal fluctuations. Figure 22 displays the pulse train of the command voltage uc.

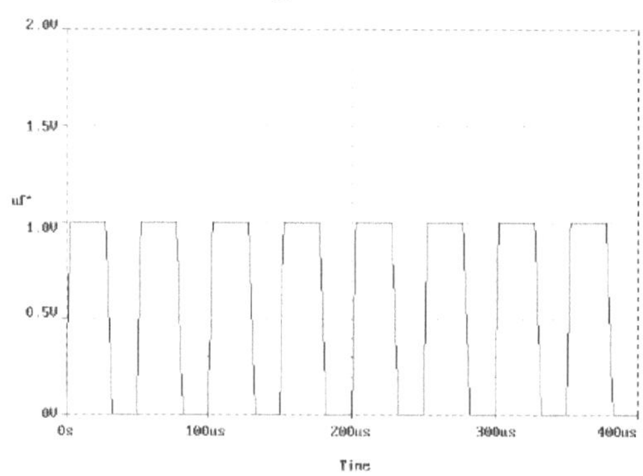


**Fig. 22. Examples of sensor impulse trains.**

The output voltage of the power amplifier is presented in Fig. 23.

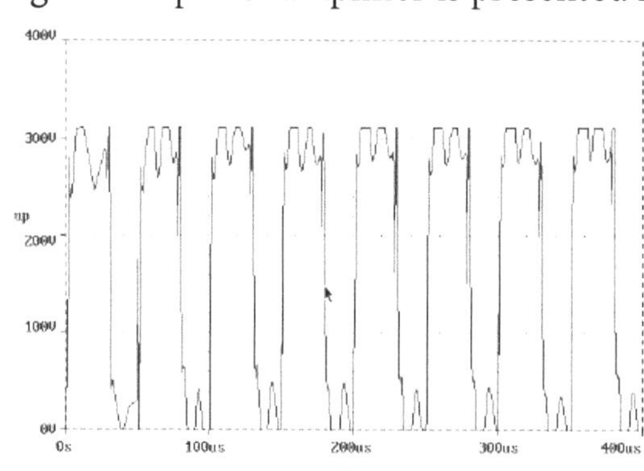


**Fig. 23. The output voltage**

The voltage $u$ over the piezoelectric transducer is presented in Fig. 24.

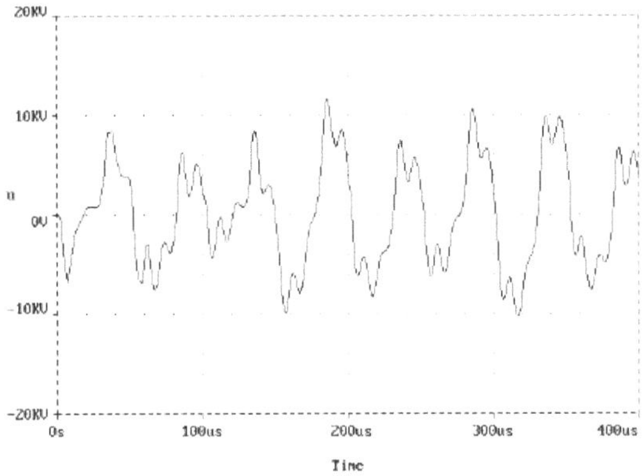


**Fig. 24. The transducer voltage**

The measured movement current *im* is presented in Fig. 25.

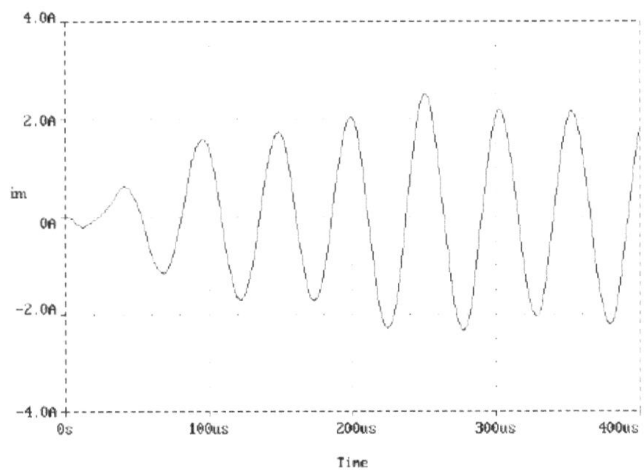


**Fig. 25. The movement current**

The control system was developed for a plastic welding machine in the range of 3000 W at 40 kHz.

## 7. CONCISE SUMMARY

This chapter presented two methods for parameter identification at the piezoelectric transducers used in high power ultrasonic applications as welding, cleaning and other. The methods are offering information about the equivalent electrical circuit: parameters and resonance frequency, efficiency and other. The first parameter estimation method is using the transformation from non-parameter model – the Nyquist hodograph - to a parameter model - the transfer function of the transducer's impedance, testing the piezoelectric transducers with a sinusoidal signal with variable frequency. The second parameter estimation method is based on an automatic measurement of piezoelectric transducer impedance using a deterministic convergence scheme with a gradient method with continuous adjustment – the method of deepest descent with a maximum slope. Some practical results are given. Some indications to implement the method using LabView are given. In the end, the paper provides a method for frequency control at ultrasonic high power piezoelectric transducers, using a feedback control systems based on the first derivative of the movement current. This method assures a higher efficiency of the energy conversion and greater frequency stability. A simulation for two kinds of transducer model is made. The control principle is implanted on a power electronic generator. Some transient characteristics are presented. The frequency control system was modelled and simulated using Matlab and Simulink. Two models for the mechanical part of the transducer are chosen. Two different regimes for the time variations of the mechanical parameters of the transducer was chosen and tested. A Simulink model and a simulation result are presented. The simulation results have proven that the control principle developed in this paper gives good quality criteria for the output frequency control. The control system is implemented using a power inverter with transistors working in commutation at high frequencies and analogue circuits for command. Transient characteristics of the control systems are presented. The frequency control system may be developed for piezoelectric transducers in a large scale of constructive types, powers and frequencies, using general usage analogue components, at a low price, with good control criteria.

## REFERENCES

1. Bose, B.K., (2000). Energy, Environment and Advances in Power Electronics, *IEEE Transactions on Power Electronics*, vol. 15, no. 4, July, 2000, pag. 688-701.
2. Chen, Y. C., Hung, L. C., Chaol, S.H. & Chien, T. H. (2008). PC-base impedance measurement system for piezoelectric transducers and its implementation on elements values extraction of lump circuit model, *WSEAS Transactions on Systems,* Volume 7, Issue 5, May 2008. Pages 521-526.
3. Eykhoff, P. (1974). *System Identification*, John Wiley and Sons, Chicester, U.K., 1974.
4. Furuichi, S. & Nose, T., (1981). Driving system for an ultrasonic piezoelectric transducer, *U.S. patent 4271371.*
5. Gallego-Juarez, J. A. (2009). Macrosonics: Sound as a Source of Energy, In: *Recent Advances in Accoustic and Music, Theory and Applications, Proceedings of the 10th WSEAS International Conference on Acoustics & Music Theory & Applications*, pag. 11-12, ISBN: 978-960-474-061-1, ISSN 1790-5095, Prague, Czech Rep., March 23-25, 2009.
6. Hulst, A.P., (1972). Macrosonics in industry 2. Ultrasonic welding of metals, *Ultrasonics*, Nov., 1972.

7. Khmelev, V.N., Barsukov, R.V., Barsukov, V., Slivin, A.N. and Tchyganok, S.N., (2001). System of phase-locked-loop frequency control of ultrasonic generators, *Proceedings of the 2nd Annual Siberian Russian Student Workshop on Electron Devices and Materials*, 2001. pag. 56-57.
8. Kirsch, M. & Berens, F., (2006). Automatic frequency control circuit, *U. S. Patent 6571088*. Marchesoni, M., (1992). High-Performance Current Control Techniques for Applications to
9. Multilevel High-Power Voltage Source Inverters, In *IEEE Trans. on Power Electronics*, Jan.
10. Mori, E., 1989. High Power Ultrasonic Wave Transmission System, In *J. Inst. Electron. Inf. Commun. Eng.*, vol. 72, no. 4, April.
11. Morris, A.S., (1986). Implementation of Mason's model on circuit analysis programs, In *IEEE Transactions on ultrasonics, ferroelectric and frequency control*, vol. UFFC-33, no. 3.
12. Neppiras, E.A., (1972). Macrosonics in industry, 1. Introduction. *Ultrasonics*, Jan., 1972.
13. Prokic, M., (2004). *Piezoelectric Converters Modelling and Characterization*, MPI Interconsulting, Le Locle, Switzerland.
14. Radmanovic, M. D. & Mancic, D. D., (2004). *Design and Modelling of Power Ultrasonic Transducers*, University of Nis, 2004, MPI Interconsulting.
15. Senchenkov, I.K., (1991). Resonance vibrations of an electromechanical rod system with automatic frequency control, *International Applied Mechanics*, Vol. 27, No. 9, Sept., Springer, N. Y.
16. Sulivan, R.A. (1983). Power supply having automatic frequency control for ultrasonic bonding, *U. S. Patent 4389601.*
17. Tertisco, M. & Stoica, P. (1980), *Identificarea si estimarea parametrilor sistemelor*, Editura Academiei Romaniei, Bucuresti, 1980.
18. Volosencu, C. (2008). Frequency Control of the Pieozoelectric Transducers, Based on the Movement Current, *Proceedings of ICINCO 2008 – Fifth International Conference on Informatics in Control, Automation and Robotics*, ISBN 978-989-8111-35-7, Funchal, Madeira, Portugal, May 11-15, 2008.

# 09 WAVE DIGITAL FILTERS DESIGN

## 1. INTRODUCTION: AN OVERVIEW

Filters are utilized in several applications within control automation and robotics, such as image processing, alarm system automation, adaptive signal processing, vibration control, mobile robot control, system identification, and speech system recognition. This chapter, titled "Design of the Wave Digital filters," presents a straightforward method for designing, analyzing, and implementing low-pass, high-pass, band-pass, and band-stop wave digital filters. These filters are derived from reference LC filters in the lattice configuration. Additionally, tables will be provided to facilitate the simple design of wave digital filters.

Wave digital filters are obtained by deriving them from LC-filters. The reference filter is composed of many elements connected in parallel and series. Due to the dependence of the load resistance R on the element or source to which the port belongs, it is not possible to directly connect the elements to a network without considering this factor. The filters' components are interconnected using serial and parallel adapters. The discrete version of these adaptors involves connecting them in one port using delay elements. Modifying the port resistance can be accomplished by use parallel and series adaptors. These adaptors include the essential components of adders, multipliers, and inverters.

## 2. SERIAL AND PARALLEL ADAPTORS.

In this chapter, we use adaptors with three ports. Blocks the serial and parallel reflection- free adapters and theirs signal-flow diagram are shown in figure 1.

The coefficient of the 3-port reflection-free serial adaptor in figure 1A) is calculated from the port resistances Ri i=1,2 by (1).

$$B = \frac{R_1}{R_1 + R_2} \tag{1}$$

The coefficient of the reflection-free parallel adaptor in figure 1B) can be calculated from the port conductance Gi i=1,2 by (2)

$$A = \frac{G_1}{G_1 + G_2} \tag{2}$$

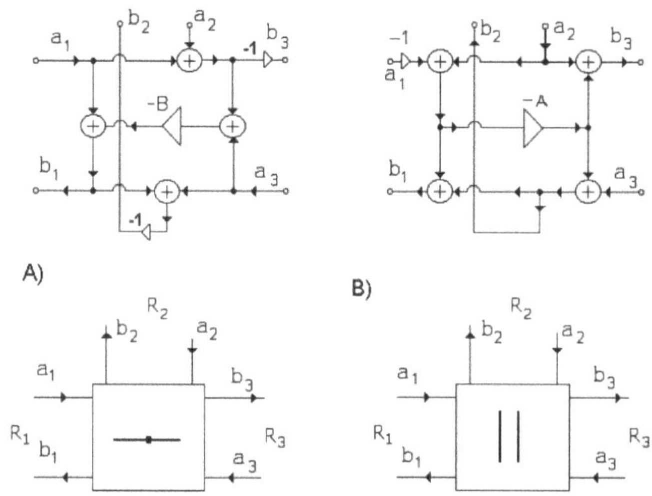


**Fig. 1. A) Three port serial adaptor whose port 3 is reflection-free and its signal flow-graph,**

B) Three port parallel adaptor whose port 3 is reflection-free and its signal flow-graph.

The coefficients of the dependent parallel adaptor in the figure 2B) can be get from port conductances Gi i=1,2,3 by (3)

$$A_1 = \frac{2G_1}{G_1 + G_2 + G_3} \qquad A_2 = \frac{2G_2}{G_1 + G_2 + G_3} \tag{3}$$

The coefficients of the dependent serial adaptor in the figure 2A) can be obtained from the port resistances Ri i=1,2,3 by (4)

$$B_1 = \frac{2R_1}{R_1 + R_2 + R_3} \qquad B_2 = \frac{2R_2}{R_1 + R_2 + R_3} \tag{4}$$

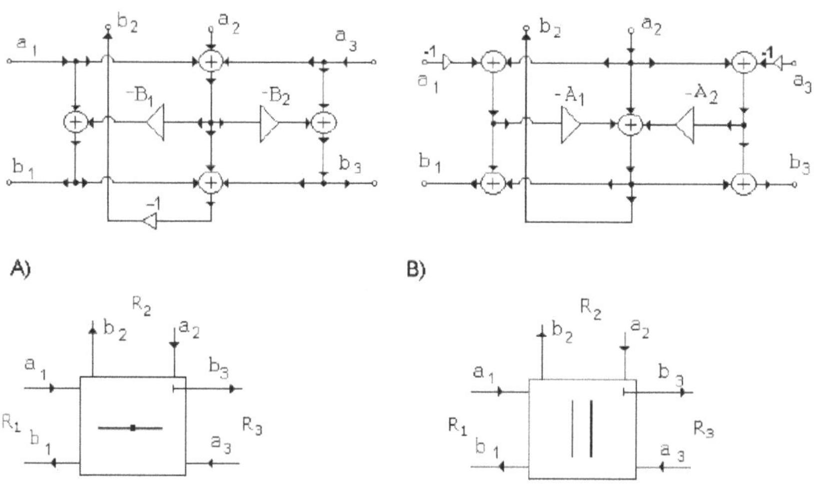


**Fig. 2. A) Three-port serial dependent adaptor and its signal-flow graph, B)Three-port parallel dependent adaptor and its signal flow graph.**

When connecting adaptors, the network must not contain any feedback loops without a delay element in order to guarantee that the structure is realizable. This means that we cannot connect the dependent adaptors from figure 2A) and 2B). The free-port dependent parallel is reflection free at port 3 if G3 = G1+G2 and three-port dependent serial adaptor is reflection free if R3=R1+R2.

## 3. EXAMPLES

By means of examples we shall demonstrate calculation of the low-pass and high-pass wave digital filters. The most important components for the realization of wave digital filters according to Fettweis procedure (Fettweis 1972) are the ladder LC filters. The tables for wave digital structures was designed for the corner frequency f1=1/(2pi)=0.159155 and sampling frequency fs=0.5

## 4. REALIZATION OF THE LOW-PAS FILTER.

In the first example we shall realize a Butterworth low-pass of the 5th order and Amax=3 dB. In the figure 3 we show the structure of a 5th order ladder LC reference Butterworth filter and its corresponding block connection in the digital form.

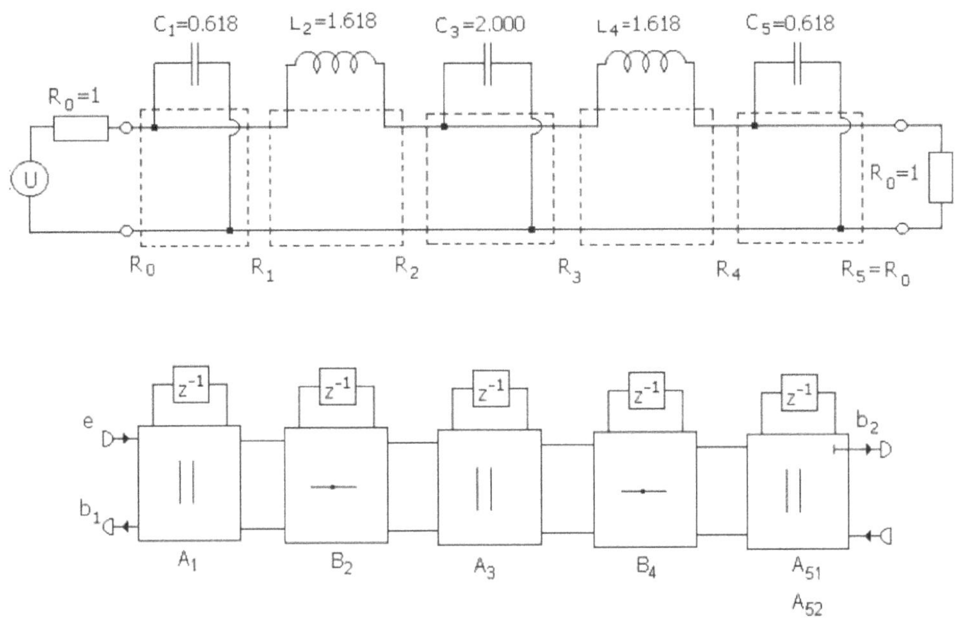


**Fig. 3. LC reference Butterworth low-pass filter and its corresponding digital block connection.**

First we must calculate from figure 3 (G0=1) wave port resistances R1,R2,R3,R4 then the coefficients of the parallel and serial adaptors A1,B2,A3,B4 and finally the coefficients of the dependent parallel adaptor A51,A52 according to (1)-(3).

$$G_1 = G_0 + C_1 = 1.618 \qquad G_3 = G_2 + C_3 = 2.447$$

$$R_1 = \frac{1}{G_1} = 0.618 \qquad R_3 = \frac{1}{G_3} = 0.408$$

$$R_2 = R_1 + L_2 = 2.236 \qquad R_4 = R_3 + L_4 = 2.026$$

$$G_2 = \frac{1}{R_2} = 0.447 \qquad G_4 = \frac{1}{R_4} = 0.493$$

$$A_1 = \frac{G_0}{G_0 + C_1} = 0.618 \qquad B_2 = \frac{R_1}{R_1 + L_2} = 0.276$$

$$A_3 = \frac{G_2}{G_2 + C_3} = 0.182 \qquad B_4 = \frac{R_3}{R_3 + L_4} = 0.201$$

$$A_{51} = \frac{2G_4}{G_4 + C_5 + G_5} = 0.467 \qquad A_{52} = \frac{2G_5}{G_4 + C_5 + G_5} = 0.947$$

The program for the analysis of the wave digital filter of the 5th order written in MATLAB follows, and the frequency response obtained is given in figure 4. The equation in the program for computing XN1-XN4, BN4-BN1,N1-N9 and YN(i) was obtained from the structure in the figure 4.

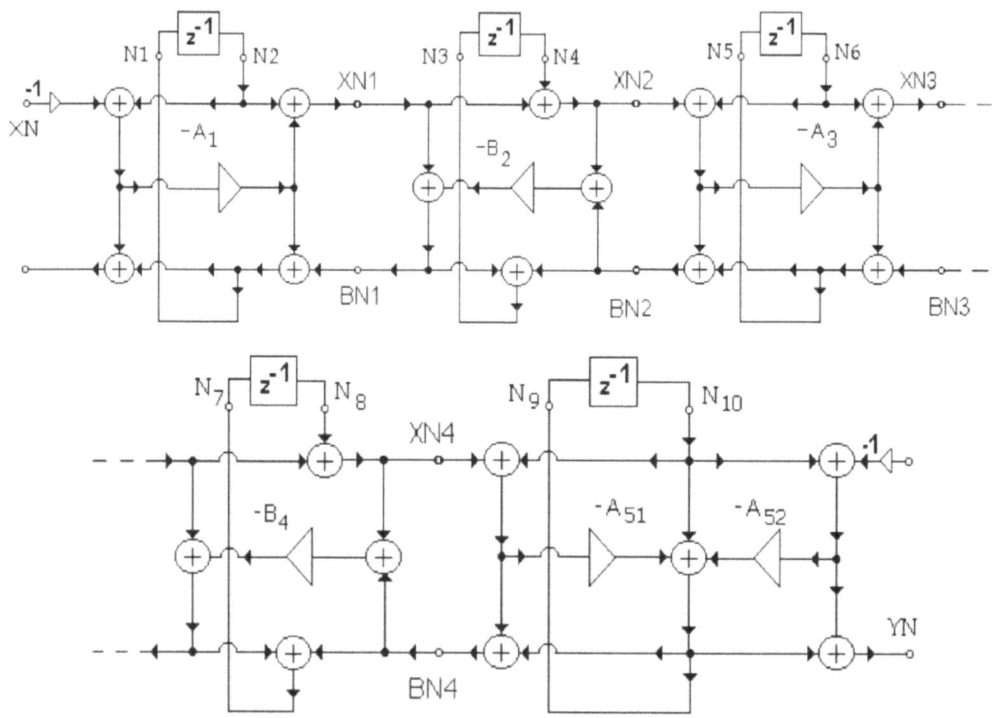


**Fig. 4. Butterworth wave digital low-pass filter of the 5th order.**

```
A1=0.618146;B2=0.276524;A3=0.182858;B4=0.201756;A51=0.467353;A52=0.947276;
N2=0; N4=0; N6=0; N8=0; N10=0; XN=1
for i=1:1:200
```

```
        XN1=A1*XN-A1*N2+N2;        XN2=XN1+N4;
         XN3=N6-A3*XN2-A3*N6;       XN4=XN3+N8;
         BN4=XN4-A51*XN4+2*N10-A51*N10-A52*N10;
         BN3=XN3-B4*XN4-B4*BN4;
         BN2=XN2-A3*XN2+BN3+N6-A3*N6;
         BN1=XN1-BN2*XN2-B2*BN2;
         N1=XN*A1-A1*N2+BN1;      N3=BN1+BN2;
         N5=BN3-A3*XN2-A3*N6;     N7=BN3+BN4;
         N9=N10-A51*XN4-A51*N10-A52*N10;
         YN(i)=2*N10-A51*XN4-A51*N10-A52*N10;
        N2=N1;N4=N3;N6=N5;N8=N7;N10=N9;XN=0;
end
                  [h,w]=freqz(YN,1,200)
                  Plot(w,20*log10(abs(h)))
```

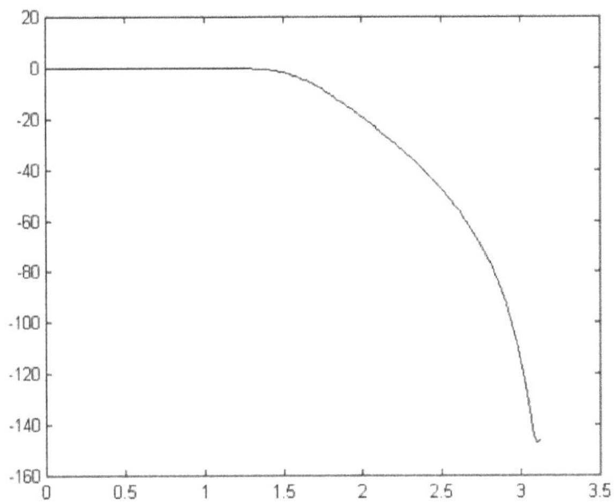


**Fig. 5. Frequency response in dB of the Butterworth low-pass filter.**

## 5. THE DESIGN OF THE LOW-PASS AND HIGH-PASS CHEBYSHEV FILTER

As a second example we shall propose low-pass and high-pass Chebychev wave digital filter for n=5, Amax=3 dB. From table 7, we get the values of the WDF A1=0.223, B2=0.226, A3=0.182, B4=0.192, A51=0.383 and A52=0.360. Using previous MATLAB program and changing the values of the coefficients we get the attenuation of Chebychev low-pass and high-pass filter in the figure 6. Additionally to get the high-pass filter from the previous program we also have to change N2= -N1; N4= -N3; N6= -N5; N8= -N7; N10= -N9.

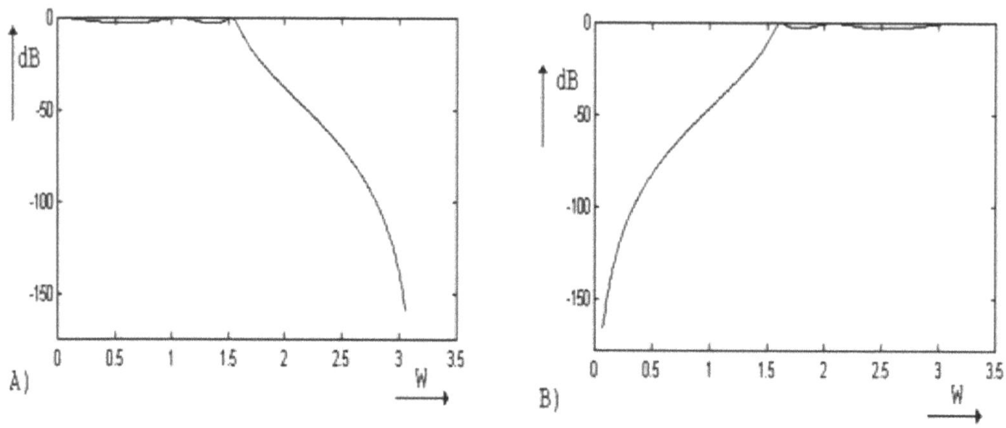


**Fig. 6. Attenuation of the Chebychev low-pass and high-pass filter**

## 6. REALIZATION OF THE LOW-PASS CAUER WDF.

In figure 7 the structure of the 5th order ladder LC reference Cauer low-pass filter is shown. The values of the LC filter was obtained from table (Rudolf Saal, p. 114, 1979) (C 0550 for Θ=30o), Amax=1.2494 dB, Amin=70.5 dB and Ωs=2.0000. (Saal 1979).

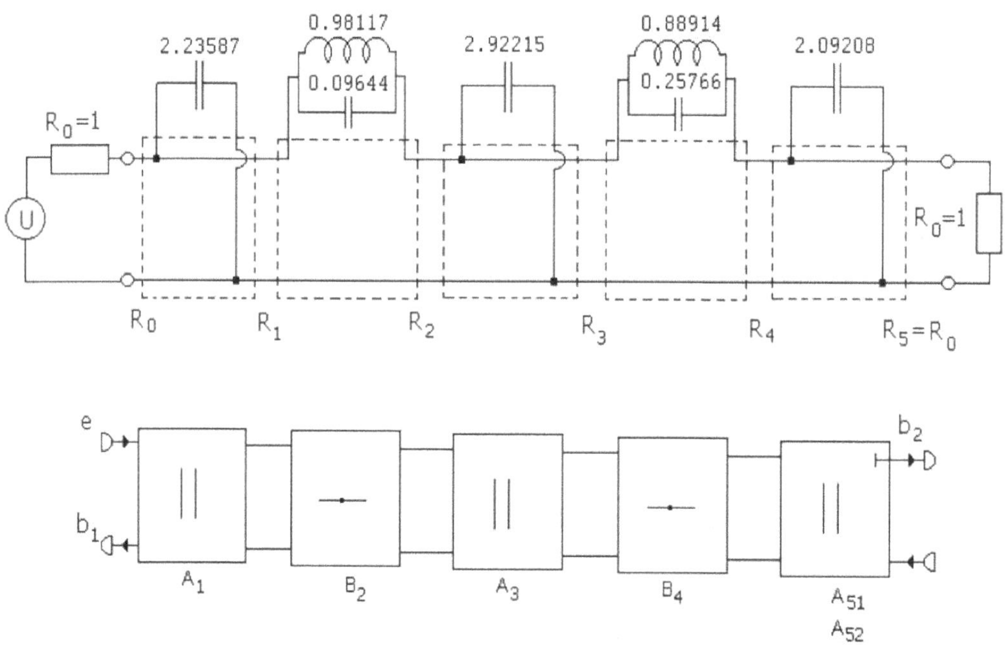


**Fig. 7. LC reference Cauer low-pass filter**

Parallel resonant LC circuit in low-pass filter figure 7 wil be realized by digital structure according to Figure 8. Structure of the wave digital Cauer low-pass filter of the 5th order is presented in the Figure 9.

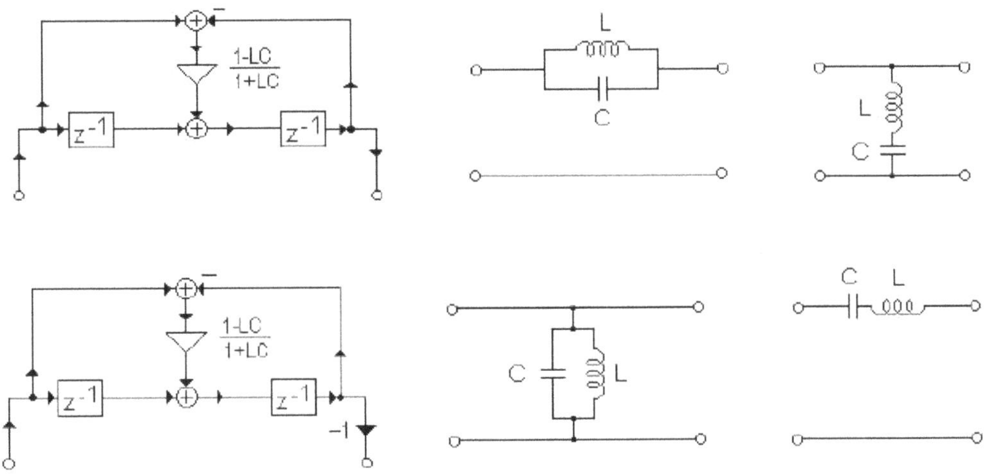


**Fig. 8. Discrete realization of parallel and serial LC circuit.**

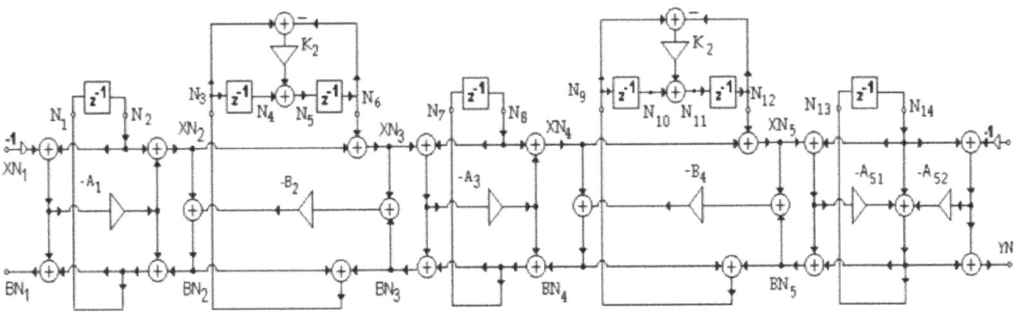


**Fig. 9. Structure of the Cauer low-pass filter n=5.**

With the assistance of the following MATLAB program we can calculate coefficients of the wave digital filter A1,B2,A3,B4,A51 and A52. The attenuation of the low-pass Cauer filter is presented in Figure 10. The program was obtained from the structure in the Figure 9. The input data of the LC filter was obtained from the catalogue of the Cauer filter (Saal 1979.) C(1)=2.235878; C(3)=2.922148; C(5)=2.092084; L(2)=0.981174; L(4)=0.889139;

```
C(1)=2.235878; C(3)=2.922148; C(5)=2.092084; L(2)=0.981174; L(4)=0.889139;
C(2)=0.096443; C(4)=0.257662; XN1=1; N2=0; N4=0; N6=0; N8=0; N10=0; N12=0;
N14=0; G(1)=1; G(2)=G(1)+C(1); R2=1/G(2);
R(3)=R(2)+1/(C(2)+1/L(2));       G(3)=1/R(3);     G(4)=G(3)+C(3);    R(4)=1/G(4);
R(5)=R(4)+1/(C(4)+1/L(4));       G(5)=1/R(5);     K(2)=(L(2)*C(2)-1)/(L(2)*C(2)+1);
K(4)=(L(4)*C(4)-1)/(L(4)*C(4)+1);  A(1)=G(1)/G(2);  B(2)=R(2)/R(3);
A(3)=G(3)/(G(3)+C(3));           B(4)=R(4)/R(5);
A(51)=2*G(5)/(G(5)+C(5)+1);      A(52)=2/(G(5)+C(5)+1);

for i=1:1:200
```

```
XN2=A(1)*XN1+N2-A(1)*N(2);        XN3=XN2+N6;
XN4=N8-A(3)*XN3-A(3)*N8;        XN5=XN4+N12;
BN5=XN5-A(51)*XN5+2*N14-A(51)*N14-A(52)*N14*;
BN4 =XN4-B(4)*BN5-B(4)*XN5;
BN3 =XN3-A(3)*XN3+BN4+N8-A(3)*N8;
BN2 =XN2-B(2)*XN3-B(2)*BN3;
N1 =A(1)*XN1-A(1)*N2+BN2;        N3=BN2*BN3;
N5=-K(2)*N3+K(2)*N6+N4:        N7=BN4-A(3)*XN3-A(3)*N8;
N9=BN4+BN5;                    N11=N10-N9*K(4)+N12*K(4);
N13=N14-A(51)*XN5-A(51)*N14-A(52)*N14;
YN(i)=2*N14-A(51)*XN5-A(52)*N14-A(51)*N14;
N2=N1;N4=N3;N6=N5;N8=N7;N10=N9;N12=N11;N14=N13;XN1=0;
end
[h,w]=freqz(YN,1,200);   plot(w,20*log10(abs(h)))
```

High-pass can be obtained by changing in the program N2=-N1; N4=-N3; N6=-N5; N8=-N7; N10=-N9; N12=-N11; N14=-N13. In Figure 10 the attenuation of low-pass and high-pass filter are presented.

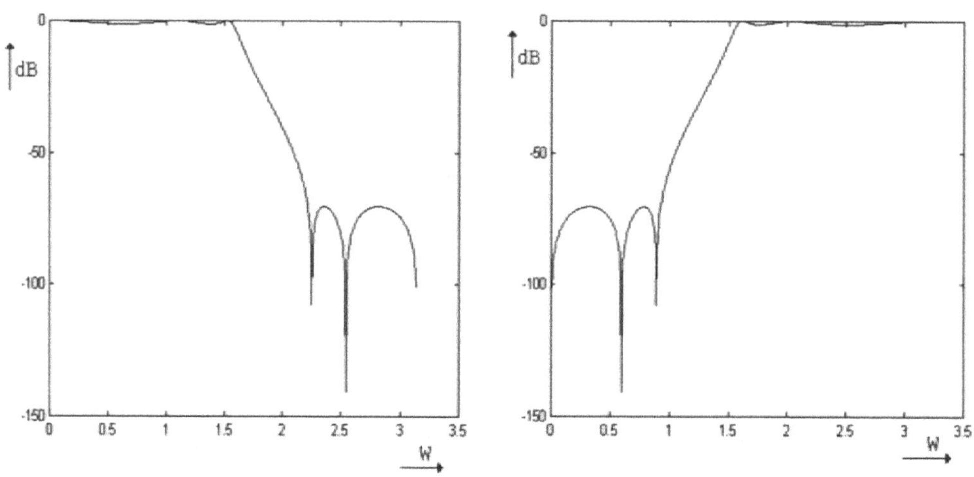


**Fig. 10. Frequency response of the Cauer low-pass and high-pass wave digital filter**

## 7. THE CAUER BAND-PASS FILTER'S DESIGN

We shall realize band-pass filter that accomplish the tolerance scheme given in Figure 11. The values of the normalized low-pass filter for C0350 θ=53 can be obtained from the table of Cauer filter (Saal 1979). The values of the normalized low-pass filter are shown in figure 12A. With the transformation of the low-pass filter to the band-pass filter can be obtained values of the band-pass filter presented in Figure 12B.

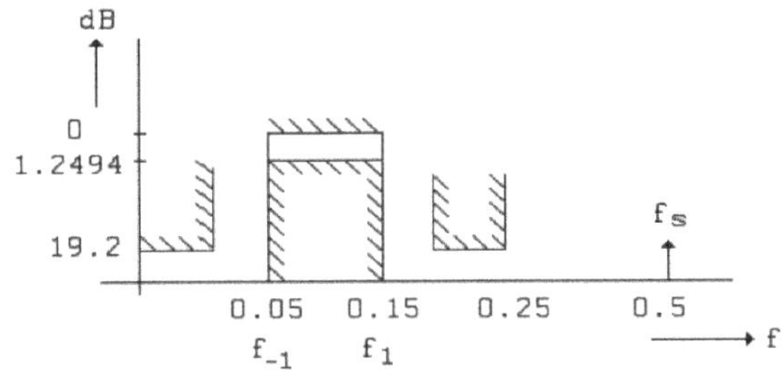


**Fig. 11.Tolerance scheme of the band-pass filter.**

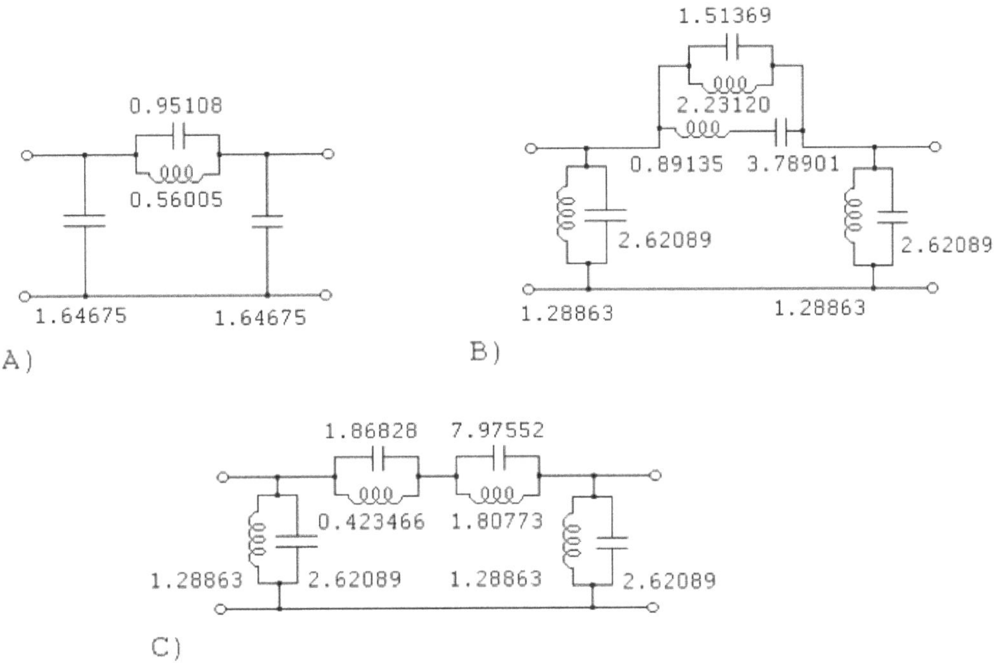


**Fig. 12. A) LC low-pass Cauer filter, B) LC band-pass Cauer filter, C) LC band-pass filter after impedance transformation.**

Serial-parallel resonant circuit in the longitudinal branch in the figure 12B) can be transformed into parallel-parallel resonant circuit, figure 13, by equations (5) up to (11).

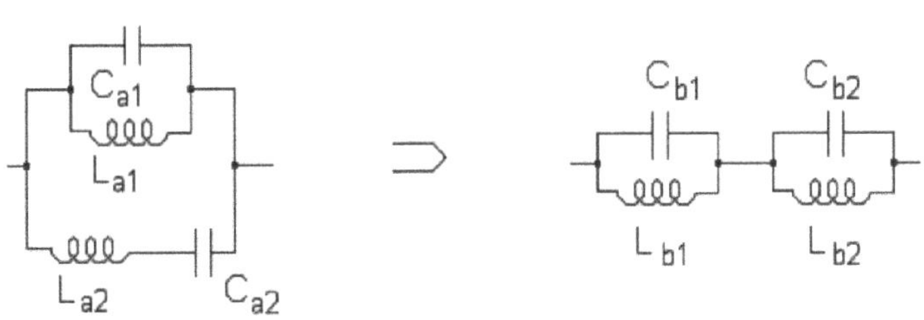


**Fig. 13. Impedance transformation of the serial-parallel resonant LC circuit to the parallel- parallel resonant circuit**

$$a = \left(1 + \frac{C_{A1}}{C_{A2}} + \frac{L_{A2}}{L_{A1}}\right)^2 - 4\frac{C_{A1}L_{A2}}{C_{A2}L_{A1}} \quad (5)$$

$$u = \sqrt{1 - \frac{4L_{A2}}{aL_{A1}}} \quad (6)$$

$$v = \sqrt{1 - \frac{4C_{A1}}{aC_{A2}}} \quad (7)$$

$$L_{B2} = L_{A1}\frac{1-u}{2} \quad (8)$$

$$L_{B1} = L_{A1}\frac{1-u}{2} \quad (9)$$

$$C_{B2} = aC_{A2}\frac{1+v}{2} \quad (10)$$

$$C_{B1} = aC_{A2}\frac{1-v}{2} \quad (11)$$

From the circuit in the figure 14 can be obtained the values Yi=Ci+1/Li (i=1, 2, 3, 4) of the parallel resonant circuit and the resistance R1, R2, R3 and R4 at the output port of each adaptors. Parallel resonant circuit can be realized by the discrete structure presented in figure 8. The coefficients Ki i=(1, 2, 3, 4) of the wave digital band-pass filter presented in figure 8 can be obtained by following equation

$$K_i = \frac{L_iC_i - 1}{L_iC_i + 1}$$

$K_1$=0.543105 $K_2$=-0.11659

$K_3$=0.870278 $K_4$=0.543105

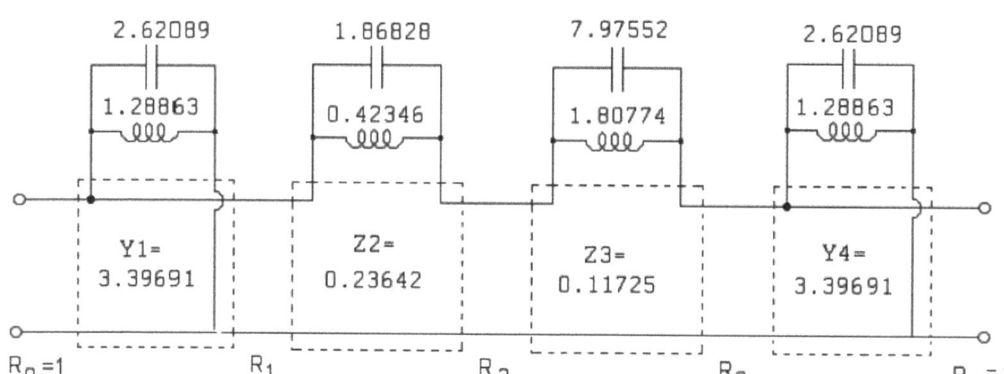


**Fig. 14. Cauer LC band-pass filter after the impedance transformation**

From the port admittances Yi and impedances Ri we get the following values,

$$G_1=G_{0}+Y_1=4.39611 \qquad R_1=0.227432$$

$$R_2=R_1+1/Y_2=0.463853$$

$$R_3=R_2+1/Y_3=0.581104 \qquad G_3=1.720862$$

and the coefficients of the parallel and serial adaptors of the wave digital band-pass in the figure 15. At the output of the band-pass filter the parallel dependent adaptor must be used in order to realize the load resistor R4=1.

$$A_1=\frac{G_0}{G_0+G_1}=\frac{1}{1+3.3969}=0.2274$$

$$B_2=\frac{R_1}{R_1+Z_2}=\frac{0.2274}{0.2274+0.2364}=0.4903$$

$$B_3=\frac{R_2}{R_2+Z_3}=\frac{0.4638}{0.4638+0.1172}=0.7982$$

$$A_{41}=\frac{2G_3}{G_3+Y_4+G_4}=\frac{2\cdot 1.7208}{1.7208+3.3969+1}=0.5625$$

$$A_{42}=\frac{2\cdot G_4}{G_3+Y_4+G_4}=\frac{2}{1.7208+3.3969+1}=0.3269$$

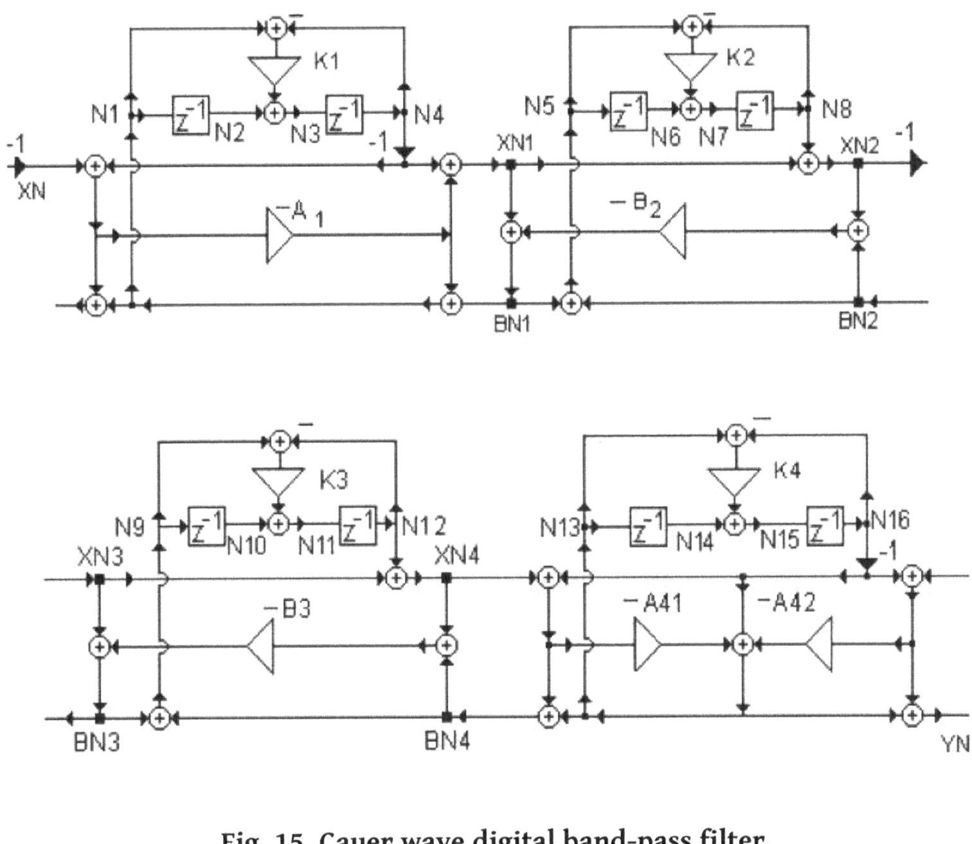


**Fig. 15. Cauer wave digital band-pass filter.**

## 8. TABLES OF THE WAVE DIGITAL FILTERS.

In the figure 16 there are frequency response and structure of the wave digital filter. The structure of Butterworth WDF is created by connection of the parallel and serial adaptors terminated at the port 3 with delay elements. At the end of the structure must be connected parallel dependent adaptor to realize for n odd load resistance RL=1. In case of n even at the end of structure must be connected serial dependent adaptor.

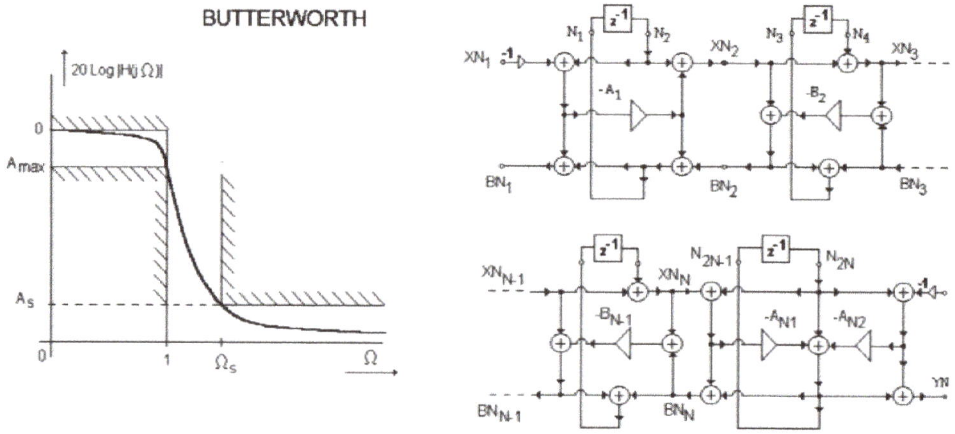


**Fig. 16. Frequency response and structure of the Butterworth wave digital filter.**

## 9. TABLES OF THE BUTTERWORTH WAVE DIGITAL FILTER

In tables 1-5 are the elements of the Butterworth wave digital filters for various attenuation Amax in the pass-band and n=3, 4, 5, 6 and 7. The tables was designed for sampling frequency fs=0.5.

**Table 1. Elements of the Butterworth WDF N=3, Amax in dB.**

| Amax | A1 | B2 | A31 | A32 |
|---|---|---|---|---|
| 0.1 | 0.6517 | 0.3788 | 0.5494 | 0.9453 |
| 0.2 | 0.6246 | 0.3419 | 0.5096 | 0.9309 |
| 0.3 | 0.6082 | 0.3206 | 0.4856 | 0.9210 |
| 0.4 | 0.5962 | 0.3056 | 0.4682 | 0.9133 |
| 0.5 | 0.5867 | 0.2941 | 0.4545 | 0.9068 |
| 0.6 | 0.5789 | 0.2846 | 0.4331 | 0.9012 |
| 0.7 | 0.5721 | 0.2767 | 0.4334 | 0.8963 |
| 0.8 | 0.5662 | 0.2698 | 0.4249 | 0.8918 |
| 0.9 | 0.5609 | 0.2637 | 0.4174 | 0.8876 |
| 1.0 | 0.5561 | 0.2583 | 0.4105 | 0.8838 |

**Table 2. Elements of the Butterworth WDF N=4, Amax in dB.**

| Amax | A1 | B2 | A3 | B41 | B42 |
|---|---|---|---|---|---|
| 0.1 | 0.6764 | 0.3694 | 0.3210 | 0.5690 | 0.9679 |
| 0.2 | 0.6568 | 0.3424 | 0.2924 | 0.5385 | 0.9599 |
| 0.3 | 0.6449 | 0.3268 | 0.2760 | 0.5200 | 0.9545 |
| 0.4 | 0.6364 | 0.3157 | 0.2645 | 0.5067 | 0.9503 |
| 0.5 | 0.6295 | 0.3071 | 0.2556 | 0.4961 | 0.9468 |
| 0.6 | 0.6239 | 0.3000 | 0.2484 | 0.4874 | 0.9437 |
| 0.7 | 0.6190 | 0.2941 | 0.2423 | 0.4798 | 0.9409 |
| 0.8 | 0.6147 | 0.2889 | 0.2369 | 0.4732 | 0.9385 |
| 0.9 | 0.6109 | 0.2843 | 0.2326 | 0.4673 | 0.9362 |
| 1.0 | 0.6074 | 0.2802 | 0.2281 | 0.4620 | 0.9431 |

**Table 3. Elements of the Butterworth WDF N=5, Amax in dB.**

| Amax | A1 | B2 | A3 | B4 | A51 | A52 |
|---|---|---|---|---|---|---|
| 0.1 | 0.7022 | 0.3784 | 0.2876 | 0.3188 | 0.6021 | 0.9815 |
| 0.2 | 0.6872 | 0.3658 | 0.2655 | 0.2951 | 0.5781 | 0.9722 |
| 0.3 | 0.6782 | 0.3531 | 0.2532 | 0.2813 | 0.5636 | 0.9742 |
| 0.4 | 0.6716 | 0.3441 | 0.2446 | 0.2717 | 0.5530 | 0.9718 |
| 0.5 | 0.6664 | 0.3371 | 0.2380 | 0.2642 | 0.5446 | 0.9699 |
| 0.6 | 0.6621 | 0.3313 | 0.2325 | 0.2580 | 0.5376 | 0.9682 |
| 0.7 | 0.6584 | 0.3264 | 0.2280 | 0.2529 | 0.5317 | 0.9667 |
| 0.8 | 0.6551 | 0.3222 | 0.2240 | 0.2484 | 0.5264 | 0.9653 |
| 0.9 | 0.6522 | 0.3184 | 0.2205 | 0.2444 | 0.5217 | 0.9641 |
| 1.0 | 0.6495 | 0.3149 | 0.2173 | 0.2408 | 0.5174 | 0.9629 |

**Table 4. Elements of the Butterworth WDF N=6, Amax in dB.**

| Amax | A1 | B2 | A3 | B4 | A5 | B61 | B62 |
|---|---|---|---|---|---|---|---|
| 0.1 | 0.7256 | 0.4125 | 0.2871 | 0.2635 | 0.3355 | 0.6363 | 0.9895 |
| 0.2 | 0.7131 | 0.3944 | 0.2696 | 0.2458 | 0.3149 | 0.6170 | 0.9870 |
| 0.3 | 0.7066 | 0.3837 | 0.2595 | 0.2356 | 0.3029 | 0.6053 | 0.9854 |
| 0.4 | 0.7014 | 0.3761 | 0.2523 | 0.2284 | 0.2948 | 0.5968 | 0.9841 |
| 0.5 | 0.6072 | 0.3702 | 0.2468 | 0.2229 | 0.2879 | 0.5901 | 0.9830 |
| 0.6 | 0.6938 | 0.3653 | 0.2422 | 0.2184 | 0.2824 | 0.5845 | 0.9821 |
| 0.7 | 0.6909 | 0.3611 | 0.2384 | 0.2145 | 0.2779 | 0.5797 | 0.9831 |
| 0.8 | 0.6883 | 0.3574 | 0.2350 | 0.2112 | 0.2739 | 0.5754 | 0.9805 |
| 0.9 | 0.6859 | 0.3542 | 0.2321 | 0.2083 | 0.2703 | 0.5716 | 0.9798 |
| 1.0 | 0.6838 | 0.3512 | 0.2294 | 0.2056 | 0.2671 | 0.5681 | 0.9792 |

**Table 5. Elements of the Butterworth WDF N=7, Amax in dB.**

| Amax | A1 | B2 | A3 | B4 | A5 | B6 | A71 | A72 |
|---|---|---|---|---|---|---|---|---|
| 0.1 | 0.7462 | 0.4392 | 0.2994 | 0.2497 | 0.2628 | 0.3597 | 0.6679 | 0.9940 |
| 0.2 | 0.7635 | 0.4236 | 0.2842 | 0.2352 | 0.2473 | 0.3415 | 0.6522 | 0.9927 |
| 0.3 | 0.7307 | 0.4144 | 0.2754 | 0.2268 | 0.2384 | 0.3308 | 0.6428 | 0.9917 |
| 0.4 | 0.7265 | 0.4078 | 0.2692 | 0.2208 | 0.2320 | 0.3232 | 0.6358 | 0.9911 |
| 0.5 | 0.7231 | 0.4027 | 0.2643 | 0.2162 | 0.2271 | 0.3173 | 0.6304 | 0.9904 |
| 0.6 | 0.7204 | 0.3984 | 0.2603 | 0.2125 | 0.2231 | 0.3125 | 0.6258 | 0.9899 |
| 0.7 | 0.7280 | 0.3948 | 0.2569 | 0.2093 | 0.2197 | 0.3083 | 0.6218 | 0.9895 |
| 0.8 | 0.7159 | 0.3916 | 0.2539 | 0.2065 | 0.2168 | 9.3047 | 0.6184 | 0.9890 |
| 0.9 | 0.7140 | 0.3887 | 0.2513 | 0.2041 | 0.2142 | 0.3015 | 0.6153 | 0.9887 |
| 1.0 | 0.7123 | 0.3862 | 0.2490 | 0.2019 | 0.2118 | 0.2986 | 0.6124 | 0.9883 |

## 10. TABLES OF THE CHEBYCHEV WAVE DIGITAL FILTER

In tables 6-9 are the elements of Chebychev wave digital filters for various attenuation Amax in the pass-band and n=3, 5, 7 and 9. The tables was designed for sampling frequency fs=0.5.

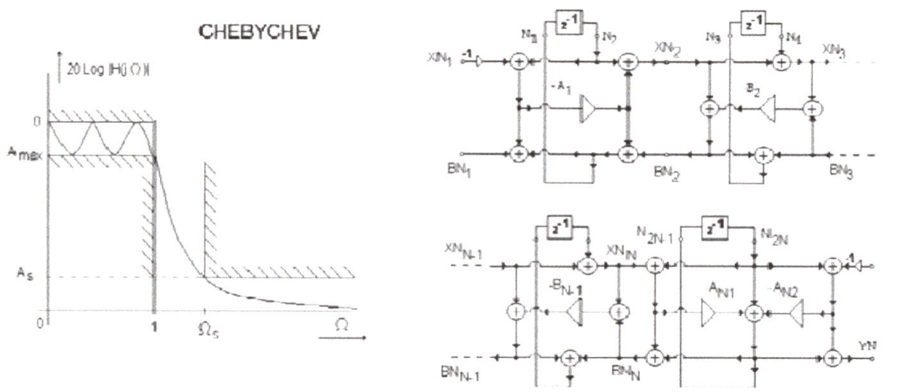


**Fig. 17. Frequency response and structure of Chebychev wave digital filter**

**Table 6. Elements of Chebychev WDF N=3, Amax in dB.**

| Amax | A1 | B2 | A31 | A32 |
|---|---|---|---|---|
| 0.10 | 0.4922 | 0.3022 | 0.4620 | 0.7574 |
| 0.25 | 0.4341 | 0.2774 | 0.4310 | 0.6812 |
| 0.50 | 0.3852 | 0.2599 | 0.4126 | 0.6114 |
| 1.00 | 0.3307 | 0.2496 | 0.3995 | 0.5293 |
| 2.00 | 0.2695 | 0.2445 | 0.3929 | 0.4331 |
| 3.00 | 0.2300 | 0.2442 | 0.3925 | 0.3696 |

**Table 7. Elements of Chebychev WDF N=5, Amax in dB.**

| Amax | A1 | B2 | A3 | B4 | A51 | A52 |
|---|---|---|---|---|---|---|
| 0.10 | 0.4658 | 0.2536 | 0.2161 | 0.2245 | 0.4179 | 0.7374 |
| 0.25 | 0.4197 | 0.2404 | 0.2059 | 0.2132 | 0.3987 | 0.6721 |
| 0.50 | 0.3696 | 0.2311 | 0.1975 | 0.2044 | 0.3869 | 0.5965 |
| 1.00 | 0.3190 | 0.2262 | 0.1911 | 0.1981 | 0.3798 | 0.5168 |
| 2.00 | 0.2610 | 0.2251 | 0.1857 | 0.1933 | 0.3797 | 0.4229 |
| 3.00 | 0.2231 | 0.2265 | 0.1828 | 0.1922 | 0.3830 | 0.3608 |

**Table 8.Elements of Chebychev WDF N=7, Amax in dB**

| *Amax* | *A1* | *B2* | *A3* | *B4* | *A5* | *B6* | *A71* | *A72* |
|---|---|---|---|---|---|---|---|---|
| *0.10* | *0.4585* | *0.2437* | *0.2022* | *0.1947* | *0.1962* | *0.2123* | *0.4049* | *0.7313* |
| *0.25* | *0.4087* | *0.2316* | *0.1944* | *0.1884* | *0.1903* | *0.2028* | *0.3875* | *0.6550* |
| *0.50.* | *0.3653* | *0.2250* | *0.1893* | *0.1861* | *0.1867* | *0.1968* | *0.3872* | *0.5925* |
| *1.00* | *0.3158* | *0.2213* | *0.1847* | *0.1834* | *0.1836* | *0.1919* | *0.3735* | *0.4206* |
| *2.00* | *0.2587* | *0.2210* | *0.1805* | *0.1814* | *0.1812* | *0.1880* | *0.3734* | *0.3921* |
| *3.00* | *0.2213* | *0.2227* | *0.1783* | *0.1806* | *0.1802* | *0.1862* | *0.3782* | *0.3589* |

**Table 9.Elements of Chebychev WDF N=9, Amax in dB**

| Amax | A1 | B2 | A3 | B4 | A5 | B6 | A7 | B8 | A91 | A92 |
|---|---|---|---|---|---|---|---|---|---|---|
| 0.10 | 0.4554 | 0.2399 | 0.1970 | 0.1886 | 0.1854 | 0.1860 | 0.1908 | 0.2081 | 0.4000 | 0.7287 |
| 0.25 | 0.4064 | 0.2287 | 0.1912 | 0.1847 | 0.1821 | 0.1825 | 0.1863 | 0.1996 | 0.3836 | 0.6569 |
| 0.50 | 0.3636 | 0.2227 | 0.1867 | 0.1823 | 0.1800 | 0.1844 | 0.1834 | 0.1943 | 0.3751 | 0.5908 |
| 1.00 | 0.3145 | 0.2193 | 0.1826 | 0.1804 | 0.1783 | 9.1787 | 0.1811 | 0.1899 | 0.3708 | 0.5123 |
| 2.00 | 0.2578 | 0.2194 | 0.1789 | 0.1790 | 0.1769 | 0.1773 | 0.1793 | 0.1864 | 0.3273 | 0.4196 |
| 3.00 | 0.2205 | 0.2213 | 0.1769 | 0.1784 | 0.1763 | 0.1604 | 0.1647 | 0.1873 | 0.3753 | 0.3583 |

## 11. TABLES OF CAUER WAVE DIGITAL FILTER

In tables 10-12 are the elements of Cauer wave digital filters for various attenuation Amax in the pass-band and N=3, 5 and 7. The tables was designed for sampling frequency fs=0.5.

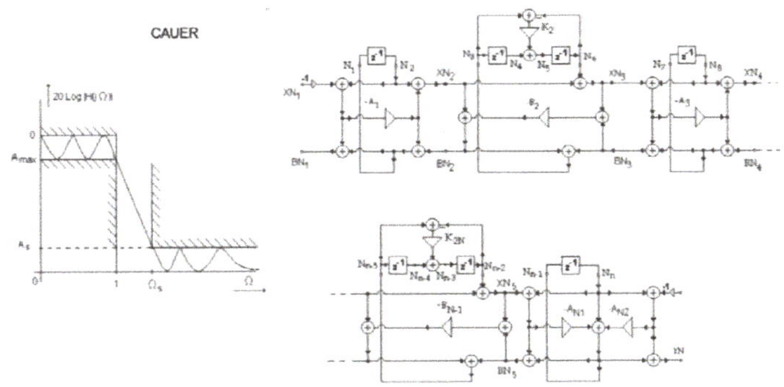


**Fig. 18. Frequency response and structure of Cauer filter**

**Table 10. Elements of Cauer WDF N=3, Ωs=4.8097, K2=-0.93686**

| C | As | A1 | B2 | A31 | A32 |
|---|---|---|---|---|---|
| 0.0004 | 24.7 | 0.7504 | 0.5766 | 0.7314 | 0.9629 |
| 0.0017 | 30.7 | 0.7209 | 0.4907 | 0.6664 | 0.9374 |
| 0.0039 | 34.3 | 0.6673 | 0.4569 | 0.6272 | 0.9160 |
| 0.0109 | 38.7 | 0.6190 | 0.4063 | 0.5778 | 0.8803 |
| 0.0279 | 42.8 | 0.5706 | 0.3612 | 0.5339 | 0.8365 |
| 0.0436 | 44.8 | 0.5461 | 0.3459 | 0.5140 | 0.8114 |
| 0.0988 | 48.3 | 0.4983 | 0.3161 | 0.4804 | 0.7573 |
| 0.1773 | 50.9 | 0.4614 | 0.2979 | 0.4590 | 0.7110 |
| 0.2803 | 53.0 | 0.4306 | 0.2855 | 0.4442 | 0.6699 |
| 1.2494 | 60.0 | 0.3147 | 0.2593 | 0.4118 | 0.4999 |

**Table 11. Elements of Cauer WDF N=5, Ωs=2.000, K2=-0.827, K4=-0.627**

| Amax | As | A1 | B2 | A3 | B4 | A51 | A52 |
|---|---|---|---|---|---|---|---|
| 0.0017 | 41.3 | 0.657 | 0.399 | 0.330 | 0.436 | 0.734 | 0.915 |
| 0.0039 | 44.8 | 0.627 | 0.373 | 0.311 | 0.404 | 0.690 | 0.893 |
| 0.0109 | 49.2 | 0.585 | 0.343 | 0.290 | 0.369 | 0.639 | 0.856 |
| 0.0279 | 53.3 | 0.543 | 0.318 | 0.272 | 0.341 | 0.596 | 0.813 |
| 0.0436 | 55.3 | 0.522 | 0.306 | 0.264 | 0.323 | 0.577 | 0.788 |
| 0.0988 | 58.8 | 0.479 | 0.288 | 0.251 | 0.310 | 0.546 | 0.736 |
| 0.1773 | 61.4 | 0.446 | 0.277 | 0.243 | 0.298 | 0.527 | 0.691 |
| 0.2803 | 63.5 | 0.418 | 0.270 | 0.237 | 0.289 | 0.514 | 0.652 |
| 1.2494 | 70.5 | 0.309 | 0.256 | 0.221 | 0.269 | 0.492 | 0.487 |

**Table 12. Elements of Cauer WDF N=7, Ωs=1.41421, K2=-0.7815, K4=-0.3491, K6=-0.4867**

| Amax | As | A1 | B2 | A3 | B4 | A5 | B6 | A71 | A72 |
|---|---|---|---|---|---|---|---|---|---|
| 0.0004 | 43.5 | 0.6901 | 0.4211 | 0.3521 | 0.4838 | 0.4446 | 0.4972 | 0.8781 | 0.9364 |
| 0.0017 | 49.5 | 0.6543 | 0.3827 | 0.3240 | 0.4464 | 0.4166 | 0.4471 | 0.7973 | 0.9071 |
| 0.0039 | 53.0 | 0.6167 | 0.3611 | 0.3093 | 0.4274 | 0.4015 | 0.4209 | 0.7502 | 0.8844 |
| 0.0109 | 57.5 | 0.5776 | 0.3353 | 0.2924 | 0.4064 | 0.3841 | 0.3912 | 0.6972 | 0.8482 |
| 0.0279 | 61.6 | 0.5375 | 0.3134 | 0.2785 | 0.3897 | 0.3698 | 0.3672 | 0.6527 | 0.8056 |
| 0.0436 | 63.5 | 0.5168 | 0.3038 | 0.2725 | 0.3826 | 0.3636 | 0.3569 | 0.6293 | 0.7766 |
| 0.0988 | 67.1 | 0.4758 | 0.2880 | 0.2624 | 0.3712 | 0.3534 | 0.3401 | 0.6015 | 0.7300 |
| 0.1773 | 69.7 | 0.4434 | 0.2784 | 0.2558 | 0.3642 | 0.3470 | 0.3295 | 0.5823 | 0.6862 |
| 0.2803 | 71.7 | 0.4158 | 0.2720 | 0.2511 | 0.3593 | 0.3425 | 0.3221 | 0.5697 | 0.6472 |
| 1.2494 | 78.7 | 0.3084 | 0.2606 | 0.2378 | 0.3475 | 0.3310 | 0.3035 | 0.5484 | 0.4845 |

## 12. REALIZATION OF WAVE DIGITAL FILTERS WITH DSP C6711 BY SIMULINK

The simulink models of the parallel and serial adaptors figure 19 was added in the simulink browse library between common used block.

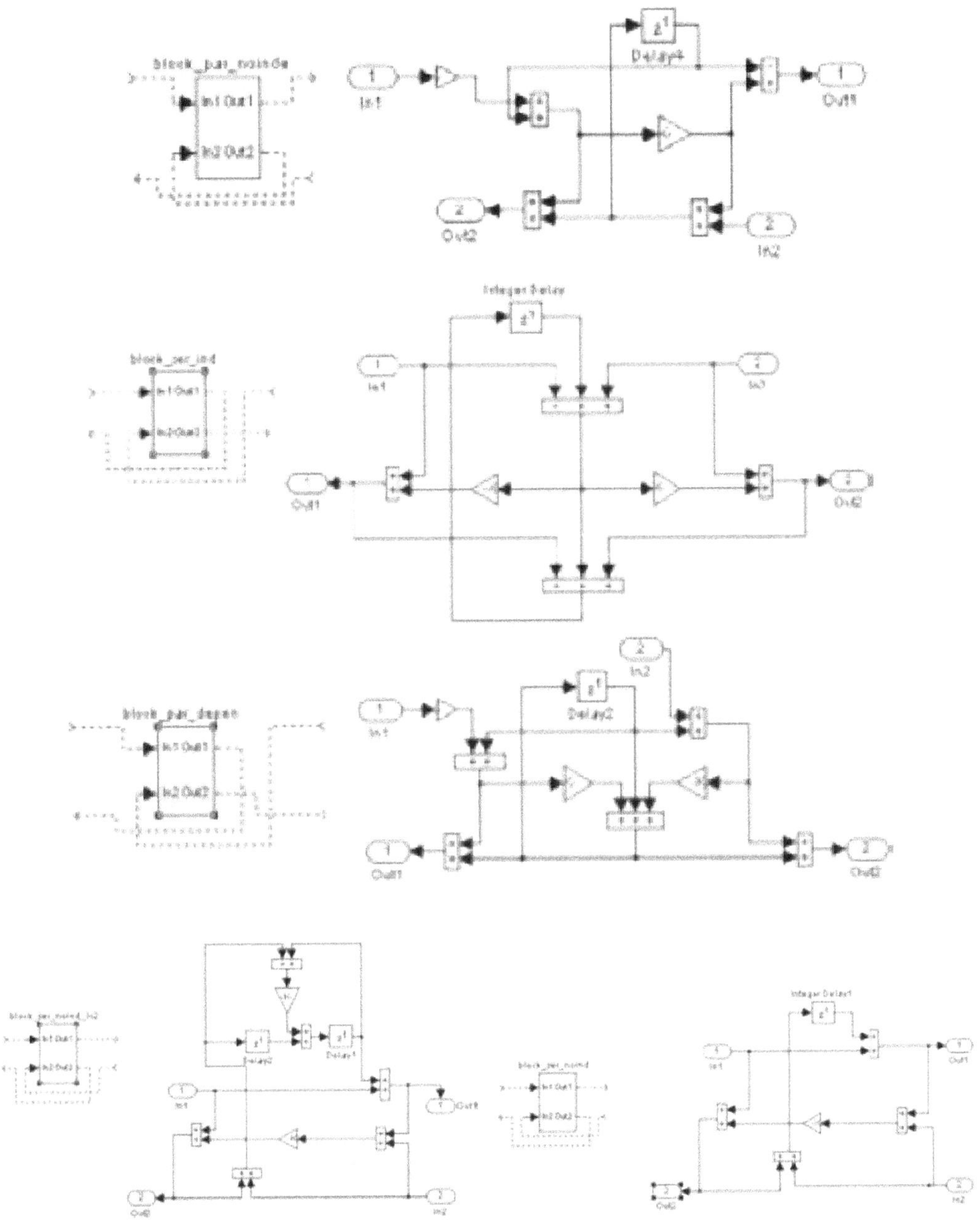

**Fig. 19. Models of the serial and parallel adaptors deposited in simulink browse library.**

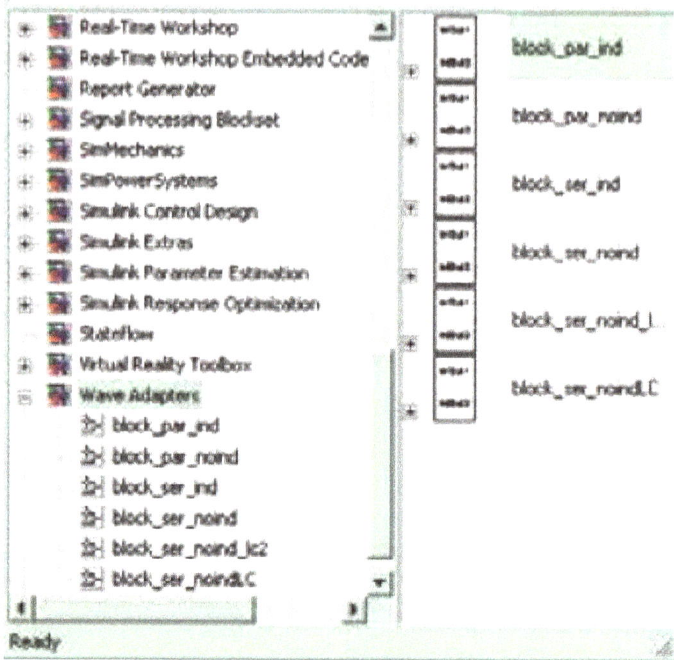


**Fig. 20. Parallel and serial adaptors deposited in simulink browse library.**

The simulink model of the fifth order filter in the figure 4 corresponds to the realization of a wave digital filter application on TMS320C6711 DSK using Embedded Target for Texas Instruments TMS320C6000 DSP Platform. The model of the filter, figure 21, was created by means of serial and parallel block that were added to the window simulink library browser. In the output and input of WDF were added ADC and DAC convertors of the TMS320C6711 that are in the simulink library browser, Embedded Target for TI C6000 DSP and C6711DSK board support.

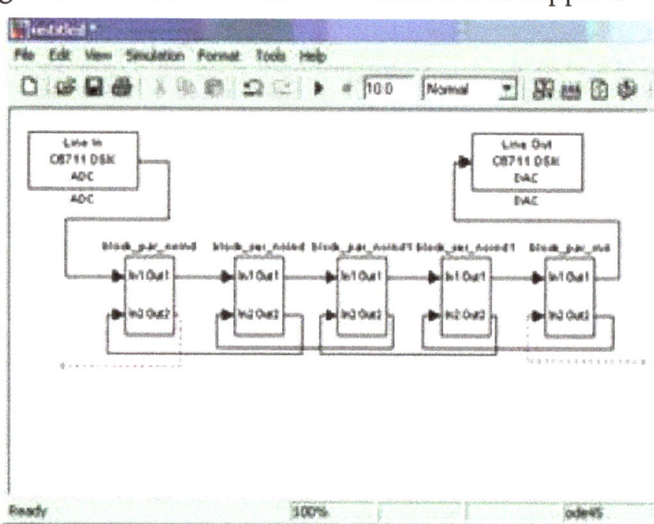


**Fig. 21. Realization of fifth order wave digital filter by TMS320C6711 simulink.**

This project created in code composer studio can be seen in figure 22 and can run on the DSPC6711.

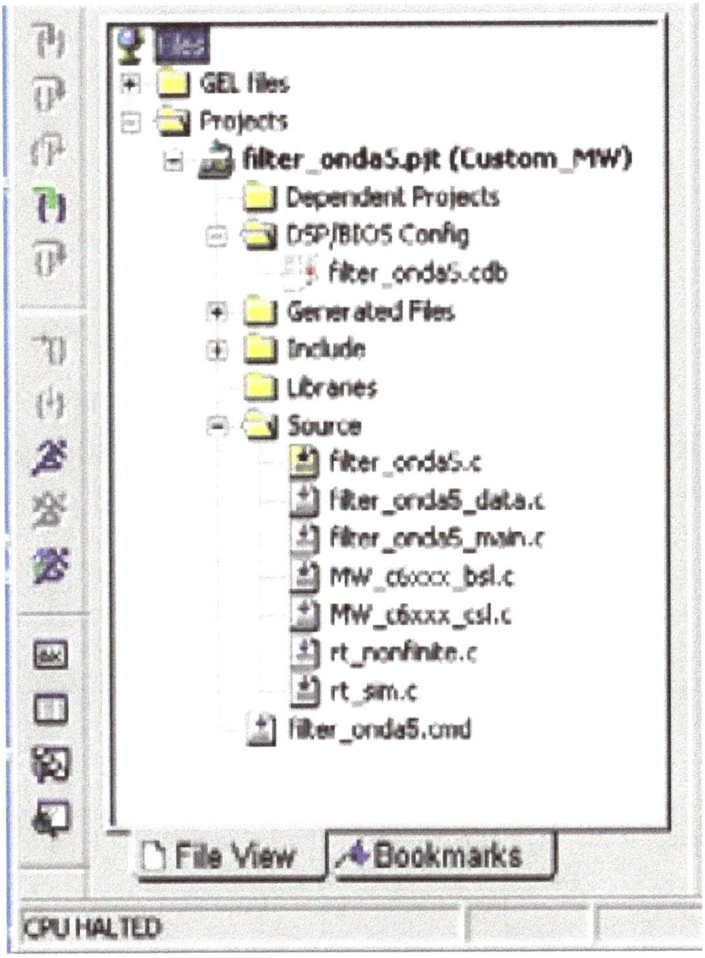


**Fig. 22. Projects of the realization of Wave digital filter by TMS320C6711**

## 13. BANK OF FILTERS WITH CHEBYCHEV AND INVERSE CHEBYCHEV WAVE DIGITAL FILTERS

In this example we shall design bank of filters with wave digital filters. We use Chebychev filter for order N=3 and Amax=0.1 dB. Transfer function of the Chebychev low-pass we obtain by matlab using the commands [a,b]=cheby1(3,0.1,0.5,'s')

$$H_0(s)=\frac{1}{2.0353s^3+2.0116s^2+2.5206s+1.0000}=\frac{1}{h_0(s)}$$

$$Fl(s)=\frac{1}{2.0353s^3+1.5265s}=\frac{1}{f(s)}$$

From these equations we can obtain input impedance Ze(s) of the circuit and the structure in figure 23.

$$Z_e(s) = \frac{h0(s)+f(s)}{h0(s)-f(s)} = \frac{4.0707s^3 + 2.0116s^2 + 4.0472\,s + 1}{2.0116s^3 + 0.9941s^2 + 1}$$

$$Z_e(s) = \cfrac{1}{2.0235s + \cfrac{1}{0.9941s + \cfrac{1}{2.0235s + 1}}}$$

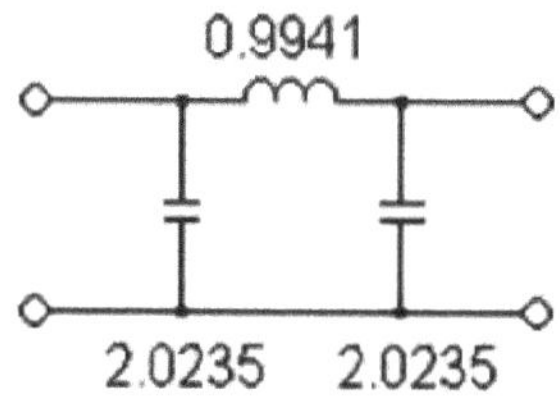


**Fig. 23. LC prototype of Chebychef low-pass filter.**

From the LC prototype of the Chebychev filter we obtain discrete realization of the wave digital filter that is demonstrate in figure 24, where

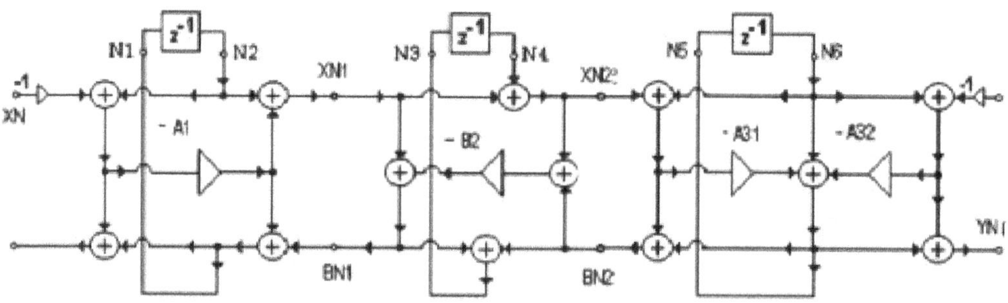


**Fig. 24. Low-pass wave digital filter with the values of the elements A1=0.3307, B2=0.2496, A31=0.3995, A32=0.5293.**

In the parte synthesis of the bank of filters we choose low-pass Chebychev inverse filter that is inverse to the low-pass in the part of analysis bank.

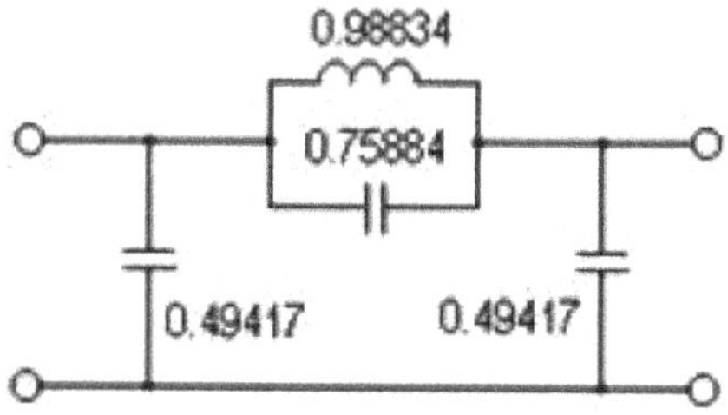


**Fig. 25. LC Chebychev inverse filter**

From the Chebychev inverse filter figure 25, the adaptor are coefficients of the parallel and Serial adaptor are

$A_1$=0.6693 $B_2$=0.2496 $A_{31}$=0.7032 $A_{32}$=0.5293 K=0.1428

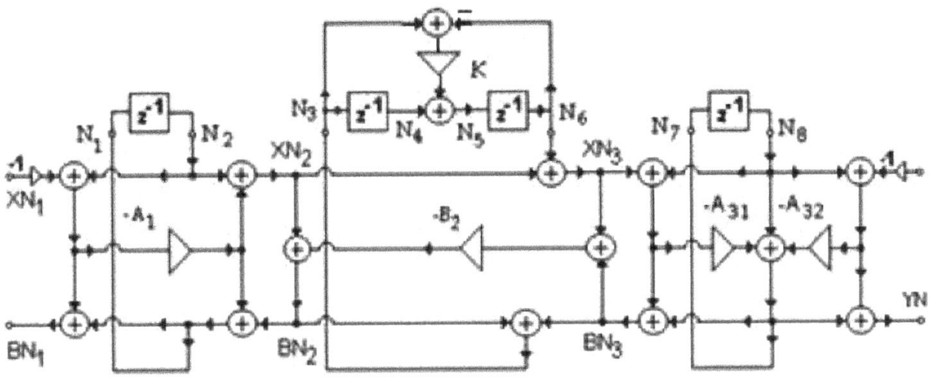


**Fig. 26. Low-pass wave digital filter Chebychev inverse with coefficients A1=0.6693 B2=0.2496 A31=0.7032 A32=0.5293 K=0.1428**

Normalized attenuation of the low-pass and high-pass filter is demonstrated in the figure 27.

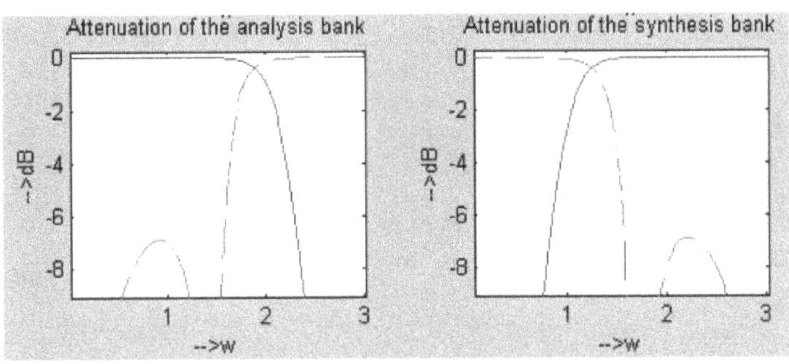


**Fig. 27. Normalized attenuation of the filter bank.**

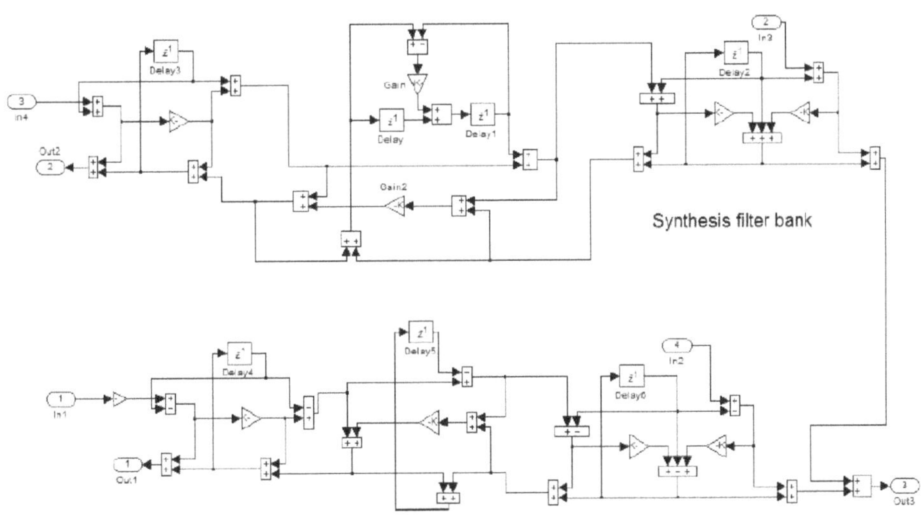


**Fig. 28. Analysis wave digital filter bank**

In the figure 28 is presented analysis part of the filter bank. This part is designed by Chebychev low-pass filter and inverse Chebychev high-pass filter. The coefficients of the chebychev low-pass filter are A1=0.3307, B2=0.2496, A31=0.3995 and A32=0.5293. The coefficients of inverse Chebychev high pass filter are A1=0.6692, B2=0.2496, A31=0.7032, A32=0.5293 and K=0.1428. In the adders connected in the output of the delay elements must be changed the signature because must be fulfill the condition H1(z)=G0(-z), where G0(z) is the transfer function of the inverse Chebychev low-pass transfer function from the Synthesis filter bank from figure 29.

In the figure 29 is presented synthesis part of the filter bank. This part is designed by Chebychev inverse low-pass filter and Chebychev high-pass filter. The coefficients of the chebychev low-pass filter are A1=0.3307, B2=0.2496, A31=0.3995 and A32=0.5293. The coefficients of inverse Chebychev high pass filter are A1=0.6692, B2=0.2496, A31=0.7032, A32=0.5293 and K=0.1428. In the adders, connected in the output of the delay elements of the Chebychev high-pass wave digital filter, must be changed the signature because must be fulfill the condition G1(z)= -H0(-z), where H0(z) is the transfer function of the Chebychev low-pass transfer function from the Analysis filter bank, figure 28.

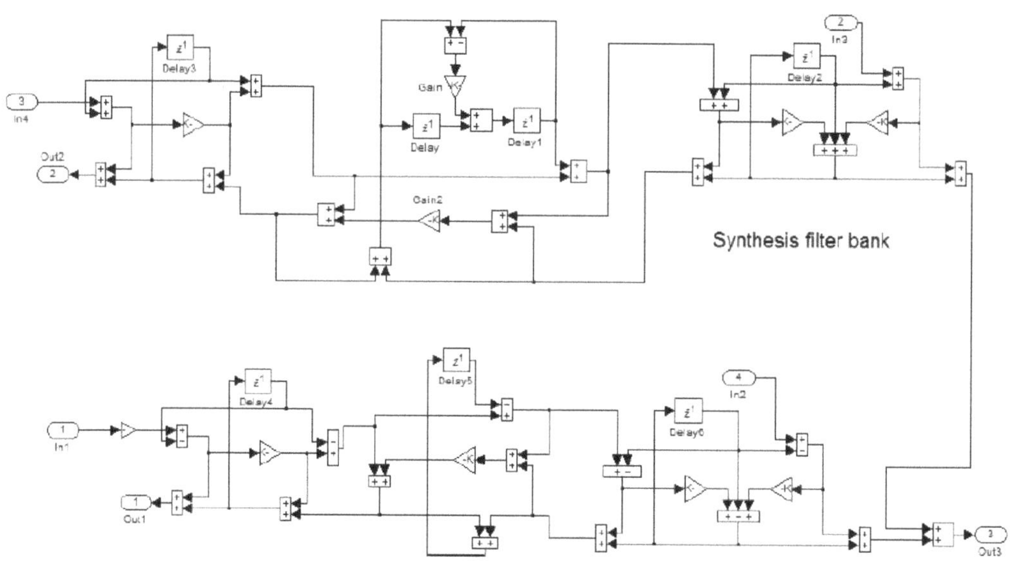


**Fig. 29. Synthesis wave digital filter bank**

In figure 30 is presented the model of the filter bank composed from analysis and synthesis parts and from down-sampler and sub-sampler. The input is connected to a speech signal saved in Workspace1. The output of the filter-bank was connected with oscilloscope. Figure 31 shows speech signal in the input and the output of the filter bank and spectrograms of a speech signal in the output and input of the filter bank. From the responses in the figure 31 we can deduce that filter bank composed from Chebychev and Inverse Chebychev filter has the good properties.

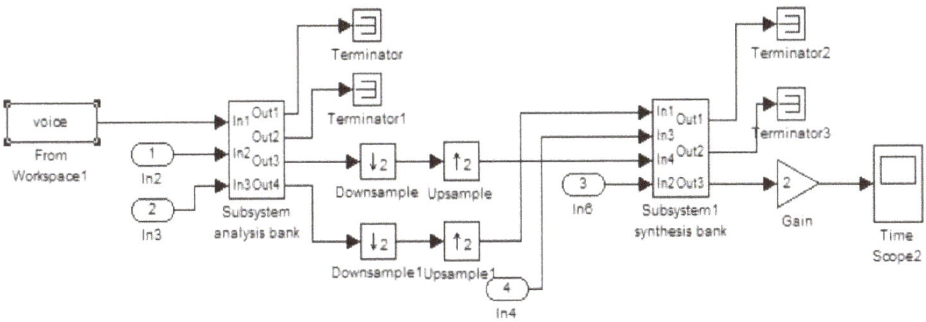


**Fig. 30. Simulink model of the wave digital filter bank**

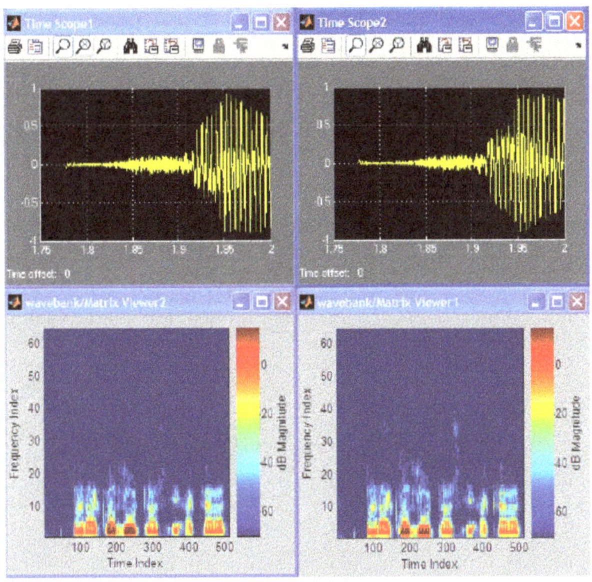


**Fig. 31. Diagrams of the voice in the input and output of the wave digital filter bank.**

## 14. CONCISE SUMMARY

While the wave digital filter's structure is more intricate compared to other structures, the method for implementing it on the DSP is very straightforward. It is also quite effortless to suggest a general algorithm for any desired order of the wave digital filter. These structures are less susceptible to quantization error compared to other types of filters. By making minor adjustments to the existing programs, it is possible to generate additional tables for the WDF. The components of the programs can be employed to create wave digital filters in digital signal processors (DSP). In this chapter, a filter-bank was built using Chebyshev and inverse Chebyshev filters. The Signal Processing toolbox in MATLAB was used to simulate this filter bank, which includes the voice signal stored in Workspace1. Figure 31 depicts the findings received from analyzing a speech signal using the Signal Processing Toolbox in MATLAB.

## REFERENCES


1. Fettweis.: *Digital filter structures related to classical filter networks.* Arch. Elektron Uebertragunstech., 1972,25, pp.78-89.
2. A.Sedlmayer and A. Fettweis.: *Digital filters with true ladder configurations* Int. J. Circuit Theory and Appl., 1973, 1, pp. 1-5.
3. F. García Ugalde, B Psenicka and M.A. Rodríguez.: *Implementation of Wave Digital Filters on a DSP Using Simulink.* IASTED, Circuits, Siognal and Systems, November 20-22, 2006, San Francisco, CA, USA
4. B. Psenicka, F. García-Ugalde and A. Romero Mier y Teran.: *Synthesis of the low-pass and high- pass Wave digital filters.ICINCO 2008. International conference on informatics in control and Robotics.* May 11-15, 2008, Madeira-Portugal.
5. Rudolf Saal.: Handbuch zum Filterentwurf AEG Telefunken, 1979.
6. Sanjit K. Mitra .: Digital Signal Processing. A computer-Based Approach. McGraw-Hill Companies, INC. New York, 1998.

# 10 NEURAL PDF CONTROL STRATEGY FOR A SIMULATOR OF A HYDROELECTRIC STATION

## 1. INTRODUCTION: AN OVERVIEW

Dinorwig is Europe's largest pumped storage hydroelectric system. Dinorwig is a power station situated in an environmentally sensitive region in North Wales. It is run by the First Hydro Company and has 16km of tunnels that have been erected deep under Elidir Mountain. The station is equipped with six turbines, each rated at 300 MW. These turbines drive synchronous generators that supply power to the British national grid. Dinorwig power plant offers swift and efficient frequency regulation during periods of high electricity demand. Its pump/turbines can achieve maximum power generation in under 16 seconds. The hydroelectric station is equipped with a solitary tunnel that extracts water from an upper reservoir, known as Lake Marchlyn, and directs it into a manifold. The manifold then divides the primary flow into six separate penstocks. Every penstock supplies a turbine to produce electricity by utilizing a guide vane to control the flow. Each unit has its own separate feedback loop to adjust the electrical supply. The input signal for the power loop is the divergence of the grid frequency from its set point of 50 Hz, which establishes an outer control loop for frequency. Dinorwig's reversible Francis turbines utilize off-peak electricity, typically during nighttime, to transfer water from the lower reservoir, Lake Peris, back to Lake Marchlyn. Mansoor et al. have developed a complex simulation model of this plant that incorporates multiple variables and nonlinearity. This model has enhanced our comprehension of the plant's properties (Mansoor, et al., 1999; Mansoor, et al., 2000). The primary characteristics of the system are non-minimum-phase dynamics, inadequately damped poles (related to water-hammer in the supply tunnel and electrical synchronization), and a non-linear correlation between flow and power. It is well-established (Kundur, 1994; Working group on prime mover energy supply, 1992) that the turbines are hydraulically connected due to their shared energy supply, resulting in a considerable coupling. Therefore, the plant is very suitable for implementing auto-tuning control.

The chapter commences with a concise examination of the non-linear mathematical model of the power plant. First, the theory of neural networks is examined. This is followed by an explanation of how neural pseudo-derivative control (PDF) is applied to the Dinorwig model, as described by Kang et al. in 1991. Subsequently, a concise overview of some principles of anti-wind-up PID is provided, followed by an examination of the software platform that is built upon Simulink© models. Ultimately, the findings demonstrate the enhanced responsiveness achieved by the implementation of neural PDF.

## 2. HYDROELECTRIC PLANT MODEL

The hydroelectric plant model has three distinct subsystems: the guide vane, nonlinear hydraulics, and the turbine/generator (refer to figure 1). Mansoor et al. (2000) constructed a complex mathematical model that incorporates a rate restriction and saturation in the dynamics of the guide vane. This model is visually represented in figure 2.

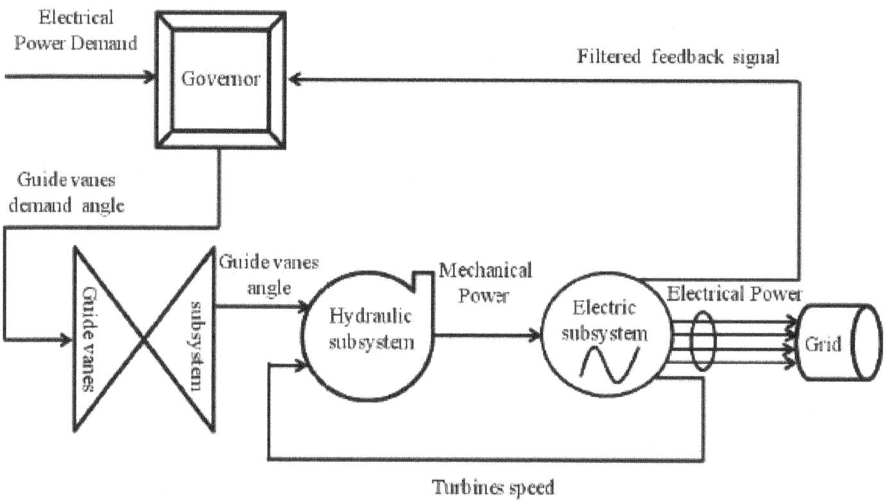


**Fig. 1. MIMO model of the hydroelectric plant with two penstocks.**

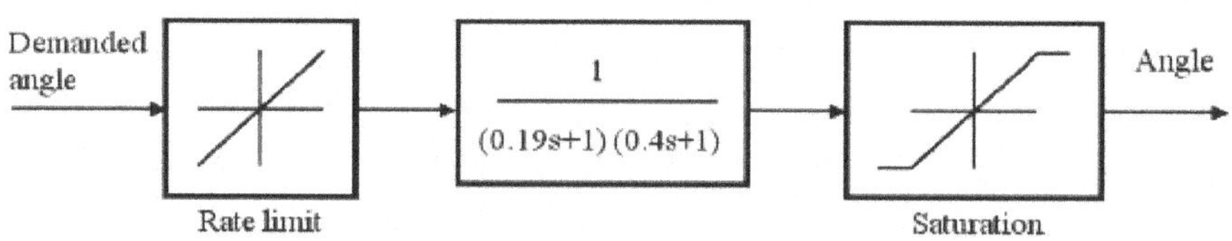


**Fig. 2. Guide vane subsystem.**

In this study a nonlinear model that takes into account the effects of the water column, including water compressibility and pipe wall elasticity, was employed (Working group on prime mover energy supply, 1992). Figure 3 shows the nonlinear elastic model of a single penstock. The coupling effect is represented with the inclusion of the main tunnel, which is modelled in the same form as a penstock. *At* is the turbine gain; its value depends directly on the turbine MW rating and inversely on the Generator MVA rating. *fp* is the head loss coefficient for the penstock. *Z0* is the surge impedance of the conduit. *Te* is the wave travel time, defined as the time taken for the pressure wave to travel the length of the penstock (*l*) to the open surface. *v* is the velocity of sound in water (Kundur, 1994).

$$T_e = \frac{l}{v} \quad (1)$$

$$Z_0 = \frac{T_W}{T_e} \quad (2)$$

*Tw* is the *water starting time* of the main tunnel and the penstocks. Kundur (1994) defines the *water starting time* as the time required for a head to accelerate the water in the penstock from standstill to a specific velocity. Its value depends directly on the constructional dimensions of the main tunnel and penstocks. In this nonlinear model, figure 3, G is the per unit gate opening and P*mech* is the mechanical power produced by a single turbine. The value of Te depends directly on the length of the penstock and inversely on the wave velocity (equation 1). Zo depends directly on the flow rate and inversely on the head of water and on the acceleration due to gravity (equation 2). The value of At depends directly on the turbine MW rating and inversely on the Generator MVA rating (Mansoor, 2000). The models are expressed in the per-unit system, normalized to 300 MW and 50 Hz.

The electrical subsystem is based on the 'swing' equations (Kundur, 1994) and includes the effect of synchronizing torque. For noise reduction a first order filter is included in the feedback loop (figure 4).

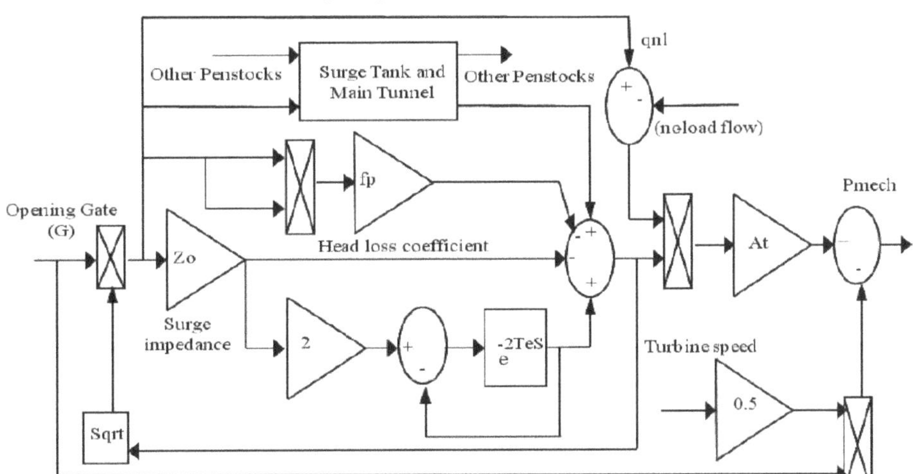


**Fig. 3. Hydraulic subsystem**

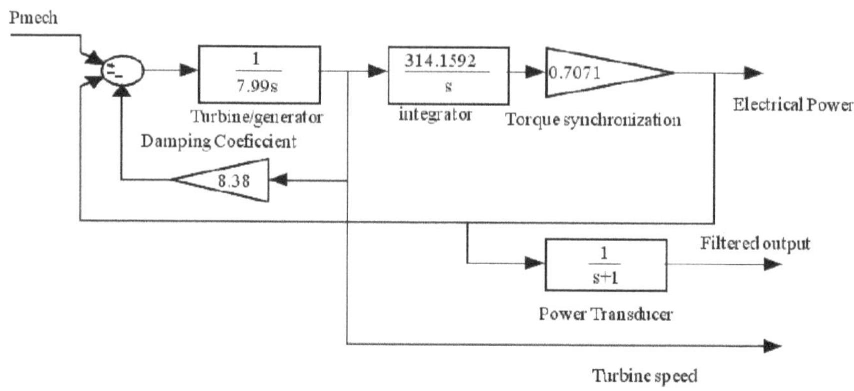


**Fig. 4. Electrical subsystem.**

## 3. ARTIFICIAL NEURAL NETWORKS USED FOR ADAPTIVE CONTROL.

### 3.1 Basic theory

Artificial Neural Networks (ANN) has been a popular research theme within the scientific community since the early 1980's. The increasing range of application of ANNs in computational modelling and evolutionary structures has led to the development of diverse types of networks, which are loosely based on the structures of biological nervous systems. Munakata (2008) reports five basic types of ANN that are focused on the solution of control design problems. These are listed below:

- **Supervised control:** This type of network tries to imitate a human or a computer program which already knows how to perform a task.
- **Direct Inverse Control (DIC):** The ANN consists of mapping the multivariable output values of the system under control versus the values of each independent variable. The desired behaviour is supplied by a human trainer or an 'expert' computer program in association with a database.
- **Neural Adaptive Control:** Here it is necessary to achieve a pre-specified target because the ANN is substituted for the (linear) mappings used in conventional adaptive control. The two schemes most often used are the Self-Tuning Regulator (STR) and Model- Reference Adaptive Control (MRAC).
- **Back-propagation Through Time (BTT):** In this case, the user specifies a utility function or performance measure to be maximized and a model of the external environment. Back-Propagation is used to calculate the derivative of the utility function *summed across all future times* with respect to *current* actions. These derivatives are then used to adapt the ANN which outputs the actions, or to adapt a schedule of actions. This type of applications can be found in pattern recognition and intelligent control of manufacturing processes (Miller, et al., 1991).

- **Adaptive Critic Methods:** This scheme is similar to BTT but the ANN, now called a critic network, evaluates the progress that the system is making.

The five ANN strategies mentioned above have gained acceptance and are extensively used in the solution of actual complex, non-linear control problems. One engineering discipline that has been enriched by the properties of the ANN is adaptive control theory. This particular theory offers the possibility of adjusting the parameters of the regulator in order to reduce the difference between the set-point and the output of the process. Narendra & Mukhopadhyay (1996) address the application in adaptive control of Back Propagation (BP), the most common type of ANN because it computes rapidly and is easy to implement in an algorithm. A typical back-propagation neural network is formed as shown in figure 5. Back-Propagation is known as a *training method* for the neural network (ANN) and many authors have reported successful application of both the basic method and its more advanced variants to practical cases (Werbos, 1974; Rumelhart, et al., 1986; Albeto Aguado Behar, 2000).

The BP training method has the capability to develop as precise an approximation as required for any non-linear function y=f(x), starting on set of xi,yi pairs of sample data. The method is in fact a generalization of the fastest descent learning law presented by Widrow and Hoff (1960). They analyzed a two layer ANN called Adaptive Linear Network (ADALINE) and developed a local descent method using the gradient direction. The method proposed by Widrow & Hoff was generalized to multiple layers and given the name "back-propagation algorithm" (Werbos, 1974; Rumelhart, et al., 1986). BP is the most widely used type of ANN for optimization problems. The user is allowed to pick any utility function, performance measure, or cost function, which can be maximized or minimized, and the method computes the impact on present or future actions. BP can be implemented using almost any Programmable Device (Gracios, et al., 2005).

The typical ADALINE structure is a linear combination of nodes interconnected to form several layers which may or may not have interactions between them. Typically, the layers are known (according to the flow of information through the Network) as the input, hidden and output layers. There is no limit to the number of internal layers contained within the hidden layer. Each layer is fully connected to the succeeding layer (figure 5). The arrows indicate the flow of information during the recall. Initially, the weights as well as the thresholds of the sigmoid functions are set with small random values. Training begins by presenting data to the input of the ANN. During training, information is propagated back through the network and used to adjust the connection weights. The training exercise will be considered satisfactory upon convergence of the weight adjustment, that is, the cost function that measures the I/O mapping of the ANN is minimized.

The number of layers in an ANN is very important. A two-layer NN can only perform well on a limited range of problems (Minsky & Papert, 1988). On the other hand, multilayer ANNs have a wide spectrum of applications and can deal with problems that are prohibitively difficult for the two-layer ANN. As discussed by Rumelhart et al (1986), the addition of internal layers allows the back propagation algorithm to develop an internal representation of dynamic processes and this feature can be crucial to finding a solution. The internal models formed in two-layer ANNs with back-propagation are linear ARX in form.

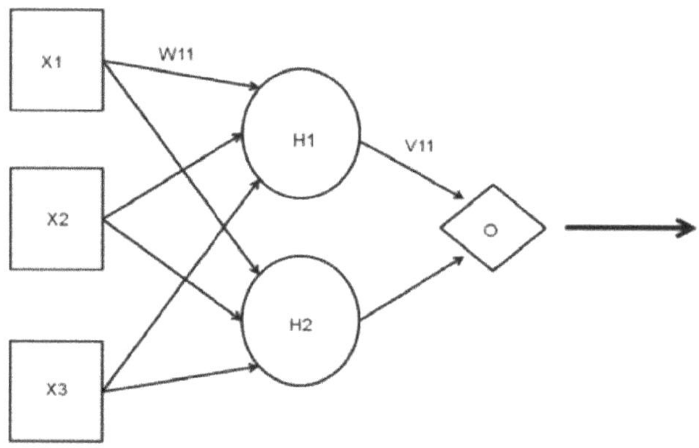


**Fig. 5. Generic structure for three layer ANN.**

The procedure to calculate the propagation values for each layer is as follows. The input layer receives a value Xi, which is propagated with a weight Wij to the intermediate layer, according to equation (3) where S is called the activation coefficient.

$$S_j^p = \sum_{i=1}^{n} W_{ij} X_i^p + W_{n+1,j}^p \tag{3}$$

This activation coefficient Sjp is propagated by an output function which represents the intermediate layer, in this case the sigmoid function, equation (4).

$$h(S_j^p) = \frac{1}{1+e^{-S_j^p}} \tag{4}$$

Propagation may be accomplished using alternative range-limited functions, such as tanh-1, selection of an appropriate function depends on the final application. As shown in figure 5, it is also necessary to define an exciting function, in order to access the output layer, equation (5), and this is obtained when the weights Vjk have been calculated.

$$r_k^p = \sum_{j=1}^{l} V_{jk}^p h_j^p + V_{l+1,k}^p \tag{5}$$

Finally, the output function of the output neuron 'k' is obtained using the same sigmoid function that was used to connect the input and hidden layers, equation (6).

$$O_k^{\,p} = \frac{1}{1+e^{-r_k^{\,p}}} \tag{6}$$

## 3.2 Neural PDF

The primary rationale for the inclusion of Artificial Neural Networks (ANNs) in the study, design, and implementation of control techniques is their adaptability. If it is feasible to obtain any prior knowledge about the structure of the model of a system to be controlled, this information can be utilized to enhance the adjustment of conventional controllers, such as PD or PID, and in this particular scenario, PDF. Several advancements have been made in the field of artificial neural networks with the goal of developing efficient and rapid methods for computing and modifying the parameters of discrete PID control systems (Narendra & Mukhopadhyay, 1996; Garcez & Garcez, 1995). This work employs a comparable approach to adjust a discrete probability density function (PDF).

ANN has been utilized in multiple studies involving hydroelectric systems. Garcez and Garcez (1995) utilized PI neural control on a linear simulator of a 20 MW hydroelectric power station. Djukanovic, et al. (1997) verified the effectiveness of an adaptive-network that utilizes fuzzy inference for controlling a low head hydropower plant. The authors Yin-Song, et al. (2000) have devised a self-learning control system that utilizes a PID fuzzy neural network and implemented it in a hydraulic turbine governor. In a recent study, Shu-Qing et al. (2005) conducted a comparison between a PID controller and a hybridized controller that combines evolutionary algorithms and fuzzy neural networks. The controllers were tested as governors for a hydroelectric power plant model.

This work utilizes a back-propagation technique to fine-tune the parameters of a discrete PDF regulator. Aguado Behar (2000) successfully employed this method to fine-tune a PID controller in real-world applications. Figure 6 depicts the Neural-PDF technique presented. The regulation can be determined by:

$$v_j(t+1) = v_j(t) + \eta sign(\frac{\partial e_y}{\partial e_u})\delta^1 h_j \tag{7}$$

$$w_{ji}(t+1) = w_{ji}(t) + \eta sign(\frac{\partial e_y}{\partial e_u})\delta^2{}_j x_i \tag{8}$$

$$\frac{\partial E(t)}{\partial v_j} = -\delta^1 h_j \frac{\partial e_y}{\partial e_u} \tag{9}$$

Equations 7 and 8 are expressed to recursively adjust the weights for each neuronal interconnection Vj and Wij. Equation 9 is used to develop the minimization of the transfer function for the error. δ is included to calculate the gradient of the function and to express the change of sign in the evolution of the process.

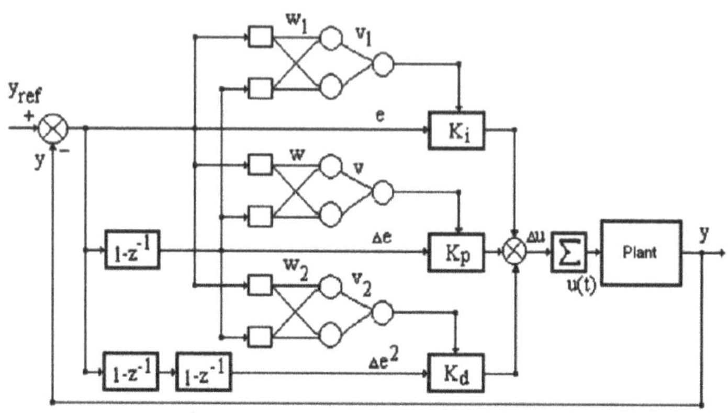


**Fig. 6. Neural PDF.**

# 4. TRADITIONAL CONTROLLERS DESIGNED FOR HYDROELECTRIC POWER STATIONS

## 4.1 Dinorwig Governor Configuration

Dinorwig is equipped with a digital Governor, the overall setup of which is illustrated in Figure 7. There are two control loops, one for power and one for frequency (Mansoor, 2000). The power control loop adjusts the turbine's guide vane based on the product of the power deviation and the speed regulation droop. The Droop gain is employed to modify the speed reference of the governor. This control utilizes a PI setup. The frequency control loop supplies a reference to the power control loop that is directly proportional to the frequency error. Additionally, the system has a derivative feed-forward loop to effectively react to fluctuations in grid frequency.

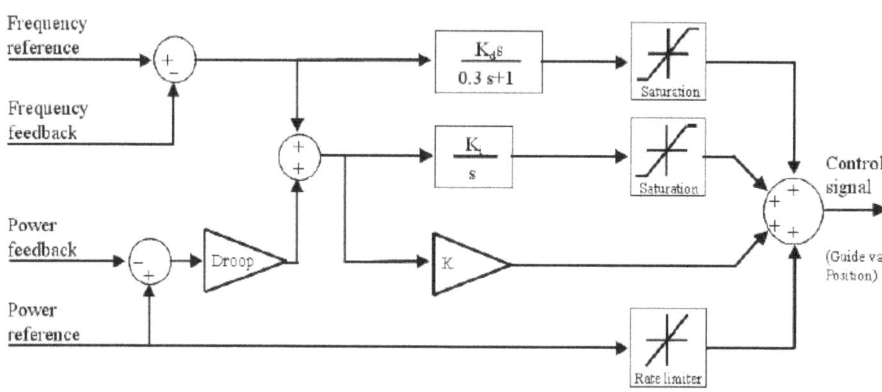


**Fig. 7. Scheme of the Dinorwig Governor.**

The generators must maintain the speed within an operational band defined by the regulatory authority. When the reference is raised the governor valve will open, thus increasing generation. On the other hand, when the output signal is lowered the reference signal to the governor valve will close, decreasing generation (Wright, 1989). At Dinorwig the governor operates with two droop settings; 1% for high regulation and 4% for low regulation (Mansoor, 2000). The power reference signal sets the reference position for guide vane opening and defines the operating point for the unit when it is working in frequency regulation mode (part load response). Changing the power reference, which also acts as a feed-forward signal, directly sets the guide vane position, in order to produce a rapid reaction when big changes in the power reference appear. The guide vane position reference signal (control signal) is produced by adding the output signals from the P, I and D parts to the feed-forward signal. The power feedback loop compensates the system for the non- linear relationship between guide vane opening and power.

## 4.2 Anti-windup PID

With careful tuning, PI control can offer good and robust performance. However, all Plant are subject to constraints and their behaviour changes when the constraints are activated. In these circumstances, the performance of a linear controller, such as PI, can deteriorate significantly (Peng, et. al., 1996). When the plant has actuator saturation the integrator value becomes excessively large compared to a linear response (an actuator without saturation), and it then "winds up". In addition, a higher integrator output and a longer settling time are caused by the saturation effect (Peng, et. al., 1996; Bohn & Atherton, 1995; Goodwin, et al., 2001). In other words, windup is produced when the control signal saturates the actuator, because increasing the control signal can no longer accelerate the response of the plant. If this behaviour persists the integrator value can become very large, without affecting the output of the plant. As a consequence, when recovering from saturation, bringing the system back to its correct steady-state value requires the control error to be of the opposite sign for a long time. The result is a large overshoot and a long settling time (Bohn & Atherton, 1995).

Figure 8 shows a general PI controller that includes a tracking anti-windup scheme (Bohn & Atherton, 1995). This controller has an internal feedback path, which drives the integrator to a negative value and forces the output of the system to be in the linear range. The internal saturation is used to reduce the integrator input. As can be seen from Figure 8 the signal to be integrated is modified by the proportional gain ($K$), therefore the values of the integral gain ($Ki$) are adjusted in order to maintain equivalence with the classic PI. The saturation limit and the dead zone depend on the constraints fixed by the operator; a value of 0.95 p. u. is commonly used. In this work, the responses of the plant when it is governed by a PI with tracking anti-windup structure will be used as a basis of comparison.

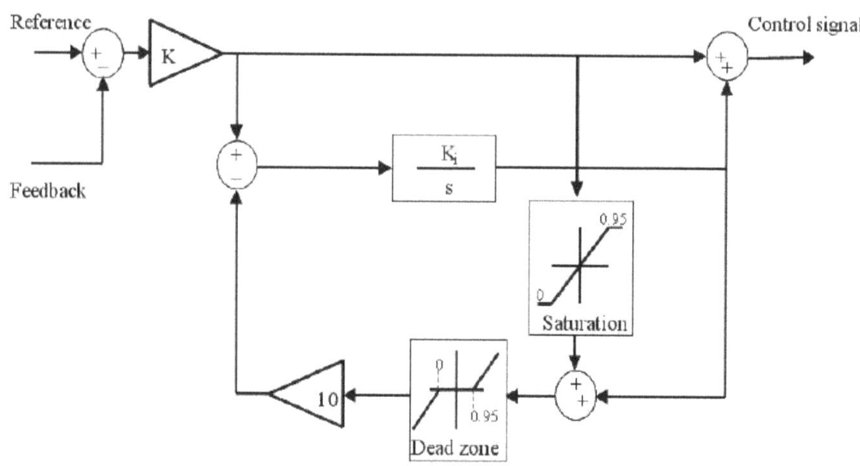


**Fig. 8. General scheme of PI anti-windup.**

## 5. SIMULINK© MODEL AND PROGRAM

The Simulink© software tool was used to facilitate studies of the power plant under different governors. This tool has libraries of specific functions (blocks) and the power plant models were constructed using these standard Simulink© functions. Using a dialog box, the parameters of a specific block can be adjusted. For example, the operating point of linear models may be changed. These models can represent the power plant as a SISO or MIMO system and linear or nonlinear behaviour may be selected. Figure 9 shows a schematic of the Simulink© power plant model. The full hydroelectric station model is constructed by combining the four sub-systems: Guide vane dynamics, hydraulic subsystem, turbine/generator and sensor filters. Each block is part of the Simulink© library developed for this study; they can be selected to represent a diversity of modes of operation. For instance, there are three models available to simulate the hydraulic subsystem - linear, nonlinear non-elastic and nonlinear elastic. The guide vane dynamics can be selected with or without rate limitation and saturation. The sensor filters block is a fixed block. The grid model can be adjusted to represent different conditions of the national grid. The governor block has the option of classical and advanced controllers.

Customised Simulink© S-functions were developed for the neural PDF algorithms. These functions can be incorporated within Simulink© models. The neural PDF block accepts η (learning parameters) and sample time. Its inputs are the reference and the output signals of the plant and its output is the control signal. The versatility of Simulink© allows the plant model to be changed easily or the control algorithm to be modified and new results to be viewed and assessed. The neural algorithm calculates the optimal values of the control law parameters.

The current criterion of optimality is quadratic error, where the error is the output deviation from the set-point; however this criterion can be changed if necessary. The algorithm takes around 10 iterations (the exact value depending on the magnitude of the change) to find the "best" range of parameter values (training time). When these ranges have been reached the parameters stay constant until the set-point or the plant model change.

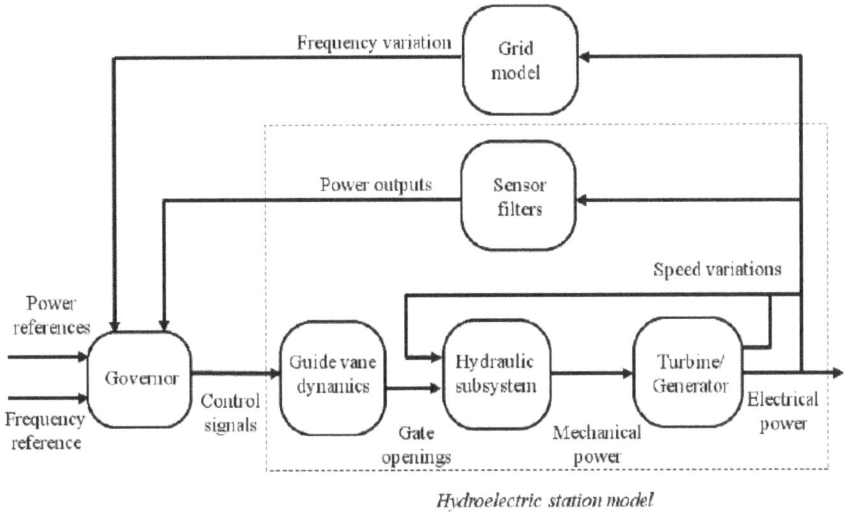


**Fig. 9. Simulink© power plant model.**

## 6. SIMULATION RESULTS

As discussed previously, the role of a hydroelectric station in frequency control mode is to provide timely and accurate supply of its demanded power contribution to the power system. The actual form of the power demand is related to Grid frequency variation but, for testing, it can be specified in terms of step, ramp and random input signals.

For all simulations, the model is expressed in the per-unit system, normalized to 300 MW and 50 Hz, and assumes a Grid system with infinite busbars. The neural PDF controller was connected to the nonlinear model of the hydroelectric power plant. A PI controller with parameters fixed at K=0.1 and Ti=0.12 (as currently implemented in practice) is used as a basis of comparison. Figure 10 shows the small step responses (0.05 p.u.) of the hydroelectric plant under anti-windup PI and neural PDF controllers for one unit operational. Figure 11 shows the small step responses (0.05 p.u.) of the power station when all six units are connected. In both cases, the hydroelectric plant performs better with the neural PDF controller, being respectively 10% and 30% faster in the one unit operational and six units operational cases. The undershoot is also reduced in both cases when a PDF controller is driving the process.

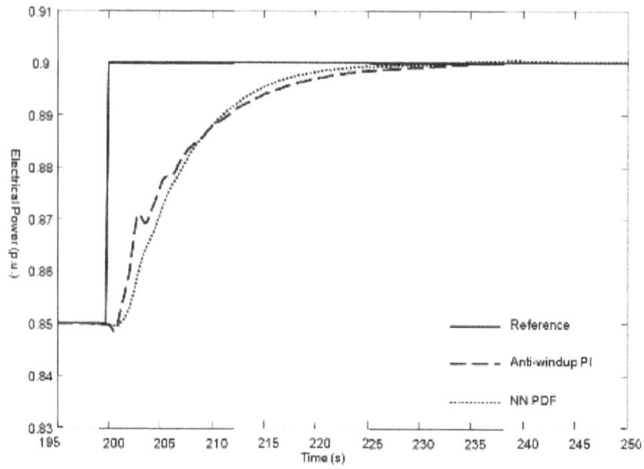


**Fig. 10. Small-step response of the hydro plant with neural PDF and PI controllers for the case of one unit in operation.**

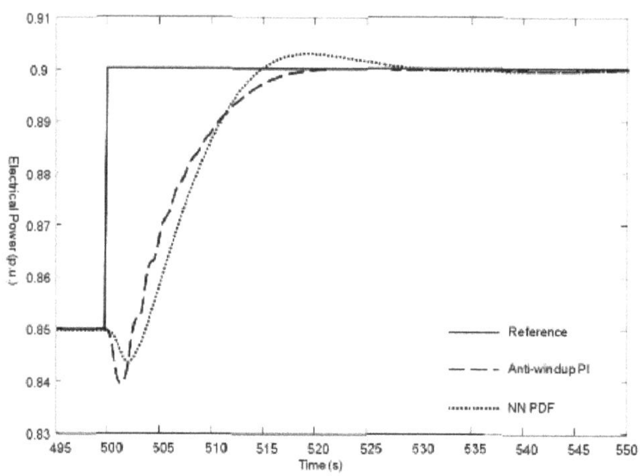


**Fig. 11. Small-step response of the hydro plant with neural PDF and PI controllers for the case of six units in operation.**

Figure 12 shows the large ramp responses (0.35 p.u.) of the hydroelectric plant with anti- windup PI and neural PDF controllers for one unit operational. Figure 13 shows large ramp responses (0.35 p.u.) of the power station when six units are generating. In both cases, the performance is better using the neural PDF controller, the response being 15% and 13% faster in, respectively, the one unit operational and six units operational cases. Again, the PDF controller reduces the undershoot.

To evaluate the cross coupling interaction a 0.8 p.u. step was applied simultaneously at t=300 to units 2-6 and the perturbation of unit 1 observed. Figure 14 shows that, although the neural PDF response has a higher overshoot, the PI response has a longer settling time and a higher undershoot.

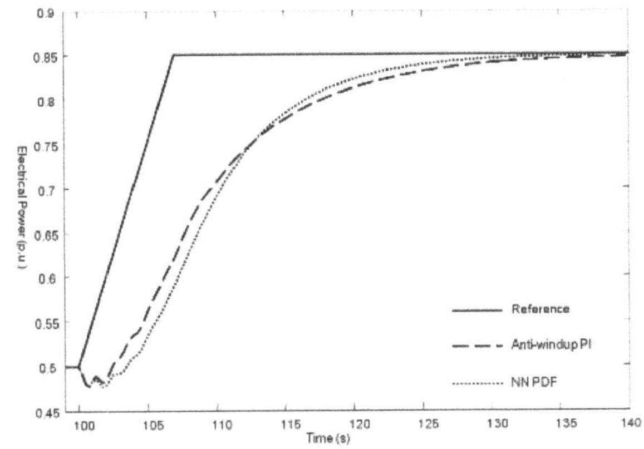


**Fig. 12. The large ramp response of the hydro plant with neural PDF and PI controllers with one unit in operation.**

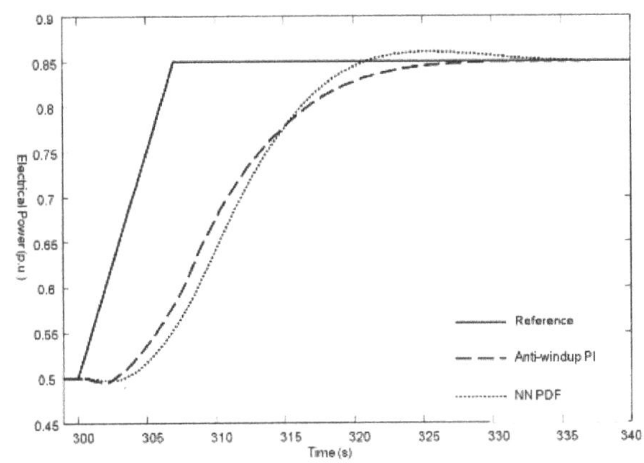


**Fig. 13. The large ramp response of the hydro plant with neural PDF and PI controllers with six units in operation.**

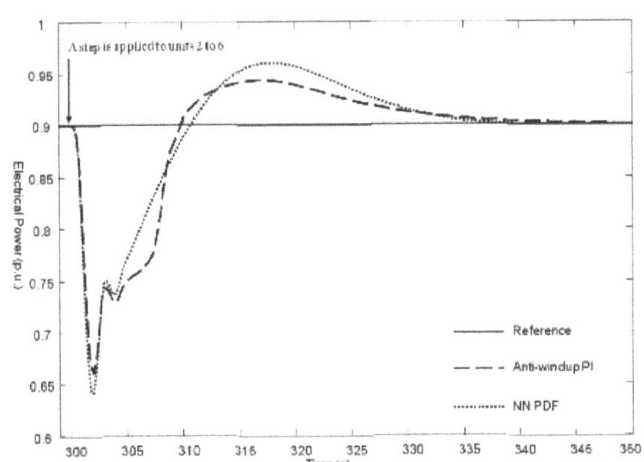


**Fig. 14. The cross coupling response of the hydro plant with PI and neural PDF controllers.**

## 7. CONCISE SUMMARY

This chapter presents a software program that simulates a hydroelectric plant and enables the evaluation of several controllers by comparing their performance. The modularity of this platform has played a crucial role in systematically enhancing the intricacy of the simulations. The open design enables the swift integration of alternative control techniques and allows for the gradual enhancement of the existing control approaches and models. The discussion has focused on the nonlinear model of pumped storage facilities. The Dinorwig power plant was subjected to this model. The model incorporates the guiding vane, hydraulic, and electrical subsystems, and encompasses the key characteristics of the plant's dynamics.

The findings have demonstrated the application of neural PDF in enhancing the dynamic response of a hydroelectric pumped-storage facility. Specifically, research has demonstrated that the system's step reaction is enhanced when utilizing neuronal probability density function (PDF). The inclusion of multivariable factors has been considered in order to accurately depict the actual plant. The connection between penstocks has been incorporated into the nonlinear model. These results show great potential for using neural PDF in this application and motivate us to focus on improving the reliability of the response in future research.

## REFERENCES

1. Aguado-Behar, A., "Topics on identification and adaptive control" (in Spanish), Book edited by ICIMAF, La Habana, Cuba. 2000.
2. Bohn, C. and Atherton, D. P., "An analysis package comparing PID anti-windup strategies", IEEE Control Systems Magazine, vol. 15, p.p. 34-40. 1995.
3. Djukanovic, M. B., Calovic, M. S., Vesovic, B. V., and Sobajic, D. J., "Neuro-fuzzy controller of low head hydropower plants using adaptive-network based fuzzy inference system", *IEEE Trans. on Energy Conversion* , 12, pp. 375-381. 1997.
4. Garcez, J. N., and Garcez, A. R., "A connectionist approach to hydroturbine speed control parameters tuning using artificial neural network", Paper presented at *38th IEEE Midwest Symposium on Circuits and Systems*, pp. 641-644. 1995.
5. Goodwin, G. C., Graebe, S. F. and Salgado, M. E., "Control system design", Prentice Hall, USA. 2001.
6. Gracios, C., Vargas, E. & Diaz-Sanchez A., "Describing an IMS by a FNRTPN definition: a VHDL Approach", *Elsevier Robotics and Computer-Integrated Manufacturing*, 21, pp. 241–247. 2005.
7. Kang, J. K., Lee, J. T., Kim, Y. M., Kwon, B. H., and Choi, K. S., "Speed controller design for induction motor drivers using a PDF control and load disturbance observer", Paper presented at *IEEE IECON*, Kobe, Japan, pp. 799-803. 1991.

8. Kundur, P., Power System Stability and Control, New York, NY: Mc Graw Hill. 1994. Mansoor, S. P., "Behaviour and Operation of Pumped Storage Hydro Plants", Bangor, U.K.: PhD. Thesis University of Wales. 2000.
9. Mansoor, S. P., Jones, D. I., Bradley, D. A., and Aris, F. C., "Stability of a pumped storage hydropower station connected to a power system", Paper presented at *IEEE Power Eng. Soc. Winter Meeting,* New York, USA, pp. 646-650. 1999.
10. Mansoor, S. P., Jones, D. I., Bradley, D. A., Aris, F. C., and Jones, G. R., "Reproducing oscillatory behaviour of a hydroelectric power station by computer simulation", *Control Engineering Practice, 8*, pp. 1261-1272. 2000.
11. Miller T., Sutton S.R. and Werbos P., *Neural Networks for Control, Cambridge Massachusetts:* The MIT Press. 1991.
12. Minsky, M. L., and Papert, S. A., Perceptrons: Introduction to Computational Geometry. Cambridge, USA: MIT Press. 1988.
13. Munakata, T., Fundamentals of the New Artificial Intelligence: Neural, Evolutionary, Fuzzy and More. London, UK: Springer-Verlag. 2008.
14. Narendra, K. S., and Mukhopadhyay, S. "Adaptive control using neural networks and approximate models", Paper presented at *American Control Conference,* Seattle, USA, pp. 355-359. 1996.
15. Peng, Y., Vrancic, D. and Hanus, R., "Anti-windup, bumpless, and conditioned transfer techniques for PID controllers", IEEE Control Systems Magazine, vol. 16, p.p. 48-57. 1996.
16. Rumelhart, D. E., McClelland, J. L., and Group, T. P., Parallel distributed processing: Explorations in the microstructure of cognition (Vol. 1). Cambridge, USA: MIT Press.1986.
17. Shu-Qing, W., Zhao-Hui, L., Zhi-Huai, X., and Zi-Peng, Z. "Application of GA-FNN hybrid control system for hydroelectric generating units", Paper presented at *International Conference on Machine Learning and Cybernetics 2*, pp. 840-845. 2005.
18. Werbos, P. J., Beyond regression: New Tools for Prediction and Analysis in the Behavioral Sciences, Cambridge, USA: PhD. Thesis Harvard University. 1974.
19. Widrow, B., and Hoff, M. E., "Adaptive switching circuits", Paper presented at *IRE WESCON Convention Record 4*, pp. 96-104. 1960.
20. Working group on prime mover energy supply, I, "Hydraulic turbine and turbine control model for system dynamic studies", *IEEE Trans.s on Power Systems , 7,* 167-179. 1992.
21. Wright, R. M., "Understanding modern generator control", IEEE Transactions on Energy Conversion, vol. 4, p.p. 453-458. 1989.
22. Yin-Song, W., Guo-Cai, S., & Ong-Xiang, "The PID-type fuzzy neural network control and it's application in the hydraulic turbine generators", Paper presented at *Power Engineering Society meeting. 1*, pp. 338-343. 2000.

# 11 ADVANCED NETWORK SYSTEM FOR PROCESS CONTROL: APPLICATIONS, CHALLENGES, APPROACHES

## 1. INTRODUCTION: AN OVERVIEW

The perpetual challenge faced by system controllers and integrators is in managing the expanding dimensions, advanced features, and intricate nature of sorting and handling systems. This arises when industries endeavor to enhance productivity and provide a wider range of options to consumers. The increasing adoption of larger capacity systems, which enhance throughput rates and offer more adaptability in handling product variation and changing supply demand, puts additional strain on system controllers to deliver timely and cost-effective results. This is often challenging due to vague specifications, resulting in the development of extensive and intricate programs. Given these requirements, it is not unexpected that control engineers opt to rely on previous experience when developing and incorporating a control system in order to reduce potential risks. Contemporary methods involve the use of distributed and modular systems such as programmable logic controllers (PLCs) to simplify intricate control processes.

The demand for modularity, real-time operation, integrated diagnostics, decentralized control, growing physical configurations, and functionality has led to the development of many types of distributed control systems (DCS). The DCS concept is currently used not only in process control and manufacturing automation, but also in various other fields such as communications engineering (Yang, 2006), physics and nuclear science (Cavineto et al., 2006; Kleines et al., 2004), power system engineering (Stewart, et al., 2003; Ioannides, 2004), and lighting system design (Alonso, et al., 2000). In the study conducted by Alonso et al. (2000), they utilized power lines as a means of communication to develop the Distributed Control System (DCS) for the lighting system. This was achieved by using a specially designed Neuron chip that integrated control and communication functions within a three microprocessor system. Industrial plant automation is another area where it can be applied. Its purpose is to decrease operational expenses, shorten product development time, and enhance the plant's ability to manufacture a variety of products. The use of decentralization in control systems has highlighted the significance of communication and coordination across different sub-control systems in the development of an optimal control system design. The research and development in the field of DCS network architecture is highly active due to advancements in the communications and information technology sector. The authors in (O'Hearn et al., 2002) present a method for integrating a distributed control system with a motor control center that includes several remote terminal units in an industrial facility.

The chapter is structured in the following manner: The next section presents a literature review of relevant work, focusing on existing methodologies linked to distributed process control and their respective applications. The conventional setting is thoroughly examined in conjunction with performance specifications. Furthermore, a selection of pertinent standards is emphasized, which establish the guidance for a unified strategy utilizing dispersed control. Section 3 outlines a series of methods that optimize the performance of the central process controller. Section 4 discusses the latest developments in communications and information technology that have shown potential for enhancing performance in process control. The text describes a collection of methods that reduce communication delay, enhance performance, and enable the compatibility and adaptability of different process devices. Section 5 presents the most current initiatives to establish standards in this field. Section 6 contains conclusions, which are then followed by references in section 7.

## 2. PROCESS CONTROL

The Distributed Control System (DCS) comprises a network of small, compact, and intelligent nodes strategically positioned throughout the environment. These nodes are interconnected via a network to the main process controller. The use of identical hardware streamlines the process of installation and the replacement of components in the field. The nodes interact with and regulate various sorts of devices, while the main controller supervises the operation. DCS has replaced the industry's old point-to-point design with multipoint and network-based process control, marking a significant shift in approach. The new architecture features a reduced wiring volume and dispersed processing, leading to efficient and cost-effective maintenance. These advancements have sparked study in two primary areas: (i) Network design, configuration, and communication protocols for distributed process control, and (ii) DCS system hardware and software architectures. In order to familiarize ourselves with the initial research direction, we examined a study conducted by Lian et al. in 2001. This study involved simulating and experimentally evaluating three control networks that are commonly used in the industry. The control networks assessed were Ethernet, ControlNet, and DeviceNet. The evaluation technique assessed the utilization of each control network as a communication infrastructure for a network-controlled system that connects sensors, actuators, and controllers. It included a comprehensive analysis of the functioning of the medium access control (MAC) protocol for each network. The performance parameters considered are access delay, transmission time, message delay, message collisions, message throughput, packet size, network utilization, and determinism boundaries. The analysis of each control network's timings focuses on blocking time, frame time, and propagation time. Based on several case studies, the authors determined that DeviceNet is superior than the other two networks when it comes to shorter or prioritized communications. On the other hand, ControlNet is more appropriate for time critical and non-time-critical messages. In summary, this research endeavor offers insight into the performance boundaries set by different networks.

Profibus is a vendor-independent communication protocol that allows devices from different manufacturers to communicate with each other without requiring any additional interface adjustments. In the study conducted by Tovar and Francisco in 1999, a thorough examination of Profibus networks was undertaken. The authors explore the utilization of Profibus fieldbus networks to facilitate real-time communication in industrial systems. The significant finding of this study is that real-time performance is ensured even without synchronous bandwidth allocation. The proposed methodologies are predicated on an assumption that ensures real-time traffic. The author conducted an experimental evaluation of Profibus-FMS in the study of Hong (2000). This study involves the development of an experimental model for an industrial application based on Profibus. The experimental environment comprises interconnected devices in a Profibus network, facilitating data interchange using the services offered by FMS. The model assesses the temporal attributes of a message after it has traversed the FMS service. The author proposes an optimal value for target token rotation and an acceptable segment size based on the testing results. This is recommended to decrease the delay caused by variable messages and in domain management service of FMS. Furthermore, it can be inferred that the performance of the Profibus-FMS service is significantly influenced by the traditional values of target rotation time and segment size.

In addition to examining network protocols and control networks with their corresponding delay characteristics, a research study conducted by Lian et al. (2002) explores the design concerns for a particular network, such as DCS. This research study focuses on the influence of network architecture in the Network Control System (NCS) category of Distributed Control Systems (DCS). Design issues encompass network characteristics, control parameters, and the performance of network-controlled systems. The primary evaluation metrics for network quality of service (QoS) are time delay statistics, network efficiency, network utilization, and the count of lost or undelivered messages. The authors have employed a simulation model and an experimental investigation to authenticate the performance features demonstrated by theoretical analysis. The authors show that the performance characteristics are valuable criteria for selecting network and control settings when developing an NCS. According to the authors, the performance characteristics rely on the sample time. Nevertheless, the research has not yet addressed the selection of a sampling interval.

Research is being conducted on the integration of wireless and cable devices in Profibus for process control applications. In a study conducted by Willig in 2003, the specifications of the wireless Profibus protocol standard were examined. The author presents and contrasts two methodologies for wireless Profibus, analyzing their real-time performance in the context of transmission faults. The study shows that the efficiency of the Profibus protocol in real-time applications is not sufficient when used in an IEEE 802.11 type PHY environment or when there are frequent bursts of faults. The author proposes a customized protocol for wireless stations in a scenario where the integration of wired and wireless stations is required in a shared Profibus Local Area Network (LAN). The customized protocol operates on the IEEE 802.11 PHY layer and provides the same link layer access to upper levels as the Profibus protocol. The author has demonstrated that the round robin protocol, when enhanced with three additional features - simple data relaying, simple poll relaying, and priority restriction - outperforms the Profibus protocol in situations characterized by bursts of activity and a stable ring. Nevertheless, it was determined that maintaining the stability of a logical ring in the presence of transmission defects is a challenging task. The authors (Almeida, et al., 2002) emphasize the significance of operational flexibility and event and time triggered paradigms in fieldbus communication systems. They offer a protocol that supports time-triggered communication in a flexible manner, employing network-based control. The authors also explore the effective integration of event and time-triggered traffic with temporal isolation, ensuring the desirable characteristics of both forms of traffic.

In addition to the previously mentioned domain, there are several methods to address the same issue. One solution discussed in (Prayati et al., 2004) focuses on developing a distributed real-time control application for job allocation in heterogeneous systems. The authors specifically address the challenge of meeting real-time requirements while ensuring interoperability among networked devices. The method employs the functional block allocation (FBALL) algorithm (International Electrotechnical Commission, 2007) to ensure the fulfillment of real-time requirements. However, this approach fails to take into account the network conditions when reconfiguring the process or its field devices in real time. Inspired by the emergence of widespread communication and computers, the authors in (Recht & D'Andrea, 2004) examine geographically interconnected distributed systems that possess arbitrary discrete symmetry groups. They explore the construction of distributed controllers that may be applied to a broader range of structural topologies. Another method has been employed to examine the application areas and tasks encountered by multi-robot systems that are becoming more complicated. In their publication, Farinelli et al. (2004) discuss different types of collaboration and synchronization achieved in the field of multi-robot systems.

Extensive research was conducted on the literature about network-based control, and various ways were discovered. However, the approaches discussed in the previous section are particularly noteworthy compared to others. Literature discusses several topics related to network considerations, such as protocol changes and delay characteristics. It also explores the interoperability of devices and the customization and reconfiguration of control systems. The methodologies that tackle the network environment examine timing characteristics or protocols to minimize communication delays at different phases of information processing.

The process being discussed is conventional, meaning it consists of devices and nodes that are currently available and widely utilized. Figure 1 provides a visual representation of the environment. The controller in this Figure serves as the primary device that will be examined in order to analyze the distributed control environment. The comprehensive system entails the connectivity of control devices, including distributed field and intelligent devices linked to Profibus, as well as a few intelligent control devices working inside a legacy environment. The control devices at the lowest level of the system may include standard Programmable Logic Controllers (PLCs) for the purpose of motion control in a typical drive, for instance. Intelligent devices are typically required for a specific application, where they handle decision-making and local information collection. This information is then transmitted to the main controller, which makes the ultimate decisions for network control. The devices are presumed to be capable of being reprogrammed and reconfigured to accommodate the evolving requirements of process control. The interface depicted in Figure 1 is designed for outdated intelligent devices that have been integrated into the primary distributed control system. The network being discussed is a standard one that consists of Profibus compatible network devices. These devices can link to a distant device or a process controller via either wired or wireless connectivity. The PC displayed is intended for managerial use, namely for the SCADA system. All of the field devices being considered adhere to the standards set forth in IEC 61804, while the PLCs conform to the guidelines outlined in IEC 61131. As mentioned earlier, the development of DCS for this process control requires careful consideration of two main factors:

reconfigurability and intelligence. Specifically, the DCS should be able to adapt and change its configuration as needed, and it should incorporate a framework that is compatible with Profibus, a widely used network protocol.

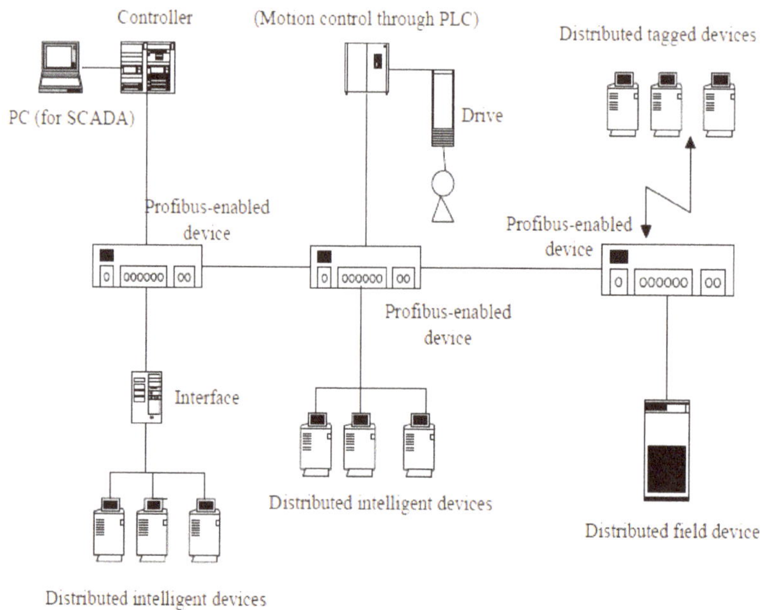


**Fig. 1. Typical Process Control Network**

For comparative purposes, we need to create a model that sets a baseline for performance. As the configuration of the network is distributed over a wide area, hence we consider a multiple input multiple output (MIMO) baseline that requires performance, which matches to that of the centralized MIMO (without network induced delays). A discrete time, causal and linear time-invariant systems with *n* states and *m* outputs can be described as (Goodwin et al., 2001):

$$X(k+1) = A\,X(k) + B\,[U(k) + D(k)];\; Y(k) = C\,X\,(k) \qquad (1)$$
$$with\ U(z) = K\,(z)\,E\,(z) = K\,(z)\ [R(z) - (C\,X\,(z) + N\,(z))]$$

where X, D, K, E, R, N are states of actual MIMO system in centralized control, disturbance, closed loop controller, error, reference, and sensor noise respectively. It should be noted that the MIMO system is assumed to be well designed meaning that its performance is exponentially stable (Goodwin et al., 2001), and that any MIMO control design technique can be used to develop an appropriate controller K. The parameters A, B, C are minimal realizations of the actual plant (Goodwin et al., 2001). Using equation (1), various states of the system can also be calculated:

$$X'(k+1) = A'X'(k) + B'\,U'(k);\; Y'(k) = C'X'(k) \qquad (2)$$
$$with\ U'(z) = K\,(z)\,E'(z) = K\,(z)[R\,(z) - C'X'(z)]$$

where A’, B’, C’ represent minimal realizations of the model. For Figure 1 (as a distributed implementation of MIMO) to achieve same performance to that of the centralized MIMO, each error or reconfiguration request needs to be communicated over the network at any sample time, and that there are no communication delays. In order to categorize time delay $d(t)$ in the network is divided into three categories:

$$d(t) = d1(t) \| d2(t) \| d3(t) \quad (3)$$

Thus, the total time delay $d(t)$ is either of the time delays, depending on the prevailing situation in the environment. The $d1$ corresponds to a normal situation (Green level), when only one field device communicates with the central process controller for a possible update or reconfiguration at any time. The $d2$ corresponds to a situation (Yellow level), when more than one device (with number set as a threshold) communicate with central process controller for possible problem getting out of hand. The $d3$ corresponds to a situation (Red level), when a set of devices (with number set as a threshold) tries to communicate with the central process controller. One obvious direction could be to minimize the delays described in equation (3) to optimize the performance to match a centralized MIMO system with no communication delays.

Since filed devices may include nodes or sensors that are non-stationary, protocol invariant and reconfigurable, the performance requirements stated by centralized MIMO would require approaches those are not based on legacy network control but should include recent advancements in sensor, information and communication technologies to match optimized performance. The combination of distributed control up to edges of the process containing sensors and tags, with recent information technology concepts like layered software and agents is an active area of research these days.

## 2.1 Relevant Standardization

The PLCs are the most widely used processing units and fieldbus is most widely used to interconnect the process controllers, sensors and actuators in a DCS system. The Profibus is the standard protocol for communication which is vendor independent so that the communication between devices from different manufacturers can be implemented without any special interface adjustment. To realize the plug and play type of operation, the DCS vendors ought to adopt certain standards both for PLCs and the communication. The International Electrotechnical Commission (IEC) and PLC manufacturing companies are actively involved in this development and establishment of standards (International Electrotechnical Commission, 2007). These standards define basic concepts for the design of modular, reusable, distributed components for the distributed control systems. Some of the most popular IEC standards are IEC 61131, IEC 61499 and IEC 61804. Because of the importance of these standards a brief introduction is presented in the next section.

**IEC 61804:** This standard addresses the need of common device description for representing the particular features of the field devices used in the DCS system.

The standard is applicable to Function Blocks (FB) for process control and specifies the Electronic Device Description Language (EDDL) thus improving the device interoperability.

**IEC 61131:** This standard has different versions ranging from 61131-1 to 61131-8. The whole range covers general information about just the PLCs, the equipment requirement and tests, defines syntax and semantics of programming languages including the function block (FB) portion of standard for PLC languages, and sets the standards for the communication. To conclude IEC 61131 was first attempt to provide common terminology and reference model about hardware (HW) and software (SW) architectures, communication, and languages for a special class of control devices i.e., the PLCs.

**IEC 61193-2:** This standard IEC applies to the inspection of electronic components, packages, and modules for use in electronic and electric equipment, manufactured under suitable process control, which prevents the outflow of nonconforming products. The objective is to provide a method for stabilizing, monitoring, and improving processes.

**IEC 61987-1:** It defines a generic structure in which product features of industrial- process measurement and control equipment with analogue or digital output should be arranged, in order to facilitate the understanding of product descriptions when they are transferred from one party to another. It applies to the production of catalogues of process measuring equipment supplied by the manufacturer of the product and helps the user to formulate requirements.

**IEC 61499:** This standard defines a generic architecture and presents guidelines for the use of function blocks as the main building block for industrial process measurement and control system applications. It extends the FB language (defined in IEC 61131-3) to more adequately meet the requirements of the distributed control in a format that is independent of implementation. The architecture is presented in terms of implement-able reference models, textual syntax and graphical representations. The standard addresses the need for modular software that can be used for distributed industrial process control by defining the basic concepts for the design of modular, reusable, distributed components for industrial control systems. It also specifies the software tool requirements.

A variety of fieldbus technologies and digital fieldbus devices have been introduced within the process industries over the last fifteen years. There has been a gradual acceptance of the fact that a variety of communication technologies are needed to fully address the application requirements of a manufacturing facility. Process control systems that use fieldbus, function blocks, and device descriptive languages already exist in the market. It is clear that the majority of these systems will continue to use function blocks into the foreseeable future, as they are modular and reusable. The function blocks hide the changing details and thus provide a stable interface for the application developers.

## 3. AUTOMATION TECHNOLOGY UTILIZES SENSORS AND TAGGING TECHNOLOGY

The RFID (Radio Frequency Identification) technology enables events that are in line with the vision of pervasive computing: using digital media to augment physical and social spaces rather than replacing them with disembodied virtual spaces. The diversification of consumer needs has made RFID and other sensor systems an important means within the manufacturing industry to achieve more efficient multi-product production. Within the RFID tag process, for example, control functions may be stored and updated as the product progresses through various stages of the operation. Specific uses of the data within the tag for process control include part routing, program operations, build orders, component needs and specific operations which may by unique to only a few products. While operating in a process control function, the RFID tag can also be used as a portable data base providing station to station data transfers, repair information, operation status, operation values, pass/fail data, host data transfers, part history and raw SPC data. One of the major advantages of using an RFID system in manufacturing is a major reduction in the size and complexity of the PC or PLC program reducing the cost associated with programming and also reduces the scan time of the controller.

The object's RFID tag may be integrated with a large object database to integrate object community into the space where social interaction occurs (Konomi et al., 2006). The Researchers at the University of Washington are curious to see what effects RFID technology could have on social networking (The RFID Ecosystem, University of Washington, 2008). To see what happens when the tags become ubiquitous, they installed two hundred antennae in and around a campus building and gave tags to twelve researchers. The result was such that their every move is recorded by computer. The system responds to nearby participants and uses algorithms to dynamically derive interconnected social clusters from a publication database - highlighting new opportunities and key design challenges. The Department of Computer Science and Engineering at the University of Washington has initiated a large scale RFID Ecosystem project to investigate user-centered RFID systems in connection with technology, business, and society and determine balance between privacy and utility (The RFID Ecosystem, University of Washington, 2008).

For vast field deployment, the authors (Bohn & Mattern, 2004) discuss distribution schemes where passive RFID tags are deployed in vast quantities and in random fashion. The authors suggest that certain deployment patterns help create novel RFID based services and applications. The density, structure of tag distributions, tag typing and clustering is also investigated by authors. For distributed business like process applications, the author (Palmer, 2007) discusses effective ways handling vast volume of RFID data for enterprise application management in real time. The steps applied together help enterprise manage volumes of raw RFID data in real time, provide reliability and accurate decisions to tackle the challenges, and exploit opportunities offered by RFID.

It is impossible to separate RFID technology from its issues and concerns surrounding its deployment, especially with respect to its potential for privacy infringement and identity theft. The RFID systems are designed to be asymmetric: tags are cheap and require low power while readers are expensive and power hungry. The various types and functionalities of these devices have been defined in an RFID class structure by the Auto-ID center, and later through Electronic Product Code (EPC) global (The EPC Global Standard, 2009), which has been subsequently refined and built-on. The number and use of standards within RFID and its associated industries is quite complex, involves a large number of bodies (ISO standard, 2009; ETSI standard, 2009; RFID Journal, 2009) and are in a continuous process of development. The various standards produced cover four key areas of RFID technology: its air interface (tag-reader communications) standards, data content and encoding, conformance and interoperability between different applications and RFID systems (RFID Journal, 2009). The various technological issues including its frequency range, standards, adoption and innovation has also been addressed at length in (Ward et al., 2006).

The RFID's have also been investigated in conjunction with software technology to optimize its usability. Integrating software agents into RFID architectures to accumulate information from tags and process them for specific use at different layers of the control plane has attracted many operational managers to investigate their deployment for real time use in industrial applications. The authors in (Naby & Giorgini, 2006) propose a possible integration between a Multi Agent framework and an RFID based application (back end). This is enabled by creating an agent for each person to develop its profile and correlating it with RFID tags movement inside a working environment. The accumulated information, after processing may be used to facilitate a concurrent mission. The RFID has also been explored for product identification in agent-based control applications to gain full control over all products and resources at shop floor level - aiming at installing a completely distributed agent-based production control system with full information control (Bratukhin & Treytl, 2006). The summary of literature survey for RFID deployment suggests investigating a multi-agent architecture as an application layer within existing RFID layered architecture that includes tag layer, RFID-reader layer, and back-end layer (Naby & Giorgini, 2006).

## 4. AGENTS FOR INTELLIGENT CONTROL

Recently there have been a number of advances in distributed artificial intelligence that provide the tools to move away from the traditional centralized, scan based programmable logic control (PLC) architecture towards a new architecture for the real-time distributed intelligent control. The Industrial control is typically implemented using large and often expensive hardware platforms that support monolithic computer control applications. The authors in (Bernnan et al., 2002) propose a general approach for dynamic and intelligent reconfiguration of real-time distributed control systems that utilizes the IEC 61499 function block model (International Electrotechnical Commission, 2007) to achieve shorter up-front commissioning time and significantly more responsive to changes, and is based on object oriented and agent based methods. Similarly, the authors in (Heck et al., 2003) discuss software technology for implementing reusable, distributed control components to make corresponding systems more flexible and dynamically re-configurable, so that they can be adapted in an easy way. Intelligent control and management of traffic and transportation systems in connected environments has been investigated in (Wang, 2005) that requires agent based technology to develop cheap, reliable and flexible control system. The authors in (Maturana et al., 2005) present distributive intelligent control architecture using agents by developing tools and methodologies for agent based solutions to a wide variety of industrial, financial and security problems, as classical control does not adapt dynamically well to the variability of the process. In another work, the authors in (Fregene et al., 2005) present a systems-and-control-oriented intelligent agent framework, as well as its composition into specific kinds of multi-agent systems to achieve the coordinated control of multiple multimode dynamical systems. The author in (Vyatkin, 2008) discusses distributed intelligent control using IEC 61499 (International Electrotechnical Commission, 2007) programming architecture for distributed automation systems in reconfigurable manufacturing systems. The testing environment consists of a network of identical software components for control of baggage handling system at the airport. The work in (Gingko Networks, 2008) provides simple decentralized ways to deal with a growing number of modern network requirements, by using distributed intelligent agents to deal with local situations in a more responsive way like sensing and observing events and changes occurring locally. In order to understand the potential use of agents in process industry, a survey has been presented in (Yang & Vyatkin, 2008) on design and validation of distributed control in process industry. The respective authors discuss distributed control benefits such as flexibility, reconfigurability and software reusability. The process control system is generally considered as a hybrid system as it usually contains both discrete and continuous dynamics. The authors highlight the importance of hybrid verification and simulation while handling validation of distributed systems.

In intelligent control design, scalability of a control system, active system reconfiguration, distributed intelligence and reduced communications are targeted benefits. In these situations, the issues like customization, decentralization and modularity at various levels of control are the main challenges in respective DCS design to facilitate distributed field operations in a process control. Intelligent network control together with active remote monitoring of a scalable distributed system has turned out to be an active area of research. The obvious choice seems to break the complex control process into two distinct processes (Memon, 2008): Local Process for simple task execution at field device level using social interaction of field process entities and generating operational parameters at the device level; Central Process for estimation of new characteristics of social entities only when entities need to be created or modified during operational stages. The independence at local process level minimizes communication between devices to avoid bottlenecks arising due to interoperability of devices, network protocol change or simple operational requirements at the device level; and provide a degree of re-configurability of field devices at the same time. Using local intelligence collected through interactions of local entities and combined with controller requirements at the central level helps solve combinatorial complexity present at any time during the operation. Thus, two areas are targeted: social interactions of entities with domain intelligence, and effective decision making set by the central process. The following is the discussion on both processes.

## 4.1 Local Process

Knowledge integration at central level requires balanced information from local entities. These entities tend to be distributed throughout the operational environment to support all operations. The job of these entities can be done effectively by distributed agents. The agents collaborate socially, learn and adjust their abilities within the constraints of the global process. Thus the agents are sophisticated software entities set by central process to execute trained intelligence at the local level. The functions within this process include: how agents collaborate and clusters are formed to accumulate intelligence and enable decision making; how access to another agent is facilitated; and the procedure for inter-cluster collaboration. To make sure that these agents form an intelligent system, the framework has to conform to four requirements (Maturana et al., 2005): that the system is decentralized; agents follow centricity; they use common language; and that the system is scalable.

### 4.1.1 Agent design & management

The agents deployed in existing infrastructures are designed to be manageable as other components with tools and procedures already in place at central process level. Each agent has software components adaptable to environmental changes. Each of these can be considered as a specialized function with some built-in capabilities, like sense→decide→act loop. The various categories of these specialized functions include:

- Producing knowledge in cooperation with other agents
- Evaluating the situation and decide to apply an appropriate action individually or through collaboration
- Acting onto the network element parameters, like fine tuning quality of service
- Forward useful information to the central process

Agents provide simple decentralized ways to deal with a growing number of modern process control network requirements. Agents can provide a set of management functions:

- Configuration functions for downloading of configuration parameters
- Control functions for monitoring the process environment and providing on-line activation and inspection of agents
- Data collection functions for enabling uploading of useful information collected by agents
- Deployment functions enable to add or remove agents dynamically

Agents can also request for additional capabilities once they discover that the task at hand cannot be fulfilled with existing ones. The programming of these agents is done at the central level where a set of heuristics is used for reasoning at the local level, and is stored as a function block diagram (like an internal script). The life cycle of the cluster is pre-set at the central process during agent design, and which can be re-negotiated upon an agent's request. The agents know about their equipment, and continuously monitor their state, as shown in Figure 2. The Figure 2 shows agents residing on tagged network elements, known as intelligent devices that can compute and communicate. Based on the intelligence, they can decide whether to participate in a mission or not, as shown in Figure 3. Thus the design of an agent is normally made open ended to add flexibility during operation.

### 4.1.2 Agent Collaboration

Each agent maintains its own view of the environment on the basis of the information obtained directly from its sensors and indirectly from the network through collaboration with its neighboring agents. As the local events are known and are properly documented in its view, the agent responds quickly and appropriately. Based on its view, agent may decide to automatically adapt its certain parameters of its own network element. The environmental view of the agent drives implicit cooperation amongst agents. This mode of cooperation is simple, robust and well situated for dynamically changing environments. The collaborating agents join (on their own will) and thus form a cluster in order to enable a decision making. Within their constraints, they accomplish their task. In addition to agents, there are other computing units that exist at the local level and which help to form a cluster. These are known as cluster directory (CD) and cluster facilitator (CF) respectively. The following steps describe the operational scene of agent collaboration in clusters:

- Agent N receives a request from central process for a task planning.
- It checks its internal scripts, if it can participate then it solves the local steps.
- For external steps, it contacts cluster directory to check for other agents if they also have external capability.
- In that case, CD provides contact details.
- Upon receipt of these details, agent N creates CFN and passes on these details.
- CFN understands coordination context.
- CFN passes the request to specified agents, and thus cluster is formed.

For efficient collaboration, CD must remain updated for recording the information of its members, such as agent name, agent locator, service name, service type, and so on. Upon joining or leaving the cluster, an agent must register or cancel registration respectively through CF. Through a query, an agent can find out other members' services and locators. Through these steps, a trust is developed amongst agents.

NE: Network Element / Device with a tag Agent

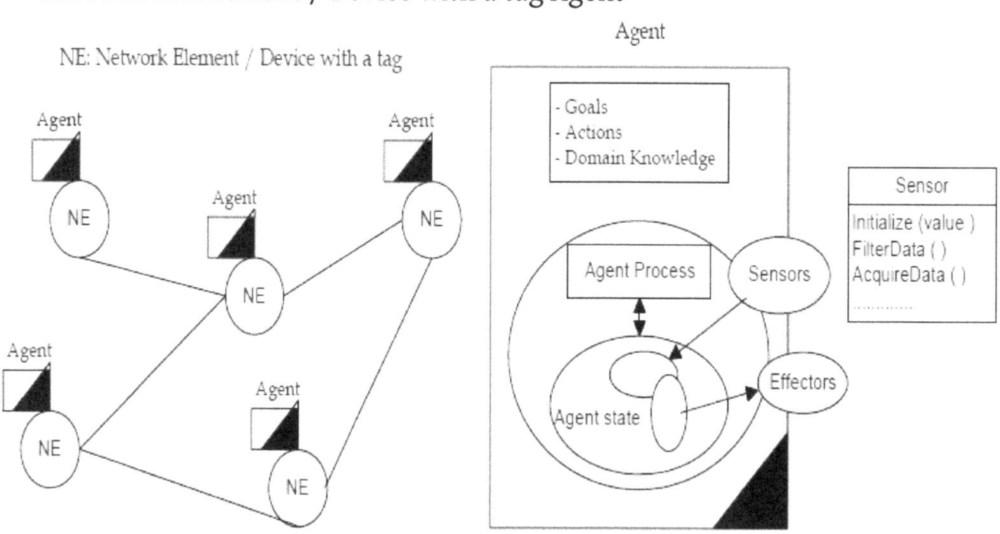


**Fig. 2. Agents in a process network**

Cluster 1: Agents (1,2,3,4); Cluster 2: Agents (3,5,6)

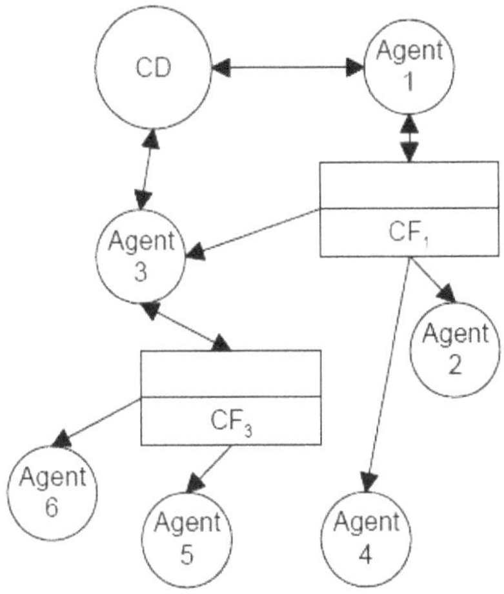


CD: Community Directory; CF: Community Facilitator

**Fig. 3. Agent Collaboration**

### 4.1.3 Agent Technology

Recently developed tools may be used to help design cluster facilitator (CF) and domain ontology, using for example DARPA Markup Language (DAML) (DAML website, 2000). The DAML extends XML (Extensible Markup Language) and RDF (Resource Description Framework) to include domain ontology. It provides rich set of constructs to create ontology and to markup information for attaining machine readability and understandability; it has capability of inference so that membership or service of cluster can be precisely defined. In case of XML and RDF, there seems to be no significant literature to manage ontology, and this is why, DAML is chosen to build cluster ontology. A number of agent standards have been in practice like Knowledge Query Manipulation Language (KQML), OMG's mobile agent system interoperability facility (MASIF) and the Foundation for Intelligent Physical Agents (FIPA). The Foundation for Intelligent Physical Agent (FIPA) Agent Management Specification (FIPA specification, 2002) is extended to develop the agent role called CF to manage cluster directory (CD) and cluster ontology. Using assistance from DAML-based ontology, the members of the cluster are able to form cluster and communicate with other agents. The interaction among domain ontology, CD and CF can be best understood using Figure 4. The Figure 4 shows how CF gets access to DAML files and facilitates the common goal of the cluster. There are tools available like Jena semantic web (HP Labs toolkit, 2003) that can be used to handle the cluster directory (CD) built using DAML, and to develop a Java class "Directory".

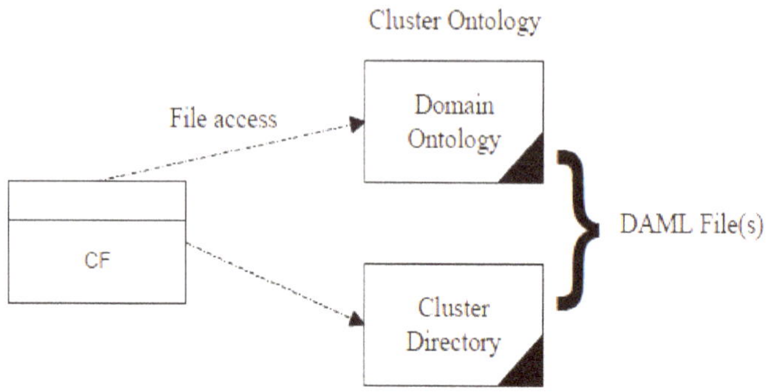


**Fig. 4. Linking CF with DAML**

From discussions above, the main functions of CD can be summarized, as:

- Add and Remove the information of an agent
- Get the list of agent names of all members
- Get the information of individual agent by name
- Get ontology used by members in the cluster
- Add external ontology if provided by an agent

Using local process mechanism and main functions of CD, the partial directory can be described as shown in Figure 5. It shows information of CF (lines 1-9) and members of cluster (lines 20-22), the cluster directory also records meta-data about cluster such as cluster name (line 12), cluster description (lines 13-15), ontology used in cluster (lines 16-18), and so on.

```
1. <cluster:CF rdf:ID="theCF">
2. <cluster:agentName>"CF"</cluster:agentName>
3. <cluster:agentDescription>
4. "DCS Cluster Facilitator"
5. </cluster:agentDescription>
6. <cluster:locator>
7. "http://dcs.ee.uaeu.ac.ae/DCS/agent/CF"
8. </cluster:locator>
9. </cluster:CF>
10.
11. <cluster:Cluster rdf:ID="DCSCluster">
12.<cluster:clusterName>"DCS"</cluster:clusterName>
13. <cluster:clusterDescription>
14. "Distributed Control System"
15. </cluster:clusterDescription>
16. <cluster ontology>
17. "http://dcs.ee.uaeu.ac.ae/DCS/ontology/dcs.daml"
18. </cluster:ontology>
19.
20. <cluster:hasCF rdf:Resource="#theCF"/>
21. <cluster:consistOf rdf:Resource="#agent1"/>
22. <cluster:consistOf rdf:Resource="#agent2"/>
23. </cluster:Cluster>
```

**Fig. 5. DCS Cluster Directory**

### 4.1.4 Example

An example can be illustrated to show how ontology may be updated (Fig. 6(b)) and that how interactions may develop in a local process. It should be noted here that basic cluster ontology (the knowledge of the local process) provided by CF remains the same but all members' domain knowledge (ontology) may not be the same. For example, user agent holds basic knowledge of the local process but does not understand the knowledge that a distributed field device holds. Through DAML-based ontology, members can communicate with each other to acquire requested service, as shown in Figure 6. It is clear from the Figure 6 that when distributed field device agent joins the cluster, it informs CF about corresponding ontology it provides (Figure 6(a)). Thus the CF maintains local process ontology plus the distributed field device ontology. When a user agent wants to perform a task, it asks CF about domain ontology and the agents that provide external capability. In response, CF informs the user agent if ontology is to be acquired (Figure 6(c)). Thus, the user agent can communicate with the distributed field device agent (Figure 6(d)).

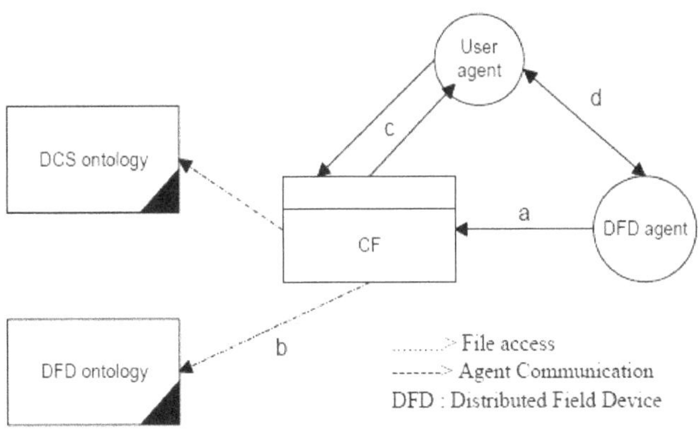


**Fig. 6. Update in ontology provided by distributed field device agent**

## 4.2 Central Process

This process handles core mechanism that glues organization's local processes to the central process. Some of the functions for example library of agents, their job description, definition of controller tasks, and domain ontology of each cluster can be defined offline before the implementation actually starts. The dynamic components are removing of agent deadlocks, security of agents, estimation of characteristics and relationships and decision making in cases of emergencies and when situation develops beyond the capabilities of agent clusters. It seems that all of these dynamic functions together may require computations, but the advantages gained are many: (i) reduced communications between central process controller and the device(s) (ii) provide simplicity to enable better interoperability (iii) intelligence gathering to build a degree of reconfigurability in a case estimated parameters exceed beyond a limit (iv) reduced human supervision. It can be also argued that complexity of this process is only a technology mismatch, and that if only small scale changes are to be decided at the central process like reconfiguration of device parameters, security of agents, then intelligence can further be distributed to the agents at the local level. Based on presented work in section 3 and in 4.1, agents can be embedded in tagged devices within a layered architecture to support business operations and services in real time. In Figure 7, the model architecture of four tiers is drawn to implement objectives of the central process. At the bottom layer (Tier 1), active readers or Profibus/Profinet enabled devices collect data, often collected on a trigger similar to a motion sensor. These readers should be controlled by one and only one edge server to avoid problems related to network partitioning. In addition, this layer supports the notion that intelligence be introduced at the edges to reduce data traffic and improve reaction at the next layer. This layer also provides hardware abstraction for various Profibus/Profinet compatible hardware and network drivers for interoperability of devices. The edge sever (Tier 2) regularly poll the readers for any update from device agents, monitors tagged devices and distributed devices through readers, performs device management, and updates integration layer. This layer may also work with system through controls and open source frameworks that provide abstraction and design layer. The integration layer (Tier 3) provides design and engineering of various objects needed for central controller as well as for field processes and for simulation levels of reconfigurability. This layer is close to business application layer (Tier 4). The monitoring of agents behavior, its parameters and cluster characteristics are done at this layer to assess the degree of reconfigurability. This layer also takes care of parameters like handling device processes, applications, security of agents, resource allocation and scheduling of processes.

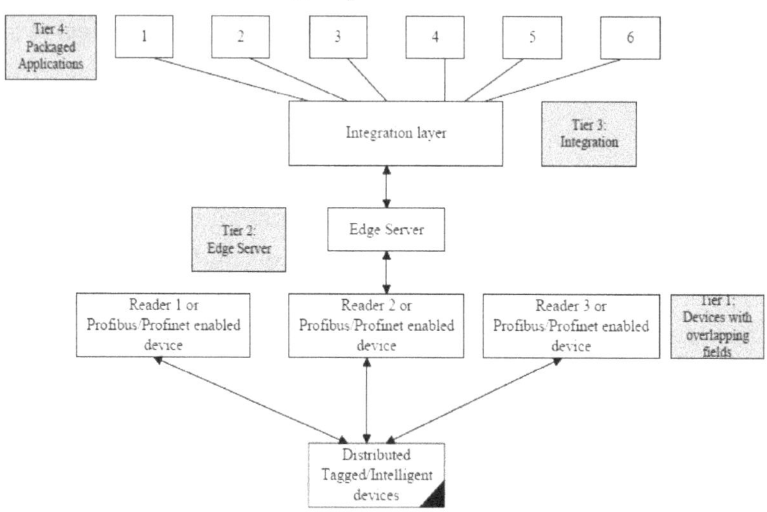


**Fig. 7. 4-Tier Reference Architecture**

The separation of edge server and integration layer improves scalability and reduces cost for operational management, as the edge is lighter and less expensive. The processing at the edge reduces data traffic to central point and improves reaction time. Similarly, the separation of integration from business applications helps in abstraction of process entities. The Tier 3 also enables it as self-healing and self-provisioning service architecture to increase availability and reduce support cost. Control messages flow into the system through business application portal to the integration layer, then on to the edge and eventually to the reader. Provisioning and configuration is done down this chain, while reader data is filtered and propagated up the chain.

The equation (3) may now be investigated again further, and evaluated using the model architecture, shown in Figure 7. The objective is to minimize the communication between field devices and the central process controller, and bring most of the local decision making intelligence at the local field level. Only when certain parameters need to be changed at local level device, then the values *d*1, *d*2, and *d*3 need to be estimated. In order to evaluate these delay parameters, it is sufficient to estimate one communication between an RFID reader and the central process controller (i.e., *d*1), as others just accumulate these delays over a number of communications. The communication between various nodes in Figure 7 may be abstracted using queuing models. In order to evaluate the performance modeling of this architecture, queuing model can be used, since communication traffic may be considered as datagram that involve traversing multiple paths, which means M/M/1 queuing system can be used. Assume that the service rate between a reader and edge server is μRE, between an edge and integration layer is μEI, and that between integration and central process at packaged application is μIC respectively. Assume also that datagram arrival rate from an RFID reader is λRE and *TIJ* is the average propagation delay between nodes *i* and *j*, where nodes *i* and *j* belong to one hop delay between any two nodes. Using these parameters, the RFID message delay from reader to central process controller may be written as accumulated delays across the entire path:

$$d_1 = d(R, \text{controller}) = \sum_{i,j}\left(\frac{\lambda ij}{\mu ij(\mu ij - \lambda ij)} + \frac{1}{\mu ij} + Tij\right) \quad (4)$$

With Giga bits per second wireless transmission rates available today, *TIJ* may be assumed negligible for one datagram traversing from one node to the other, along the entire path. The only terms left are the arrival rates and service rates along all node hops. Since 100 Mbps system (i.e., 0.014ms/message) is commonly available for RFID switches, network switches and servers (with exponential service waiting time), we may assume the corresponding values in equation (4) very easily. It may also be assumed that central process controller server message receiving rate at is 0.014ms/message, based on the same criterion. Thus, *d*1 may be estimated once we insert λIJ in the equation (4). It turns out that for given typical value of λ from reader as (say) 0.2, the estimated delay from a reader to central process controller is less than 0.1msec, which is acceptable for a central process controller that is waiting to update a set of parameters for agents down the local process. For a set of RFID tags (say 50) generating communication signals to central process controller (i.e., Yellow level situation), the estimated delay is still few milliseconds. For a situation involving 1000 tags generating messages (i.e., Red level), the total estimated delay is still less than a second.

### 4.3 Performance Gains

As presented in section 4.2, the communication delay has largely been reduced at the cost of increased intelligence at the local level. In fact, if we look at equation (3) we see that $d1(t)$, $d2(t)$ and $d3(t)$ minimize to a level when problem of the node device exceeds the threshold level of the agent intelligence. Thus this approach sets practical performance limits. However, again this is just a technology mismatch. If agent design technology reaches its maturity i.e., if the collaborative intelligence within agents exceeds combinatorial complexity of device problems then there is no need of communication between devices and the controller. Thus the requirements of the central process reduce to that of the customized design of the agents only, and its performance matches to that of the centralized MIMO system. The mechanism set at the local process provides self-healing, reliability and scalability. If a reader or service goes down, additional units can take up the workload automatically. If bottlenecks develop, the RFID system software can dynamically provision new service agents to manage increased requirements. The scalability is assured by a design at the central process that grows horizontally and vertically – like a single-CPU, tag-and- ship pilot through N-way and multi-purpose device deployments, smoothing the growth path. At the central process, design and reconfigurability can help introduce features in agents to thwart external and intrusive agents, and thus help boost security of operational devices and processes during real time. This set of gains has not been addressed in either of the approaches described in (Konomi et al., 2006; Lian et al., 2002; Maturana et al., 2005; Prayati et al., 2004). The combination of agents and tagging technology uses programming and standardized components, which adds versatility to the process control. This type of process control is suited to a wide range of applications that need wide area sensing, and control points. The exploitation of agents is expected to rise over time as other enabling technologies grow in prominence.

## 5. RECENT STANDARDIZATION

There has been a large standardization effort conducted towards process control communications and systems, covering a range of industries. It is not possible to describe all of them here, but most recent, relevant to this work is presented below (International Electrotechnical Commission standard, 2006-2009; Hart Communication Foundation standard, 2009; Aim Global RFID Guideline, 2009):

**IEC 60770-3 (2006):** The standard specifies the methods for reviewing the functionality and the degree of intelligence in intelligent transmitters, for testing the operational behavior and dynamic performance of an intelligent transmitter as well as methodologies for determining the reliability and diagnostic features used to detect malfunctions; and determining the communication capabilities of the intelligent transmitters in a communication network.

**IEC 69870-5-104 (2006):** The standard defines telecontrol companion standard that enables interoperability among compatible telecontrol equipments. It applies to telecontrol equipment and systems with coded bit serial data transmission for monitoring and controlling geographically widespread processes.

**IEC 61784-1-3 (2007):** It defines a set of protocol specific communication profiles based primarily on the IEC 61158 series, to be used in the design of devices involved in communications in factory manufacturing and process control. It contains a minimal set of required services at the application layer and specification of options in intermediate layers defined through references.

**IEC 62264-3 (2007):** It defines activity models of manufacturing operations management that enable enterprise system to control system integration. The activities defined are consistent with the object models definitions given in IEC 62264-1. The modeled activities operate between business planning and logistics functions, defined as the Level 4 functions and the process control functions, defined as the Level 2 functions of IEC 62264-1. The scope of this standard is limited to: - a model of the activities associated with manufacturing operations management, Level 3 functions; - an identification of some of the data exchanged between Level 3 activities.

**Hart 7.0 (2007):** The Hart Communication Foundation (HCF) has released the HART 7 specification, enabling more capabilities for communication with intelligent field devices, and targeting wireless communication in industrial plant environment. The specification allows building on established and field-proven international standards including IEC 61158, IEC 61804-3, IEEE 802.15.4 radio and frequency hopping, spread spectrum and mesh networking technologies.

**IEC 61298-1-4 (2008):** The specification defines general methods and procedures for conducting tests, and reporting on the functional and performance characteristics of process measurement and control devices. The methods and procedures specified in this standard are applicable to any type of process measurement and control device. The tests are applicable to any such devices characterized by their own specific input and output variables, and by the specific relationship (transfer function) between the inputs and outputs, and include analogue and digital devices.

**IEC 62424 (2008)E:** It specifies how process control engineering requests are represented in a P&ID for automatic transferring data between P&ID and PCE tool and to avoid misinterpretation of graphical P&ID symbols for PCE. It also defines the exchange of process control engineering data between a process control engineering tool and a P&ID tool by means of a data transfer language (called CAEX). These provisions apply to the export/import applications of such tools.

**IEC/PAS 62443-3 (2008):** It establishes a framework for securing information and communication technology aspects of industrial process measurement and control systems including its networks and devices on those networks, during the operational phase of the plant's life cycle. It provides guidance on a plant's operational security requirements and is primarily intended for automation system owners/operators (responsible for ICS operation).

**IEC 61850 (2009):** This is a standard for the design of electrical substation automation. Multiple protocols exist for substation automation, which include many proprietary protocols with custom communication links. The objectives set for the standard are: a single protocol for complete substation, definition of basic services required to transfer data, promotion of high interoperability between systems from different vendors, a common method/format for storing complete data, and define complete testing required for the equipments which confirms to the standard.

**IEC/PAS 62601 (2009):** It specifies WIA-PA system architecture and communication protocol for process automation based on IEEE 802.15.4. WIA-PA network is used for industrial monitoring, measurement and control applications.

**AIM Global RFID Guideline 396 (2008):** This guideline provides an overview of RFID chips and transponders, as well as the process of verifying and qualifying the design and fabrication of these chips. This guideline focuses on tagging items at the individual level, where the RFID tag can be found in different formats. These formats include a label that is attached to the inner or outer surface of a tire, or a patch that is permanently affixed to the tire. The tag can also be incorporated into the tire during the manufacturing process, becoming an integral part of the tire's structure.

## 6. CONCISE SUMMARY

The primary concept behind two processes is decentralization. The reduction in transmission delay comes at the expense of heightened intelligence at the local level. Upon examining equation (1), it becomes evident that d1(t), d2(t), or d3(t) reach their minimum level when the problem of the node device above the threshold level of the agent intelligence. If the level of collaborative intelligence surpasses the level of combinatorial complexity, then communication between devices and the controller becomes unnecessary, and the central process simply needs to focus on designing the agents. Therefore, the performance is equivalent to that of the centralized MIMO system. The central level's four-tier modular architecture facilitates the application of distributed intelligence at the field level and the design of agents. The relevant functionality for each layer has been integrated into the individual tiers at the central level. In addition, the incorporation of design and reconfigurability can facilitate the implementation of features in agents to counteract the actions of intrusive agents in real-time. Additionally, it was demonstrated that the projected latency resulting from communication between a tagged device and a central process controller is under one second when one thousand marked devices transmit their communication signal to the central process controller simultaneously. These specific benefits have not been asserted in either of the frequently explored methodologies for distributed control systems as described in the literature.

## REFERENCES


1. Almeida, L., Pedreiras, P., & Fonseca, J. (2002). The FFT-CAN Protocol: Why and How, *IEEE Transactions on Industrial Electronics*, Vol. 49, No. 6, pp. 1189-1201, December, 2002
2. Alonso, J., Ribas, J., Coz, J., Calleja, A., & Corominas, E. (2000). Development of a Distributive Control Scheme for Fluorescent Lighting based on LonWorks Technology, *IEEE Transactions on Industrial Electronic*, Vol. 47, No. 6, pp. 1253-1262, December, 2000
3. Association for Automatic Identification and Mobility, AIM Global radio frequency identification (RFID) Guideline REG 396, www.aimglobal.org/, [last accessed 04/25/2009]
4. Bernnan, Fletcher, M., & Norrie, D. (2002). An Agent-Based Approach to Reconfiguration of Real-time Distributed Control Systems, *IEEE Transactions on Robotics and Automation*, Vol. 18, No. 4, pp. 444-451, August, 2002
5. Bohn, J., & Mattern, F. (2004). Super-Distributed RFID Infrastructures, *Lecture Notes in Computer Science (LNCS)* No. 3295, pp. 1-12, Eindhoven, Netherlands, November 8- 10, Springer-Verlag, 2004
6. Bratukhin, A., & Treytl, A. (2006). Applicability of RFID and Agent-Based Control for Product Identification in Distributed Production, *Proceedings of IEEE Conference on Emerging Technologies and Factory Automation*, Vol. 20, Issue 22, pp. 1198-1205, Prague, 2006

7. Cavinato, M, Manduchi, G., Luchetta, A., & Taliercio, C. (2006). General-Purpose Framework for Real Time Control in Nuclear Fusion Experiments, *IEEE Transactions on Nuclear Science*, Vol. 53, No. 3, pp. 1002-1008, June, 2006

8. DAML website, DARPA Agent Markup Language (2000) [Online], http://www.daml.org/; [last accessed 04/22/2006]

9. ETSI RFID standards by ETSI, http://www.etsi.org/WebSite/Standards/Standard.aspx, Available online [last accessed on 04/25/2009]

10. Farinelli, A., Iocchi, L., & Nardie, D. (2004). Multirobot Systems: A Classification Focused on Coordination, *IEEE Transactions on Systems, Man, and Cybernetics*, Vol. 34, No. 5, pp. 2015-2028, October, 2004

11. FIPA website, Foundation for Intelligent Physical Agents (FIPA) Agent Management Specification (2002) [Online], http://www.fipa.org/specs/fipa00023/; [last accessed 09/21/2006]

12. Fregene, K., Kennedy, D., & Wang, D. (2005). Toward a Systems– and Control-Oriented Agent Framework, *IEEE Transactions on Systems, Man, and Cybernetics,* Vol. 35, No. 5, pp. 999-1012, October, 2005

13. Function Blocks (FB) for industrial-process measurement and control systems, IEC 61804, IEC 61131, and IEC 61499. http://www.iec.ch/searchpub/cur_fut.htm, [last accessed 02/26/2009]

14. Gingko Networks (2008). White Paper: Gingko Distributed Network Piloting System, *Gingo Networks*, September, 2008

15. Goodwin, G., Graebe, S., Salgado, M., (2001), Control System Design, New Jersey, Prentice Hall, 2001

16. Hart Communication Foundation website, Hart 7.0 Wireless Communication Protocol, http://www.hartcomm2.org/hart_protocol/wireless_hart/hart7_overview.html, [last accessed 04/27/2009]

17. Heck, B., Wills, L., & Vachtsevanos, G. (2003). Software Technology for Implementing Reusable, Distributed Control Systems, *IEEE Control Systems Magazine*, pp. 21-35, February, 2003

18. Hong, S. (2000). Experimental Performance Evaluation of Profibus-FMS, *IEEE Robotics and Automation Magazine*, pp. 64-72, December, 2000

19. HP Labs website, HP Labs (2003), [Online], Jena Semantic Web Toolkit Available:

20. http://www.hpl.hp.com/semweb/jena.htm/; [last accessed September 2006]

21. International Electrotechnical Commission (IEC, 2007) Webstore, http://webstore. iec.ch/webstore/webstore.nsf/artnum/ [last accessed 04/28/2009]

22. Ioannides, M. (2004). Design and Implementation of PLC based Monitoring Control System for Induction Motor, *IEEE Transactions on Energy Conversion*, Vol. 19, No. 3, pp. 469- 476, September, 2004
23. ISO RFID standards, http://www.iso.org/rfid, Available online [last accessed on 04/25/2009]
24. Kleines, H., Sarkadi, J., Suxdorf, F., & Zwoll, K. (2004). Measurement of Real Time Aspects of Simatic PLC Operation in the Context of Physics Experiments, *IEEE Transactions on Nuclear Science*, Vol. 51, No. 3, pp. 489-494, June, 2004
25. Konomi, S., Inoue, S., Kobayashai, T., Tsuchida, M., & Kitsuregawa, M. (2006). Supporting Colocated Interactions Using RFID and Social Network Displays, *IEEE Pervasive Computing Magazine*, Vol. 5, Issue 3, pp. 48-56, July-September, 2006
26. Lian, F., Moyne, J., & Tilbury, D. (2001). Performance Evaluation of Control Networks, *IEEE Control Systems Magazine*, pp. 66-83, February, 2001
27. Lian, F., Moyne, J., & Tilbury, D. (2002). Network Design Consideration for Distributed Control Systems, *IEEE Transactions on Control Systems Technology*, Vol. 10, No. 2, pp. 297-307, March 2002
28. Maturana, F., Staron, R., & Hall, K. (2005). Methodologies and Tools for Intelligent Agents in Distributed Control, *IEEE Intelligent Systems*, pp. 42-49, February, 2005
29. Memon, Q. (2008). A Framework for Distributed and Intelligent Process Control, *Proceedings of 5th* International Conference on Informatics in Control, Automation and Robotics, pp. 240-243, May 12-15, Madeira, Portugal, 2008
30. Naby, A. & Giorgini, S. (2006). Locating Agents in RFID Architectures, Technical Report No. DIT-06-095, University of Trento, Italy, 2006
31. O'Hearn, T., Cerff, J., & Miller, S. (2002). Integrating Process and Motor Control, *IEEE Industry Applications Magazine*, pp. 61-65, August, 2002
32. Palmer, M. (2007). Seven Principles of Effective RFID Data Management, *A Technical Primer*, Real Time Division, Progress Software, Inc., http://www.progress.com/ [last accessed 03/21/2009]
33. Prayati, A., Koulamas, C., Koubias, S., & Papadopoulos, G. (2004). A Methodology for the Development of Distributed Real-time Control Applications with Focus on Task
34. Allocation in Heterogeneous Systems, *IEEE Transactions on Industrial Electronics,* Vol. 51, No. 6, pp. 1194-1206, December, 2004
35. Recht, B. & D'Andrea, R. (2004). Distributed Control of Systems over Discrete Groups, *IEEE Transactions on Automatic Control*, Vol. 49, No. 9, pp. 1446-1452, September, 2004

36. RFID Journal, A Summary of RFID Standards, http://www.rfidjournal.com/article/articleview/609/1/1/ [last accessed on 04/25/2009]
37. Stewart, G., Gorinevsky, D., & Dumont, G. (2003). Feedback Controller Design for a Spatially Distributed Systems: The Paper Machine Problem, *IEEE Transactions on Control Systems Technology*, Vol. 11, No. 5, pp. 612-628, September, 2003
38. The EPC global, http://www.epcglobalinc.org, Available online [last accessed on 04/25/2009]
39. The RFID Ecosystem, http://rfid.cs.washington.edu/, [last accessed on 04/24/2009]
40. Tovar, E., & Francisco, F. (1999). Real-time field bus communications using Profibus Networks, *IEEE Transactions on Industrial Electronics*, Vol. 46, No. 6, pp. 1241-1251, December, 1999
41. Vyatkin, V. (2008). Distributed IEC 61499 Intelligent Control of Reconfigurable Manufacturing Systems, *Technical Report*, University of Auckland, 2008
42. Wang, F. (2005). Agent-Based Control for Networked Traffic Management Systems, *IEEE Intelligent Systems*, pp. 92-96. October, 2005
43. Ward, M., Kranenburg, R., & Backhouse, G. (2006). RFID: Frequency, Standards, Adoption and Innovation, *Technical Report in JISC Technology and Standards Watch,* May, 2006
44. Willig, A. (2003). Polling based MAC protocols for improving real-time performance in a wireless Profibus, *IEEE Transactions on Industrial Electronics,* Vol. 50, No. 4, pp. 806- 817, August, 2003
45. Yang, C., Vyatkin, V. (2008). Design and validation of distributed control with decentralized intelligence in process industries: A survey, *IEEE International Conference on Industrial Informatics*, pp. 1395 – 1400, Daejeon, July, 2008
46. Yang, Y. (2006). Attenuation Splice Control in the Manufacture of Fiber Optical Communication System, *IEEE Transactions on Control Technology*, Vol. 14, No. 1, pp. 170-175, January, 2006

# 12 INDUSTRIAL PROCESSES: NEURAL GENERALIZED PREDICTIVE CONTROL

## 1. INTRODUCTION: AN OVERVIEW

The manufacturing industry necessitates high-speed, rapid-response, and highly precise performances as crucial requirements. Model predictive control (MPC), created in the late 1970s, is a type of computer control system that uses a process model to forecast the future behavior of the plant (Qin & Badgwell, 2004). Over the past twenty years, Model Predictive Control (MPC) has gained significant recognition for its effectiveness in achieving set point monitoring and addressing model mismatch in many industries such as refining, petrochemical, chemical, pulp and paper manufacture, and food processing (Rossiter, 2006). The utilization of model predictive control is also implemented in the positioning control of an ultra-precision stage that is controlled by a linear actuator (Hashimoto, Goko, et al., 2008). Dynamic Matrix Control (DMC), Model Algorithmic Control (MAC), Predictive Functional Control (PFC), Extended Prediction Self Adaptive Control (EPSAC), Extended Horizon Adaptive Control (EHAC), and Generalized Predictive Control (GPC) are among the most popular MPC algorithms that have gained widespread acceptance in industry (Sorensen, Norgaard, et al., 1999). Typically, in the majority of controllers, the disturbances caused by the manipulated variable are addressed only after they have already impacted the output of the process. Therefore, it is imperative to create a controller that can anticipate and enhance process efficiency. Model Predictive Control (MPC) is a widely adopted control technique in the process industries. It utilizes an optimizer to calculate the control trajectory for a future time period, based on a dynamic model of the processes. Linear models are commonly employed in most applications of model predictive approaches to forecast the behavior of a process within the desired time frame. However, as the majority of actual processes exhibit nonlinear characteristics, it is necessary to make efforts to enhance predictive control methods in order to account for the nonlinearities of the system. The derivation of the mathematical model is the costliest aspect of implementing a nonlinear predictive control scheme. Obtaining a physically established process model can be challenging, and in certain situations, it may even be impossible. This is often owing to the complexity of the underlying process or a lack of understanding regarding essential parameters of the model. An effective approach to address these issues is to utilize neural networks as nonlinear models capable of efficiently approximating the dynamic behavior of the process.

Generalized Predictive Control (GPC) is a distinct subset of digital control systems called Model Predictive Control (MPC) that has been developed independently. The method developed by Clarke, Mohtadi, et al. in 1987 has gained widespread popularity in both industry and academics as one of the most widely used MPC methods. It has been effectively utilized in numerous industrial applications, demonstrating strong performance and a notable level of resilience. This system is capable of addressing various control issues for a diverse set of plants, using a manageable amount of design factors. The user must specify these variables based on their prior understanding of the plant and control objectives. GPC has the ability to regulate non-minimum phase plants, plants that are unstable in open loop, and plants with varying or unknown dead time. The Generalized Predictive Control (GPC) algorithm is very resilient to inaccuracies in the model and disturbances caused by sensor noise. Using a neural network to understand the dynamics of the plant can boost the ability of Generalized Predictive govern (GPC) to govern nonlinear plants and generate accurate predictions. This chapter has covered the topic of using a neural network to create a control technique called Neural Generalized Predictive Control (NGPC) (Rao, Murthy, et al., 2006). The NGPC algorithm functions in two distinct modes, namely prediction and control. It produces a series of upcoming control signals during each sampling period to maximize the efficiency of controlling the systems being managed. The control vector calculations in NGPC are performed at regular intervals and rely on both the control and prediction horizon. A computational comparison of generalized predictive control (GPC) and neural generalized predictive control (NGPC) schemes is shown in the study conducted by Rao, Murthy, et al. in 2007.The study conducted by Pitche, Sayyer-Rodsari, et al. (2000) addresses the impact of a reduced output horizon in neural generalized predictive control. The use of nonlinear model predictive control utilizing neural network was also pioneered by Chen, Yuan, et al. in 2002. The paper by Raff and Sinz (2008) introduces two different ways to model predictive control (MPC) for continuous-time systems with uncertain constraints and piecewise constant control input. These approaches are referred to as on-line MPC and off-line MPC.

There has been a significant amount of scholarly literature and conference papers discussing the application of neural network models in the implementation of Model Predictive Control (MPC) with finite prediction horizons. The majority of publications focus on the topics pertaining to the construction of neural network models. There is a lack of focus on stability and closed-loop performance, despite the fact that these concerns are still open and unsolved. The paper by Sun, Chang, et al. (2002) proposes a predictive control technique that utilizes an enhanced back propagation neural network to address real-time control in nonlinear systems with time delays.Predictive control would not be satisfactory for nonlinear processes. Similar to neural networks, fuzzy logic has also garnered significant attention for controlling nonlinear processes. Controlling nonlinear systems offers numerous benefits due to their capacity for approximation utilizing nonlinear mappings. Typically, they avoid utilizing parametric models like as transfer functions or state space equations. Hence, the outcome of modeling or regulating nonlinear systems does not follow a predictable pattern, and we can only ascertain if their performance is sufficient. If the controller necessitates the parametric representation of the nonlinear system, there are no existing means to connect the controller with the fuzzy modeling method. The fuzzy model-based prediction is obtained by considering the output operating point, and the optimum control is computed using the fuzzy prediction model and optimization techniques described in the work of Kim, Ansung, et al. (1998).

This chapter demonstrates the efficacy of a novel algorithm called Generalized Predictive Control (GPC) for controlling industrial processes. The algorithm's capability is assessed across a range of systems. This paper presents a very effective implementation of Generalized Predictive Control (GPC) by utilizing a multi-layer feed-forward neural network as the nonlinear model for the plant. This implementation, known as NGPC, significantly enhances the capabilities of GPC by efficiently controlling both linear and nonlinear processes. The typical GPC utilizes a brain model of the plant, referred to as a neural generalized predictive control (NGPC). For the purpose of verifying the effectiveness of both GPC and NGPC algorithms, we will use Duffing's nonlinear equation as a well-known example. The output of the trained neural network is utilized as the expected output of the plant. The forecasted result is utilized in the algorithm for minimizing the cost function. The GPC criterion is minimized using two distinct schemes: a Quasi Newton method and a Levenberg Marquardt algorithm. GPC and NGPC are utilized to assess the proficiency of both linear and nonlinear systems. The performance of these combinations has been evaluated based on the Integral Square Error (ISE) and Integral Absolute Error (IAE). Only a few additional steps were needed for the GPC system compared to the NGPC system in order to stabilize the output. However, the most significant observation is that there is no indication of instability. The performance of the NGPC is also evaluated using a highly nonlinear process, namely the continuous stirred tank reactor (CSTR), as well as a linear process including a DC motor.

The ideas appearing in greater or lesser degree in all the predictive control family are basically:

- Explicit use of a model to predict the process output at future time instants (horizon).
- Calculation of a control sequence minimizing an objective function.
- Receding strategy, so that at each instant the horizon is displaced towards the future, which involves the application of the first control signal of the sequence calculated at each step.

## 2. MPC STRATEGY

The methodology of all the controllers belonging to the MPC family is characterized by the following strategy, as represented in Fig.1:

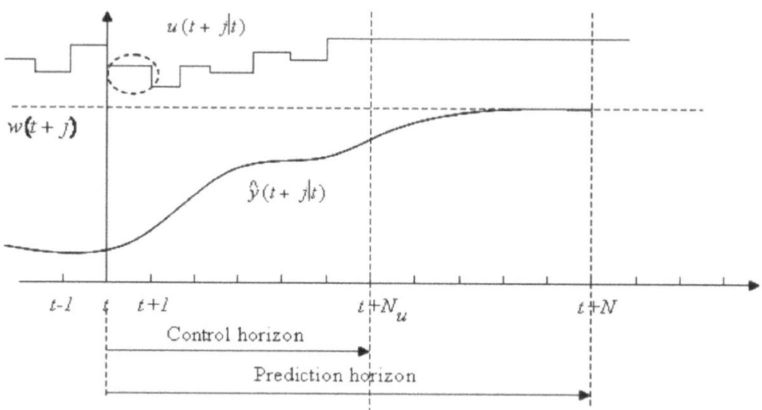


**Fig. 1. MPC Strategy**

- The future outputs for a determined horizon N, called the prediction horizon, are predicted at each instant k using the process model. These predicted outputs y(t+j/t) for j = 1 ... N depend on the known values up to instant t (past inputs and outputs) and on the future control signals u(t+j/t), j = 0 ... N-1, which are those to be sent to the system and to be calculated.
- The set of future control signals is calculated by optimizing a determined criterion in order to keep the process as close as possible to the reference trajectory w(t+j) (which can be the set point itself or a close). This criterion usually takes the form of a quadratic function of the errors between the predicted output signal and the predicted reference trajectory. The control effort is included in the objective function in most cases. An explicit solution can be obtained if the criterion is quadratic, the model is linear and there are no constraints; otherwise an iterative optimization method has to be used. Some assumptions about the structure of the future control law are made in some cases, such as that it will be constant from a given instant.

- The control signal u(t/t) is sent to the process whilst the next control signal calculated are rejected, because at the next sampling instant y(t+1) is already known and step1 is repeated with this new value and all the sequences are brought up to date. Thus the u(t+1|t) is calculated (which in principle will be different to the u(t+1|t) because of the new information available) using receding horizon control.

## 3. GENERALIZED PREDICTIVE CONTROLLER (GPC)

### 3.1 Introduction

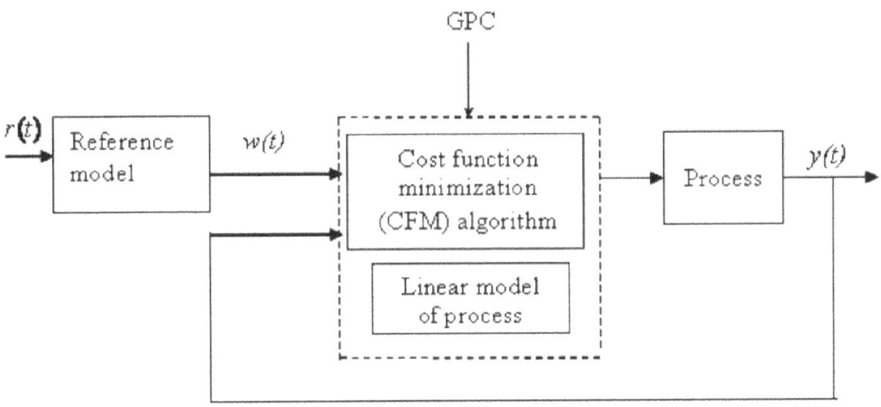


**Fig. 2. Basic Structure of GPC**

The fundamental concept of GPC is to compute a series of forthcoming control signals in a manner that minimizes a cost function spanning multiple stages, as defined over a prediction horizon. The index to be optimized is the expected value of a quadratic function that calculates the difference between the predicted output of the system and a certain reference sequence during a specific time period, in addition to a quadratic function that calculates the amount of control effort required. Generalized Predictive Control has many characteristics with other predictive controllers, as it is built on similar principles. However, it also exhibits certain distinctions. Later on, it will become evident that this solution offers an analytical approach without any limitations. It is capable of handling plants that are unstable and non-minimum phase. Additionally, it includes the idea of control horizon and takes into account the weighting of control increments in the cost function. The wide range of options offered by GPC results in a more diverse set of control objectives compared to alternative methods, some of which can be seen as smaller parts or specific instances of GPC. Figure 2 displays the GPC scheme. The system comprises of the target plant, a reference model defining the desired plant performance, a linear model of the plant, and the Cost Function Minimization (CFM) algorithm that calculates the necessary input to achieve the desired plant performance. The GPC algorithm comprises the CFM block. The GPC system begins with the input signal, r(t), which is then fed into the reference model. The tracking reference signal, denoted as w(t), is generated by this model and serves as an input to the CFM block. The CFM algorithm generates an output that serves as an input to the plant. The CFM algorithm utilizes this model to compute the subsequent control input, u(t+1), based on predictions of the plant's model response, during intervals between samples. After the cost function has been minimized, the input is then transmitted to the plant.

### 3.2 Formulation of Generalized Predictive Control

Most single-input single-output (SISO) plants, when considering operation around particular set-points and after linearization, can be described by the following:

$$A(z^{-1})y(t) = z^{-d}B(z^{-1})u(t-1) + C(z^{-1})e(t) \tag{1}$$

where *u(t)* and *y(t)* are the control and output sequence of the plant and *e(t)* is a zero mean white noise. A, B and C are the following polynomials in the backward shift operator $z^{-1}$:

$$A(z^{-1}) = 1 + a_1 z^{-1} + a_2 z^{-2} + \ldots\ldots + a_{na} z^{-na}$$

$$B(z^{-1}) = b_0 + b_1 z^{-1} + b_2 z^{-2} + \ldots\ldots + b_{nb} z^{-nb}$$

$$C(z^{-1}) = 1 + c_1 z^{-1} + c_2 z^{-2} + \ldots\ldots + c_{nc} z^{-nc}$$

where, d is the dead time of the system. This model is known as a Controller Auto- Regressive Moving-Average (CARMA) model. It has been argued that for many industrial applications in which disturbances are non-stationary an integrated CARMA (CARIMA) model is more appropriate. A CARIMA model is given by,

$$A(z^{-1})y(t) = z^{-d}B(z^{-1})u(t-1) + C(z^{-1})\frac{e(t)}{\Delta} \tag{2}$$

with $\nabla$=1 $z^{-1}$

For simplicity, C polynomial in (2) is chosen to be 1. Notice that if $C^{-1}$ can be truncated it can be absorbed into A and B.

### 3.3 Cost Function

The GPC algorithm consists of applying a control sequence that minimizes a multistage cost function,

$$J(N_1, N_2, N_u) = \sum_{j=N_1}^{N_2} \delta(j)[\hat{y}(t+j|t) - w(t+j)]^2 + \sum_{j=1}^{N_u} \lambda(j)[\Delta u(t+j-1)]^2 \tag{3}$$

Where $\hat{y}$ (t + j | t) is an optimum j-step ahead prediction of the system output on data up to time k, $N_1$ and o$N$ and are the are the minimum and maximum costing horizons, $N_u$ control horizon, δ ( j )and λ ( j ) is the future reference trajectory, which can considered to be the constant.

The objective of predictive control is to compute the future control sequence u (t ), u (t +1) ,... u (t + $N_u$) in such a way that the future plant output y ( t + j ) is driven close to w ( t + j ) . This is accomplished by minimizing J ( $N_1$, $N_2$, $N_u$ ).

### 3.4 Cost Function Minimization Algorithm

In order to optimize the cost function the optimal prediction of y (t + j ) for j ≥ $N_1$ and j ≤ $N_2$ is required. To compute the predicted output, consider the following Diophantine equation,

$$1 = E_j(z^{-1})\tilde{A}(z^{-1}) + z^{-j}F_j(z^{-1}) \quad \text{with } \tilde{A}(z^{-1}) = \Delta A(z^{-1}) \tag{4}$$

The polynomials $E_j$ and j $F_j$ are uniquely defined with degrees j-1 and na respectively. They can be obtained dividing 1 by $\tilde{A}$% ( $z^{-1}$ ) until the remainder can be factorized as $z^{-j}$ $F_j$ ($z^{-1}$). The quotient of the division is the polynomial $E_j$ ($z^{-1}$)- . An example demonstrating calculation of $E_j$ and $F_j$ coefficients in Diophantine equation (4) is shown in Example 1 given below:

. The quotient of the division is the polynomial

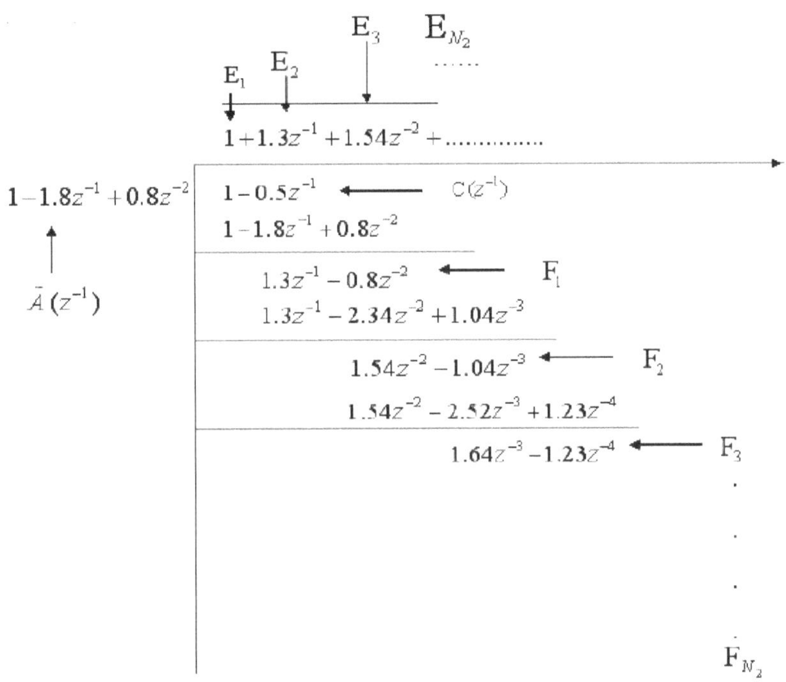


Example 1. Diophantine equation demonstration example

If (2) is multiplied by Δ $E_j$ ($z^{-1}$)$z^j$ we get,

$$\tilde{A}(z^{-1})E_j(z^{-1})y(t+j) = E_j(z^{-1})B(z^{-1})\Delta u(t+j-d-1) + E_j(z^{-1})e(t+j)$$

Substituting (4), in (5) we get,

$$(1 - z^{-j}F_j(z^{-1}))y(t+j) = E_j(z^{-1})B(z^{-1})\Delta u(t+j-d-1)) + F_j(z^{-1})y(t) + E_j(z^{-1})e(t+j)$$

which can be rewritten as:

$$y(t+j) = F_j(z^{-1})y(t) + E_j(z^{-1})B(z^{-1})\Delta u(t+j-d-1)) + F_j(z^{-1})y(t) + E_j(z^{-1})e(t+j) \quad (6)$$

degree of polynomial is E ( $z^{-1}$ ) = j - 1 the noise terms in equation are all in the future. The best prediction of

$$\hat{y}(t+j \mid t) = G_j(z^{-1})\Delta u(t+j-d-1) + F_j(z^{-1})y(t) \quad (7)$$

were $G_j(Z^{-1}) = E_j(z^{-1})\, B\, (z^{-1})$

Consider that polynomials $E_j$ and $F_j$ have been obtained by dividing 1 by $\tilde{A}$ ($z^{-1}$) until the remainder of the division can be factorized as expressed as:

$$F_j(z^{-1}) = f_{j,0} + f_{j,1}z^{-1} + f_{j,2}z^{-2} + \ldots\ldots + f_{j,na}z^{-na} \quad (8)$$

$$E_j(z^{-1}) = e_{j,0} + e_{j,1}z^{-1} + e_{j,2}z^{-2} + \ldots\ldots + e_{j,j-1}z^{-(j-1)} \quad (9)$$

Suppose that the same procedure is used to obtain $E_{j+1}$ and $F_{j+1}$, that is, dividing 1 by Ã ($z^{-1}$) until the remainder of the division can be factorized as $z^{-(j-1)}F_{j+1}(Z^{-1})$ with

$$F_{j+1}(z^{-1}) = f_{j+1,0} + f_{j+1,1}z^{-1} + \ldots\ldots + f_{j+1,na}z^{-na} \quad (10)$$

It is clear that only another step of the division is performed to obtain the polynomials $E_{j+1}$ and $F_{j+1}$. The polynomial $E_{j+1}$ will be given by:

$$E_{j+1}(z^{-1}) = E_j(z^{-1}) + e_{j+1,j}z^{-j} \quad (11)$$

with $e_{j+1,j} = f_{j,0}$. The coefficients of polynomial $F_{j+1}$ can then be expressed as:

$$f_{j+1,i} = f_{j+1,i+1} - f_{j,0}\tilde{a}_{i+1} \qquad i = 0 \ldots\, na-1 \quad (12)$$

The Polynomial $G_{j+1}$ can be obtained recursively as follows:

$$G_{j+1} = E_{j+1}B = (E_j + f_{j,0}z^{-1})B \quad (13)$$

$$G_{j+1} = G_j + f_{j,0}z^{-1}B \quad (14)$$

That is, the first j coefficient of coefficients $G_{j+1}$ will be identical to those of $G_j$ and the remaining

coefficients will be given by:

$$g_{j+1,j+i} = g_{j,j+i} + f_{j,0}b_i \qquad \text{for } i = 0 \ldots\ldots nb \quad (15)$$

To solve the GPC problem the set of control signals u (t) , u (t+1) ,..., u (t+N) has to be obtained in order to optimize expression. As the system considered has a dead time d sampling periods, the output of the system will be influenced by signal u (t) after sampling period d+1 . The values $N_1$, $N_2$ and $N_u$ defining the horizon can be defined by N2 = d+N, N2 = d+N and Nu = N. Notice that there is no point in making $N_1$ > d+1 as terms added to expression will only depend on the past control signals. On the other hand, if N1 > d + 1 , the first point in the reference sequence, being the ones guessed with most certainly, will not be taken into account.

Now consider the following set of j ahead optimal predictions given below:

$$\begin{aligned}\hat{y}(t+d+1\mid t) &= G_{d+1}\Delta u(t) + F_{d+1}y(t)\\ \hat{y}(t+d+2\mid t) &= G_{d+2}\Delta u(t+1) + F_{d+2}y(t)\\ \hat{y}(t+d+N\mid t) &= G_{d+N}\Delta u(t+N-1) + F_{d+N}y(t)\end{aligned} \tag{16}$$

which can be re-written as:

$$y = \mathbf{G}u + \mathbf{F}(z^{-1})y(t) + \mathbf{G}'(z^{-1})\Delta u(t-1) \tag{17}$$

where,

$$y = \begin{bmatrix}\hat{y}(t+d+1\mid t)\\ \hat{y}(t+d+2\mid t)\\ \dots\\ \hat{y}(t+d+N\mid t)\end{bmatrix};\ u = \begin{bmatrix}\Delta u(t)\\ \Delta u(t+1)\\ \dots\\ \Delta u(t+N-1)\end{bmatrix};\ \mathbf{G} = \begin{bmatrix} g_0 & 0 & \dots & 0\\ g_1 & g_0 & \dots & 0\\ . & . & . & .\\ g_{N-1} & g_{N-2} & \dots & g_0\end{bmatrix}$$

$$\mathbf{G}'(z^{-1}) = \begin{bmatrix}(\mathbf{G}_{d+1}(z^{-1}) - g_0)z\\ (\mathbf{G}_{d+2}(z^{-1}) - g_0 - g_1 z^{-1})z^2\\ \dots\dots\\ (\mathbf{G}_{d+N}(z^{-1}) - g_0 - g_1 z^{-1}\dots\dots g_{N-1}z^{-(N-1)})z^N\end{bmatrix};\ \mathbf{F}(z^{-1}) = \begin{bmatrix}F_{d+1}(z^{-1})\\ F_{d+2}(z^{-1})\\ \dots\\ F_{d+N}(z^{-1})\end{bmatrix}$$

Note that if the plant dead time is d > 1 the first d - 1 rows of G will be null, but if instead $N_1$ is assumed to be equal to d the leading element is non-zero. However, as d will not in general be known in the self-tuning case one key feature of the GPC approach is that a stable solution is possible even if the leading rows of G are zero.

Notice that the last two terms in (17) only depends on the past and can be grouped into f leading to:

$$\hat{y} = \mathbf{G}\,\mathrm{u} + \mathrm{f}$$

Notice that if all initial conditions are zero, the free response f is also zero. If a unit step is applied to the input at time t; that is

$$\Delta u(t) = 1,\ \Delta u(t+1) = 0,\ \dots,\ \Delta u(t+N-1) = 0$$

then expected output sequence $[\hat{y}(t+1), \hat{y}(t+2), \dots\dots \hat{y}(t+N)]^T$ is equal to the first column of matrix G. That is, the first column of matrix G can be calculated as the step response of the plant when a unit step is applied to the manipulated variable. The free response term can be calculated recursively by:

$$\mathbf{f}_{j-1} = z(1 - \tilde{A}(z^{-1}))\mathbf{f}_j + B(z^{-1})\Delta u(t-d+j) \tag{18}$$

with $\mathrm{f}_0 = y(t)$ and $\Delta u(t+j) = 0$ for $j \geq 0$.

The equation (3) can be written as:

$$J = (Gu + f - w)^T(Gu + f - w) + \lambda u^T u \tag{19}$$

where,

$$w = [w(t+d+1)\quad w(t+d+2)\quad \dots\dots\quad w(t+d+N)]^T$$

It has been considered that the future reference trajectory keeps constant along the horizon or its evolution is unknown and therefore w(t + i) = w(t ).

The equation (19) can be written as:

$$J = \frac{1}{2} u^T H u + b^T u + f_0 \tag{20}$$

where,

$$\mathbf{H} = 2(\mathbf{G}^T\mathbf{G} + \mathrm{RI})$$
$$b^T = 2(\mathrm{f}-\mathrm{w})^T\mathbf{G}$$
$$\mathrm{f}_0 = (\mathrm{f}-\mathrm{w})^T(\mathrm{f}-\mathrm{w})$$

The minimum of J, assuming there are no constraints on the control signals, can be found by making the gradient of J equal to zero, which leads to:

$$u = -H^{-1}b = (G^T G + \lambda I)^{-1} G^T (w - f) \tag{21}$$

The dimension of the matrix involved in equation (19) is N x N. Although in the non- adaptive case the inversion need be performed only once. In a self-tuning version the computational load of inverting at each sample would be excessive. Moreover, if the wrong value for dead-time is assumed, $G^T G$ is singular and hence a finite non-zero value of weighting λ would be required for a realizable control law, which is inconvenient because the accurate value for λ would not be known a priori. Notice that the control signal that is actually sent to the process is the first element of vector *U*, which is given by:

$$\Delta u = K(w - f) \tag{22}$$

where K is the first row of matrix $( G^T G + \lambda I )^{-1} G^T$. If there are no future predicted errors, that is, if ( w-f ) = 0 , then there is no control move, since the objective will be fulfilled with the free evolution of the process. However, in the other case, there will be an increment in the control action proportional (with a factor K) to that future error. Notice that the action is taken with respect to future errors, not past errors, as is the case in conventional feedback controllers.

Also notice that when only the first element of *U* is applied, then at the next sampling instants, new data are acquired and a new set of control moves is calculated. Once again, only the first control move is implemented. These activities repeated at each sampling instant, and the strategy is referred to as a receding horizon approach. It may strange to calculate an N u -step control policy and then only implement the first move. The important advantage of this *receding horizon approach* is that new information in the form of the most recent measurements y(k) is utilized immediately instead of being ignored for the next $N_u$ sampling instants. Otherwise, the multi-step predictions and control moves would be based on old information and thus be adversely affected by unmeasured disturbances.

## 4. AN OVERVIEW OF NEURAL GENERALIZED PREDICTIVE CONTROL

The Generalized Predictive Control (GPC), introduced in above section, belongs to a class of digital control methods called Model-Based Predictive Control (MBPC). GPC is known to control a non-minimum phase plants, open-loop unstable plants and plants with variable or unknown dead time. GPC had been originally developed with linear plant predictor models which, leads to a formulation that can be solved analytically. But most of the real processes show nonlinear behavior. Some work has to be done to extend the predictive control techniques to incorporate nonlinear models. Developing adequate nonlinear empirical models is very difficult and there is no model form that is clearly suitable to represent general nonlinear processes. Part of the success of standard model based predictive techniques was due to the relative ease with which step and impulse responses or low order transfer functions could be obtained. A major mathematical obstacle to complete theory of nonlinear processes is the lack of superposition principal for nonlinear systems. The selection of the minimization algorithm affects the computational efficiency of the algorithm. Explicit solution for it can be obtained if the criterion is quadratic, the model is linear and there are no constraints; otherwise an iterative optimization method has to be used. In this work a Newton-Raphson method is used as the optimization algorithm. The main cost of the Newton-Raphson algorithm is in the calculation of the Hessian, but even with this overhead the low iteration numbers make Newton-Raphson a faster algorithm for real-time control (Soloway & Haley,1997).

The Neural Generalized Predictive Control (NGPC) scheme is shown in Fig. 3. It consists of four components, the plant to be controlled, a reference model that specifies the desired performance of the plant, a neural network that models the plant, and the Cost Function Minimization (CFM) algorithm that determines the input needed to produce the plant's desired performance. The NGPC algorithm consists of the CFM block and the neural net block.

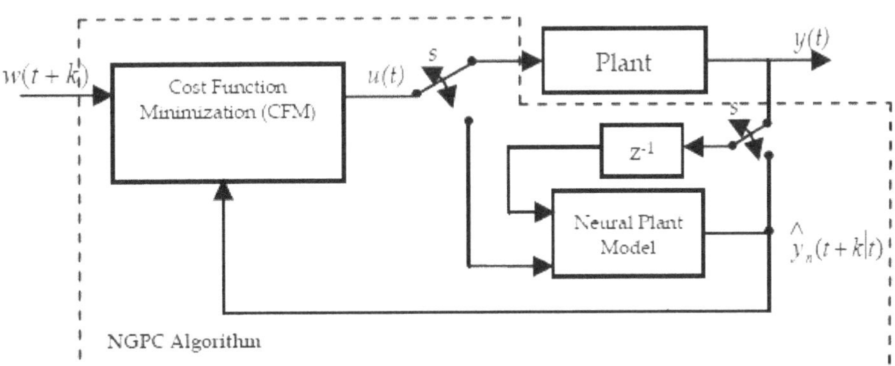


**Fig. 3. Block Diagram of NGPC System**

The NGPC system starts with the input signal, w(t), which is applied to the reference model. This model produces a tracking reference signal, w(t+k), that is used as an input to the CFM block. The CFM algorithm produces an output which is either used as an input to the plant or the plant's model. The double pole double throw switch, S, is set to the plant when the CFM algorithm has solved for the best input, u(n), that will minimize a specified cost function. Between samples, the switch is set to the plant's model where the CFM algorithm uses this model to calculate the next control input, u(n+1), from predictions of the response from the plant's model. Once the cost function is minimized, this input is passed to the plant. The computational performance of a GPC implementation is largely based on the minimization algorithm chosen for the CFM block. The selection of a minimization method can be based on several criteria such as: number of iterations to a solution, computational costs and accuracy of the solution. In general these approaches are iteration intensive thus making real-time control difficult. In this work Newton-Raphson as an optimization technique is used. Newton-Raphson is a quadratically converging. The improved convergence rate of Newton-Raphson is computationally costly, but is justified by the high convergence rate of Newton-Raphson. The quality of the plant's model affects the accuracy of a prediction. A reasonable model of the plant is required to implement GPC. With a linear plant there are tools and techniques available to make modeling easier, but when the plant is nonlinear this task is more difficult. Currently there are two techniques used to model nonlinear plants. One is to linearize the plant about a set of operating points. If the plant is highly nonlinear the set of operating points can be very large. The second technique involves developing a nonlinear model which depends on making assumptions about the dynamics of the nonlinear plant. If these assumptions are incorrect the accuracy of the model will be reduced. Models using neural networks have been shown to have the capability to capture nonlinear dynamics. For nonlinear plants, the ability of the GPC to make accurate predictions can be enhanced if a neural network is used to learn the dynamics of the plant instead of standard modeling techniques. Improved predictions affect rise time, over-shoot, and the energy content of the control signal.

$$J(N_1, N_2, N_u) = \sum_{j=N_1}^{N_2} \delta(j)\left[\hat{y}_n(t+j|t) - w(t+j)\right]^2 + \sum_{j=1}^{N_u} \lambda(j)\left[\Delta u(t+j-1)\right]^2 \quad (23)$$

Formulation of NGPC

Cost Function

As mentioned earlier, the NGPC algorithm is based on minimizing a cost function over a finite prediction horizon. The cost function of interest to this application is,

$N_1$ = Minimum costing prediction horizon

$N_2$ = Maximum costing prediction horizon Nu= Length of control horizon

$\hat{y}(t+k|t)$ = Predicted output from neural network

$u(t+k|t)$ = Manipulated input

w ( t + k ) = Reference trajectory

δ and λ = Weighing factor

This cost function minimizes not only the mean squared error between the reference signal and the plant's model, but also the weighted squared rate of change of the control input with it's constraints. When this cost function is minimized, a control input that meets the constraints is generated that allows the plant to track the reference trajectory within some tolerance. There are four tuning parameters in the cost function, $N_1$, $N_2$, $N_u$, and λ. The predictions of the plant will run from $N_1$ to $N_2$ future time steps. The bound on the control horizon is $N_u$. The only constraint on the values of $N_u$ and $N_1$ is that these bounds must be less than or equal to $N_2$. The second summation contains a weighting factor, λ that is introduced to control the balance between the first two summations. The weighting factor acts as a damper on the predicted *u(n+1)*.

## 5. COST FUNCTION MINIMIZATION ALGORITHM

The objective of the CFM algorithm is to minimize J in equation (24) with respect to [u(n+l), u(n+2), ..., u(n+$N_u$)]T, denoted as U. This is accomplished by setting the Jacobian of equation (23) to zero and solving for U. With Newton-Rhapson used as the CFM algorithm, J is minimized iteratively to determine the best U. An iterative process yields intermediate values for J denoted J(k). For each iteration of J(k) an intermediate control input vector is also generated and is denoted as:

$$U(k) \stackrel{\Delta}{=} \begin{bmatrix} u(t+1) \\ u(t+2) \\ . \\ . \\ . \\ u(t+N_u) \end{bmatrix} \qquad k = 1, \ldots .N_u \qquad (24)$$

The Newton-Raphson method is one of the most widely used of all root-locating formula. If the initial guess at the root is xi, a tangent can be extended from the point [xi, f(xi)]. The point where this tangent crosses the x axis usually represents an improved estimate of the root.

So the first derivative at x on rearranging can be given as: $x_{(i+1)} = x_{(i)} - \dfrac{f(x_i)}{f'(x_i)}$

Using this Newton-Raphson update rule, U ( k + 1) is given by,

$$U(k+1) = U(k) - \left(\frac{\partial^2 J}{\partial U^2}(k)\right)^{-1} \frac{\partial J}{\partial U}(k)\text{, where } f(x) = \frac{\partial J}{\partial U} \qquad (25)$$

and the Jacobian is denoted as,

$$\frac{\partial J}{\partial U}(k) \equiv \begin{bmatrix} \dfrac{\partial J}{\partial u(t+1)} \\ \cdot \\ \cdot \\ \cdot \\ \dfrac{\partial J}{\partial u(t+N_u)} \end{bmatrix} \tag{26}$$

Also the Hessian is given by,

$$\frac{\partial^2 J}{\partial U^2}(k) \equiv \begin{bmatrix} \dfrac{\partial^2 J}{\partial u(t+1)^2} & \cdot\ \cdot & \dfrac{\partial^2 J}{\partial u(t+1)\partial u(t+N_u)} \\ \cdot & \cdot\ \cdot & \cdot \\ \cdot & \cdot\ \cdot & \cdot \\ \dfrac{\partial^2 J}{\partial u(t+N_u)\partial u_k(t+1)} & \cdot\ \cdot & \dfrac{\partial^2 J}{\partial u(t+N_u)^2} \end{bmatrix} \tag{27}$$

Each element of the Jacobian is calculated by partially differentiating equation (23) with respect to vector U.

$$\frac{\partial J}{\partial u(t+h)} = 2\sum_{j=N_1}^{N_2} \delta(j)\left[\hat{y}_n(t+j) - w(t+j)\right]\frac{\partial \hat{y}_n(t+j)}{\partial u(t+h)} + 2\sum_{j=1}^{N_u} \lambda(j)\left[\Delta u(t+j)\right]\frac{\partial \Delta u(t+j)}{\partial u(t+h)} \tag{28}$$

where, h = 1, ...., N u .

Once again equation (28) is partially differentiated with respect to vector U to get each element of the Hessian.

$$\frac{\partial^2 J}{\partial u(t+m)\partial u(t+h)} = 2\sum_{j=N_1}^{N_2} \delta(j)\left\{\frac{\partial^2 \hat{y}(t+j)}{\partial u(t+m)\partial u(t+h)}\left[\hat{y}(t+j) - w(t+j)\right] - \frac{\partial \hat{y}(t+j)}{\partial u(t+m)}\frac{\partial \hat{y}(t+j)}{\partial u(t+h)}\right\}$$

$$+2\sum_{j=N_1}^{N_2} \lambda(j)\left\{\frac{\partial \Delta u(t+j)}{\partial u(t+m)}\frac{\partial \Delta u(t+j)}{\partial u(t+h)} + \frac{\partial^2 \Delta u(t+j)}{\partial u(t+m)\partial u(t+h)}\right\} \tag{29}$$

The $m^{th}$, $h^{th}$ elements of the Hessian matrix in equation (27) are, h=1, . . . . . , $N_u$ and m=1 , . . . . . . . . . , $N_{u.}$ The last computation needed to evaluate U (k-1) ) is the calculation of the predicted output of the plant, $\hat{y}(t+j)$, and it's derivatives. The next sections define the equation of a multilayer feed forward neural network, and define the derivative equations of the neural network.

## 6. NEURAL NETWORK FOR PREDICTION

In NGPC the model of the plant is a neural network. This neural model is constructed and trained using MATLAB Neural Network System Identification Toolbox commands (Norgaard, 2000). The output of trained neural network is used as the predicted output of the plant. This predicted output is used in the Cost Function Minimization Algorithm. If yn(t) is the neural network's output then it is nothing but plant's predicted output $\hat{y}_n(t+k|t)$. The initial training of the neural network is typically done off-line before control is attempted. The block configuration for training a neural network to model the plant is shown in Fig. 4. The network and the plant receive the same input, *u(t)*. The network has an additional input that either comes from the output of the plant, *y(t)*, or the neural network's, yn(t). The one that is selected depends on the plant and the application. This input assists the network with capturing the plant's dynamics and stabilization of unstable systems. To train the network, its weights are adjusted such that a set of inputs produces the desired set of outputs. An error is formed between the responses of the network, yn(t), and the plant, y(t). This error is then used to update the weights of the network through gradient descent learning. In this work, a Levenberg-Marquardt method is used as gradient descent learning algorithm for updating the weights. This is standard method for minimization of mean-square error criteria, due to its rapid convergence properties and robustness. This process is repeated until the error is reduced to an acceptable level. Since a neural network is used to model the plant, the configuration of the network architecture should be considered. This implementation of NGPC adopts input/output models.

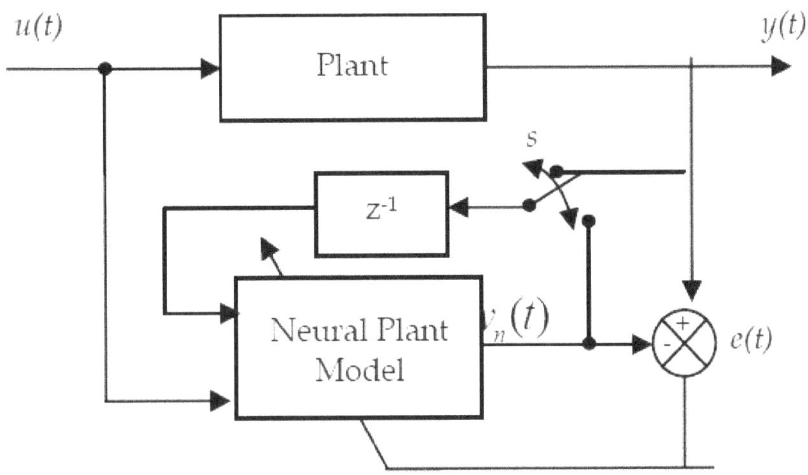


**Fig. 4. Block Diagram of Off-line Neural Network Training**

The diagram shown in Fig. 5, depicts a multi-layer feed-forward neural network with a time delayed structure. For this example, the inputs to this network consists of two external inputs, *u(t)* and two outputs *y(t-1)*, with their corresponding delay nodes, *u(t), u(t-1)* and *y(t- 1), y(t-2)*. The network has one hidden layer containing five hidden nodes that uses bi-polar sigmoidal activation output function. There is a single output node which uses a linear output function, of one for scaling the output.

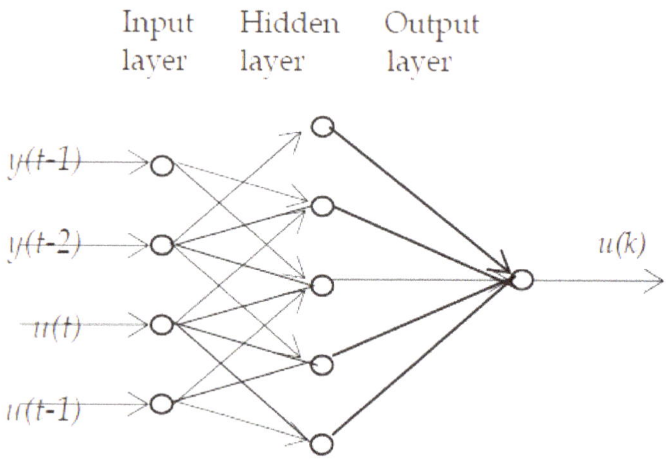


**Fig. 5.Neural Network Architecture**

The equation describing this network architecture is:

$$y_n(t) = \sum_{j=1}^{hid} w_j f_j(net(t)) \tag{30}$$

and

$$net_j(t) = \sum_{j=1}^{n_d} w_{j,i+1} u(t-i) + \sum_{j=1}^{d_d} w_{j,n_d+i+1} y(t-i) \tag{31}$$

where,

$y_n(t)$ is the output of the neural network

f (.) is the output function for the $j^{th}$ node of the hidden layer

$net_j(t)$ is the activation level of the $j^{th}$ node's output function

hid is the number of hidden nodes in the hidden layer

$n_d$ is the number of input nodes associated with u(.)

$d_d$ is the number of input nodes associated with y(.)

$w_j$ is the weight connecting the $j^{th}$ hidden node to the output node

$w_{j,i}$ is the weight connecting the $i_{th}$ hidden input node to the j th hidden node

y ((t-i)) is the delayed output of the plant used as input to the network

u ((t-i)) is the input to the network and its delays

This neural network is trained in offline condition with plants input/output data.

**Prediction Using Neural Network:** The NGPC algorithm uses the output of the plant's model to predict the plant's dynamics to an arbitrary input from the current time, t, to some future time, t+k. This is accomplished by time shifting equations equation (30) and (31), by k, resulting in the following equations given by,

$$y_n(t+k) = \sum_{j=1}^{hid} w_j f_j\left(net_j(t+k)\right) \tag{32}$$

and

$$net_j(t+k) = \sum_{i=1}^{hid} w_{j,i+1} \begin{cases} u(n+k-i), k - N_u < i \\ u(n+N_u), k - N_u \geq i \end{cases} + \sum_{j=1}^{\min(k,d_d)} \left(w_{j,n_d+i+1} y_n(t+k-i)\right)$$
$$+ \sum_{i=k+1}^{d_d} \left(w_{j,n_d+i+1} y(t+k-i)\right) \tag{33}$$

The first summation in equation (33) breaks the input into two parts represented by the conditional. The condition where k - $N_u$ < i handles the previous future values of the u up to u(t+$N_u$-1).The condition where k - $N_u$ > i sets the input from u (t + $N_u$) to u (t + k) equal to u (t + $N_u$). The second condition will only occur if $N_2$ > $N_u$. The next summation of equation (33) handles the recursive part of prediction. This feeds back the network output, $y_n$, for k or $d_d$ times, which ever is smaller. The last summation of equation (33) handles the previous values of y. The following section derives the derivatives of equation (32) and (33) with respect to the input u (t + h).

## 6.1 Neural Network Derivative Equations

To evaluate the Jacobian and the Hessian in equation (26) and (27) the network's first and second derivative with respect to the control input vector are needed.

**Jacobian Element Calculation:** The elements of the Jacobian are obtained by differentiating $y_n$(t+k) in equation (32) with respect to u(t+h) resulting in

$$\frac{\partial yn(t+k)}{\partial u(t+h)} = \sum_{j=1}^{hid} w_j \frac{\partial f_j\left(net_j(t+k)\right)}{\partial u(t+h)} \tag{34}$$

Applying chain rule to $\partial f_j\left(net_j(t+k)\right)/\partial u(t+h)$ result in

$$\frac{\partial f_j\left(net_j(t+k)\right)}{\partial u(t+h)} = \frac{\partial f_j\left(net_j(t+k)\right)}{\partial net_j(t+k)} \frac{\partial net_j(t+k)}{\partial u(t+h)} \tag{35}$$

Where $\partial f_j\left(net_j(t+k)\right)/\partial net_j(t+k)$ is the output function's derivative which will become zero as we are using a linear (constant value) output activation function and

$$\frac{\partial net_j(t+k)}{\partial u(t+h)} = \sum_{i=0}^{n_d} w_{j,i+1} \begin{cases} \delta(k-i,h), k - N_u < i \\ \delta(N_u,h), k - N_u \geq i \end{cases} \tag{36}$$
$$+ \sum_{i=0}^{\min(k,d_d)} w_{j,i+n_d+1} \frac{\partial yn(t+k-i)}{\partial u(t+h)} \delta_1(k-i-1)$$

Note that in the last summation of equation (36) the step function, δ, was introduced. This was added to point out that this summation evaluates to zero for k-i<l, thus the partial does not need to be calculated for this condition.

## 6.2 Hessian Element Calculation

The Hessian elements are obtained by once again differentiating equations (34) by u(t+m), resulting in equation (37):

$$\frac{\partial^2 yn(t+k)}{\partial u(t+h)\partial u(t+m)} = \sum_{j=0}^{n_d} w_j \frac{\partial^2 f_j(net_j(t+k))}{\partial u(t+h)\partial u(t+m)} \tag{37}$$

where,

$$\frac{\partial^2 f_j(net_j(t+k))}{\partial u(t+h)\partial u(t+m)} = \frac{\partial f_j(net_j(t+k))}{\partial net_j(t+k)} \frac{\partial^2 net_j(t+k)}{\partial u(t+h)\partial u(t+m)} + \frac{\partial^2 f_j(net_j(t+k))}{\partial net_j(t+k)^2} \frac{\partial net_j(t+k)}{\partial u(t+h)} \frac{\partial net_j(t+k)}{\partial u(t+m)} \tag{38}$$

The equation (38) is the result of applying the chain rule twice.

# 7. SIMULATION RESULTS

The objective of this study is to show how GPC and NGPC implementation can cope with linear as well as nonlinear systems. GPC is applied to the systems with changes in system order. The Neural based GPC is implemented using MATLAB Neural Network Based System Design Toolbox (Norgaard, 2000)

## 7.1 GPC and NGPC for Linear Systems

The above derived GPC and NGPC algorithm is applied to the different linear models with varying system order, to test its capability. This is done by carrying out simulation in MATLAB 7.0.1 (Mathworks Natic, 2007). Different systems with large dynamic differences are considered for simulation. GPC and NGPC are showing robust performance for these systems. In below figures, for every individual system the systems output with GPC and NGPC is plotted in single figure for comparison purpose. Also the control efforts taken by the both controllers are plotted in consequent figures for every individual figure.

In this simulation, neural network architecture considered is as follows. The inputs to this network consists of two external inputs, u(t) and two outputs y(t-1), with their corresponding delay nodes, u(t), u(t-1) and y(t-1), y(t-2). The network has one hidden layer containing five hidden nodes that uses bi-polar sigmoidal activation output function. There is a single output node which uses a linear output function, of one for scaling the output.

For all the systems Prediction Horicon N1 =1, N2 =7 and Control Horizon (Nu) is 2. The weighing factor $\lambda$ for control signal is kept to 0.3 and $\delta$ for reference trajectory is set to 0. The same controller setting is used for all the systems simulation. The following simulation results are obtained showing the plant output when GPC and NGPC are applied. Also the required control action for different systems is shown.

System I: The GPC and NGPC algorithms are applied to a second order system given below.

$$G(s) = \frac{1}{1 + 10s + 40s^2} \tag{39}$$

The Fig.6. Shows the plant output with GPC and NGPC for setpoint tracking. The Fig. 7 shows the control efforts taken by both controllers. The simulation results reveal that performance of NGPC is better than GPC.

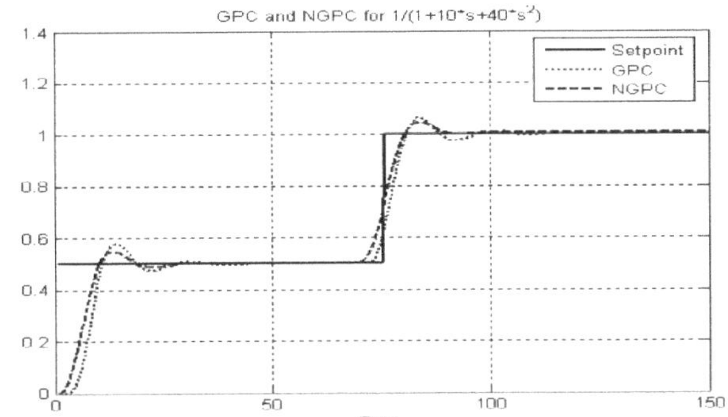


**Fig. 6. System I Output using GPC and NGPC**

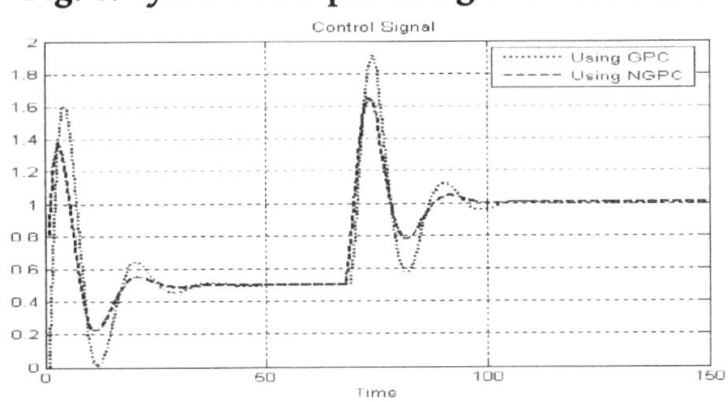


**Fig. 7. Control Signal for System I**

System II A simple first order system given below is to be controlled by GPC and NGPC.

$$G(s) = \frac{1}{1 + 10s} \tag{40}$$

Fig. 8 and Fig. 9 show the system output and control signal.

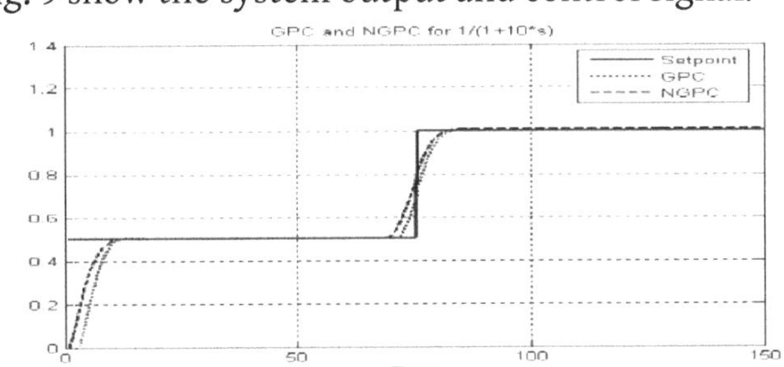


**Fig. 8. System II Output using GPC and NGPC**

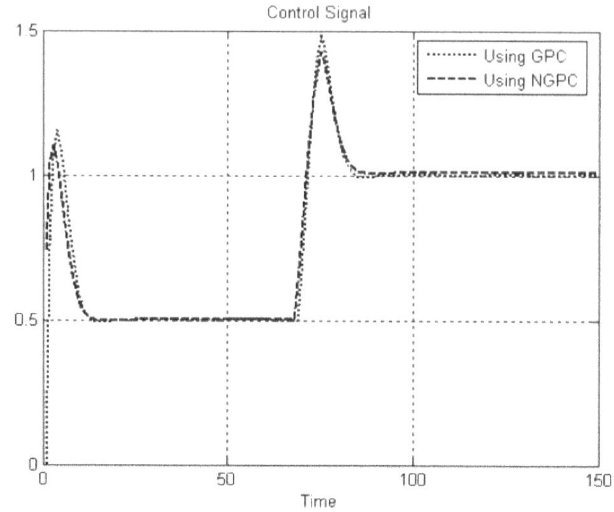


**Fig. 9. Control Signal for System II**

System III: A second order system given below is controlled using GPC and NGPC.

$$G(s) = \frac{1}{10s(1 + 2.5s)} \tag{41}$$

Fig. 10 and Fig. 11 show the predicted output and control signal.

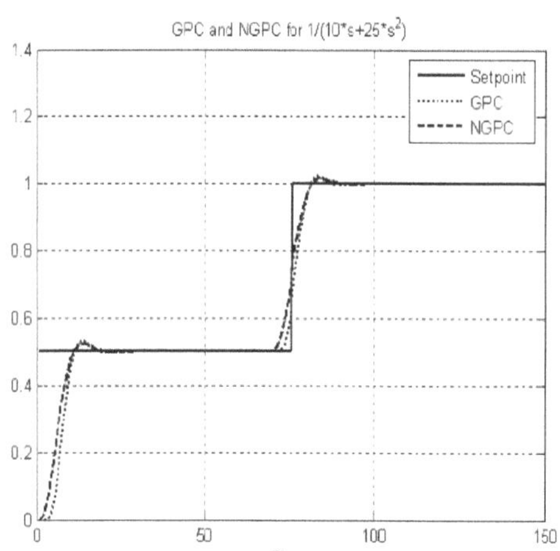


**Fig. 10. System III Output using GPC and NGPC**

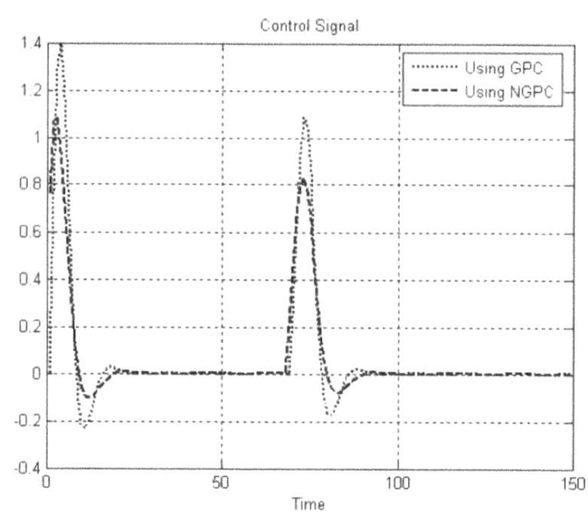


**Fig. 11. Control Signal for System III**

Before applying NGPC to the all above systems it is initially trained using Levenberg- Marquardt learning algorithm. Fig. 12 (a) shows input data applied to the neural network for offline training purpose. Fig. 12 (b) shows the corresponding neural network output.

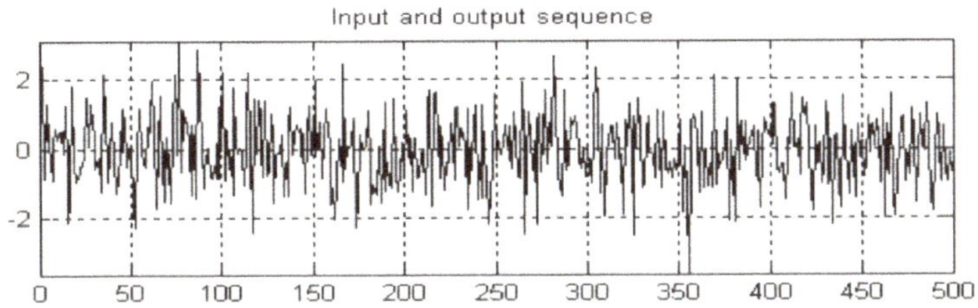


**Fig. 12. (a). Input Data for Neural Network Training**

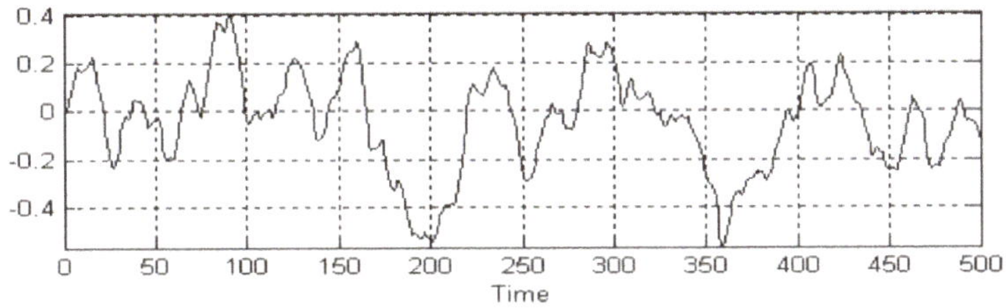


**Fig. 12. (b). Neural Network Response for Random Input**

To check whether this neural network is trained to replicate it as a perfect model or not, common input is applied to the trained neural network and plant. Fig. 13 (a) shows the trained neural networks output and predicted output for common input. Also the error between these two responses is shown in Fig. 13 (b).

The performance evaluation of both the controller is carried out using ISE and IAE criteria given by the following equations:

$$ISE = \int_0^t e^2 dt; \quad IAE = \int_0^t |e|\, dt \tag{42}$$

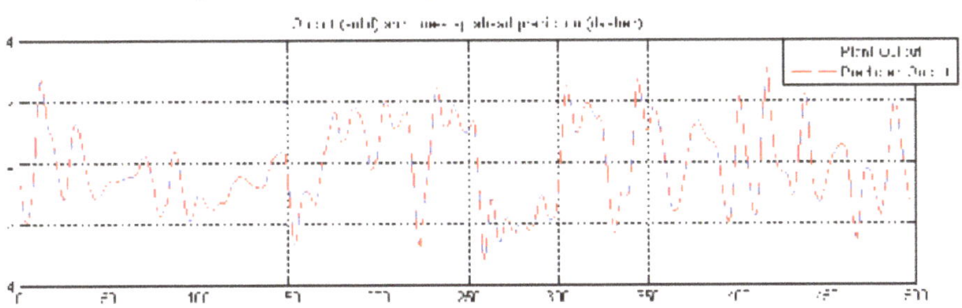

**Fig. 13. (a) Neural network and plant output**

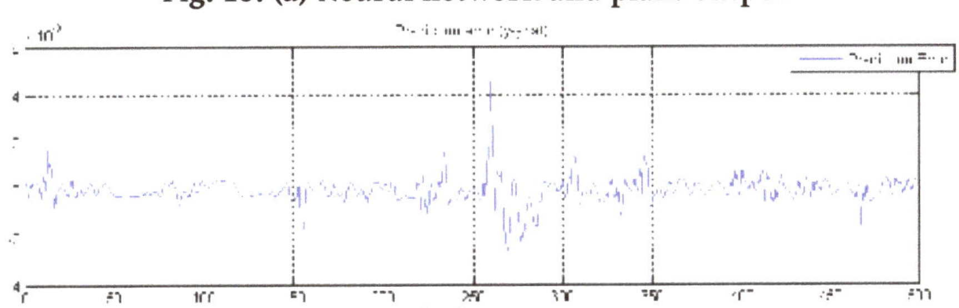

**Fig. 13. (b) Error between neural network and plant output**

The Table 1 gives ISE and IAE values for both GPC and NGPC implementation for all the linear systems given by equation (39) to equation (41). We can find that, for each system ISE and IAE for NGPC is smaller or equal to GPC. So using GPC with neural network i.e. NGPC control configuration for linear application, is also a better choice.

**Table 1. ISE and IAE Performance Comparison of GPC and NGPC for Linear System**

| Systems | Setpoint | GPC | | NGPC | |
|---|---|---|---|---|---|
| | | ISE | IAE | ISE | IAE |
| System I | 0.5 | 1.6055 | 4.4107 | 1.827 | 3.6351 |
| | 1 | 0.2567 | 1.4492 | 0.1186 | 1.4312 |
| System II | 0.5 | 1.1803 | 3.217 | 0.7896 | 2.6894 |
| | 1 | 0.1311 | 0.767 | 0.063 | 1.017 |
| System III | 0.5 | 1.4639 | 3.7625 | 1.1021 | 3.3424 |
| | 1 | 0.1759 | 0.9065 | 0.0957 | 0.7062 |

## 7.2 GPC and NGPC for Nonlinear System

In above Section GPC and NGPC are applied to the linear systems. Fig. 6 to Fig. 11. show the excellent behavior achieved in all cases by the GPC and NGPC algorithm. For each system only few more steps in setpoint were required for GPC than NGPC to settle down the output, but more importantly there is no sign of instability. In this Section, GPC and NGPC is applied to the nonlinear systems to test its capability. A well known Duffing's nonlinear equation is used for simulation. It is given by,

$$\ddot{y}(t) + \dot{y}(t) + y(t) + y^3(t) = u(t) \tag{43}$$

This differential equation is modeled in MATLAB 7.0.1 (Maths work Natic USA, 2007). Then using linearization technique ('linmod' function) available in MATLAB a linear model of the above system is obtained. This function returns a linear model in State-Space format which is then converted in transfer function. This is given by,

$$\frac{y(s)}{u(s)} = \frac{1}{s^2 + s + 1} \tag{44}$$

This linear model of the system is used in GPC algorithm for prediction. In both the controllers configuration, Prediction Horizon N1 =1, N2 =7 and Control Horizon (Nu) is 2 is set. The weighing factor $\lambda$ for control signal is kept to 0.03 and $\delta$ for reference trajectory is set to 0. The sampling period for this simulation is kept at 0.1.

In this simulation, neural network architecture considered is as follows. The inputs to this network consists of two external inputs, u(t) and two outputs y(t-1), with their corresponding delay nodes, u(t), u(t-1) and y(t-1), y(t-2). The network has one hidden layer containing five hidden nodes that uses bi-polar sigmoid activation output function. There is a single output node, which uses a linear output function, of one for scaling the output.

Fig. 14 shows the predicted and actual plant output for the system given in equation (43) when controlled using GPC and NGPC techniques. Fig.15. shows the control efforts taken by both the controller.

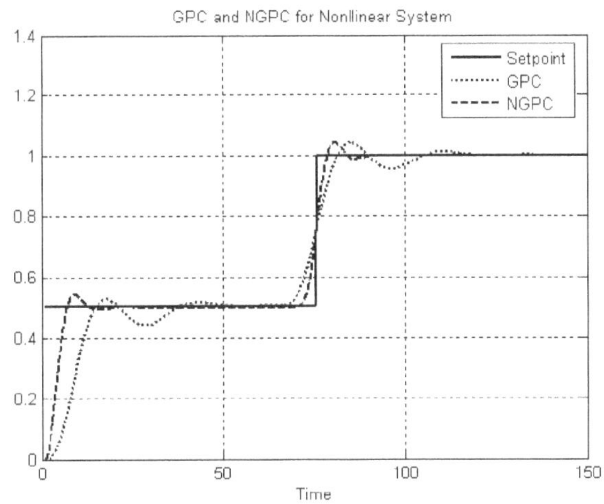


**Fig. 14.Predicted Output and Actual Plant Output for Nonlinear System**

The Fig.14, shows that, for set point changes the response of GPC is sluggish whereas for NGPC it is fast. The overshoot is also less and response also settles down earlier in NGPC as compared to GPC for nonlinear systems. This shows that performance of NGPC is better than GPC for nonlinear system. The control effort is also smooth in NGPC as shown in Fig. 15

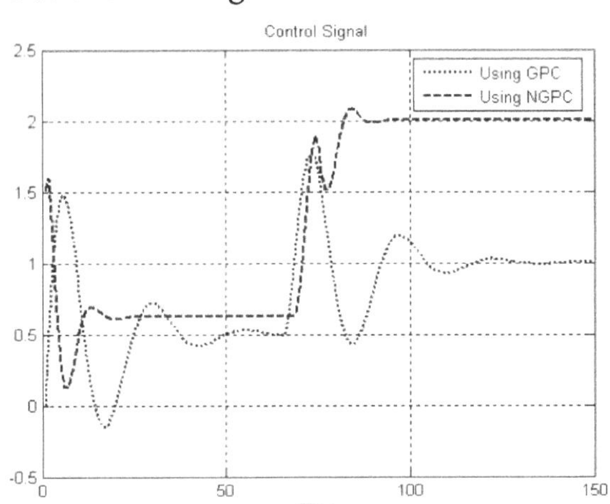


**Fig. 15. Control Signal for Nonlinear System**

Fig. 16 (a) shows input data applied to the neural network for offline training purpose. Fig. 16 (b) shows the corresponding neural network output.

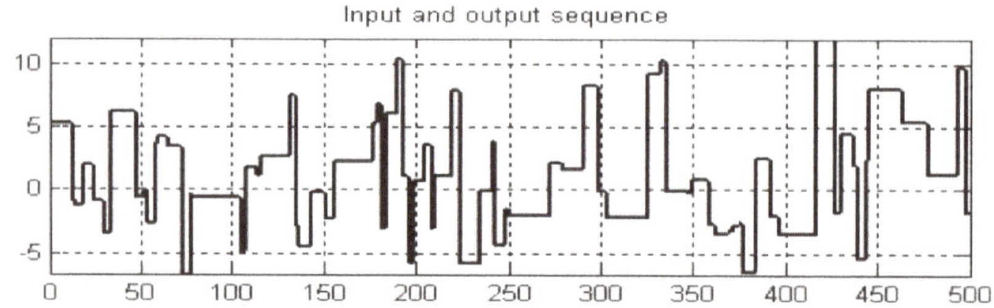


**Fig. 16. (a) Input Data for Neural Network Training**

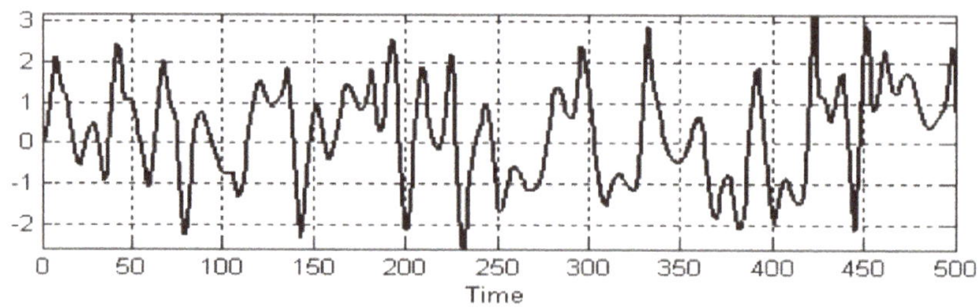


**Fig. 16. (b) Neural Network Response for Random Input**

The Table 2 gives ISE and IAE values for both GPC and NGPC implementation for the nonlinear system given by equation (43). Here a cubic nonlinearity is present. The NGPC control configuration for nonlinear application is better choice. Same results are also observed for set point equals to 1.

**Table 2. ISE and IAE Performance Comparison of GPC and NGPC for Nonlinear System**

| Setpoint | GPC | | NGPC | |
|---|---|---|---|---|
| | ISE | IAE | ISE | IAE |
| 0.5 | 1.8014 | 5.8806 | 0.8066 | 2.5482 |
| 1 | 0.1199 | 1.4294 | 0.0566 | 0.5628 |

## 7.3 Industrial processes

In order to assess the suitability of the suggested controller, an analysis of its performance has been conducted on certain industrial processes.

**Example 1: NGPC for highly nonlinear process (Continues Stirred Tank Reactor)**

Further to evaluate the performance of the Neural generalized predictive control (NGPC) we consider highly nonlinear process continuous stirred tank reactor (CSTR) (Nahas,Henson,et al.,1992) .Many aspects of nonlinearity can be found in this reactor, for instance, strong parametric sensitivity, multiple equilibrium points and nonlinear oscillations. The CSTR system, which can be found in many chemical industries, has evoked a lot of interest for the control community due to its challenging theoretical aspects as well as the crucial problem of controlling the production rate. A schematic of the CSTR system is shown in Fig.17. A single irreversible, exothermic reaction A→B is assumed to occur in the reactor.

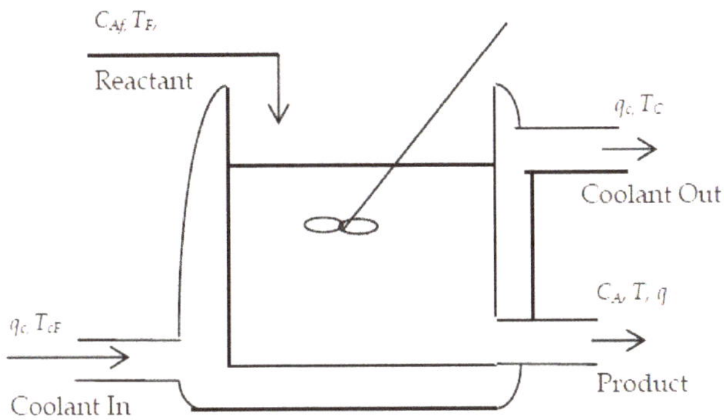


**Fig. 17. Continuous Stirred Tank Reactor**

The objective is to control the effluent concentration by manipulating coolant flow rate in the jacket. The process model consists of two nonlinear ordinary differential equations,

$$\frac{dC_A}{dt} = \frac{q}{V}(C_{Af} - C_A) - k_0 C_A e^{\left(-\frac{E}{RT}\right)}$$

$$\frac{dT}{dt} = \frac{q}{V}(T_f - T) + \frac{(-\Delta H) k_0 C_A}{\rho C_p} e^{\left(-\frac{E}{RT}\right)} + \frac{\rho_c C_{pc}}{\rho C_p V} q_c \left(1 - e^{\frac{-hA}{q_c \rho_c C_{pc}}}\right)(T_{cf} - T) \quad (45)$$

where CAf is feed concentration, CA is the effluent concentration of component A, TF, T and Tc are feed, product and coolant temperature respectively. q and qc are feed and coolant flow rate. Here temperature T is controlled by manipulating coolant flow rate qc. The nominal operating conditions are shown in Table 3.

**Table 3. Nominal CSTR operating conditions**

| | |
|---|---|
| $q = 100l\ min^{-1}$ | $E/R = 9.95\times10^3$ K |
| $C_{Af} = 1mol^{-1}$ | $\Delta$-H = $2 \times10^5$ $calmol^{-1}$ |
| $T_f = 350K$ | $\rho, \rho^c = 1000gl^{-1}$ |
| $T_{cf} = 350K$ | $C_p, C_{cp} = 1cal\ g^{-1}\ K^{-1}$ |
| $V = 100l$ | $q_c = 103.41l\ min^{-1}$ |
| $h_A = 7 \times105\ cal\ min^{-1}\ K^{-1}$ | T = 440.2K |
| $k_o = 7.2 \times1010\ min^{-1}$ | $C_A = 8.36 \times10^{-2}\ mol^{-1}$ |

The operating point in Table 3 corresponds to the lower steady state. For these conditions, there are three (two stable and one unstable) steady states. The objective is to control CA by manipulating coolant flow rate qc.

The corresponding model under certain assumptions is converted into transfer function form as,

$$\frac{y(s)}{u(s)} = \frac{0.42 * e^{-0.75s}}{(13.4 * s + 1)} \tag{46}$$

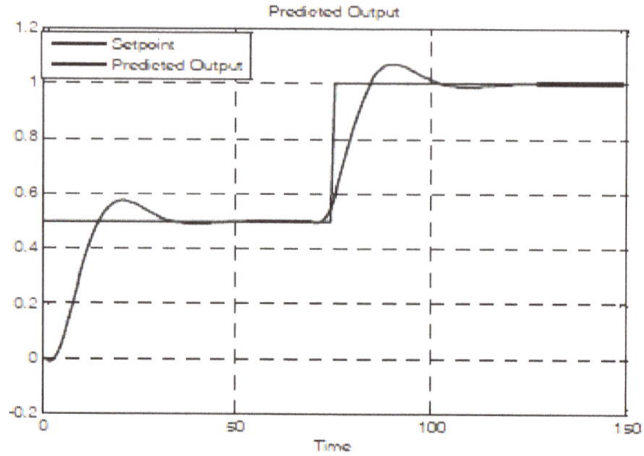


**Fig. 18. System output using NGPC**

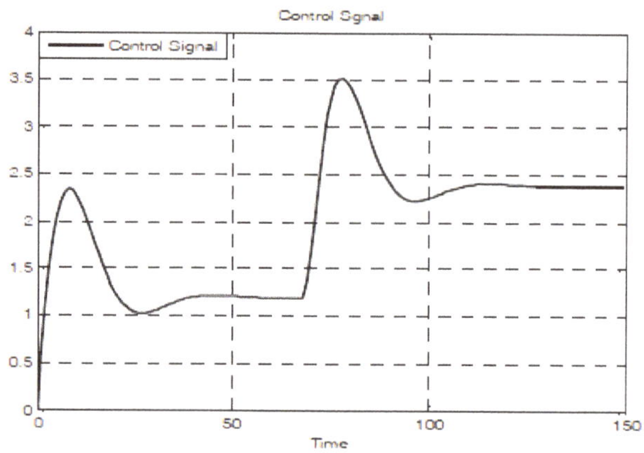


**Fig. 19.Control signal for system**

Fig. 18 shows the plant output for NGPC and Fig.19 shows the control efforts taken by controller. Performance evaluation of the controller is carried out using ISE and IAE criteria. Table 4 gives ISE and IAE values for NGPC implementation for nonlinear systems given by equation (46).

**Table 4. ISE and IAE Performance Comparison of NGPC for CSTR**

| Systems | Setpoint | NGPC | |
|---|---|---|---|
| | | ISE | IAE |
| System I | 0.5 | 1.827 | 3.6351 |
| | 1 | 0.1186 | 1.4312 |

**Example 2: NGPC for** highly linear system **(dc motor)**

Here a DC motor is considered as a linear system from (Dorf & Bishop,1998). A simple model of a DC motor driving an inertial load shows the angular rate of the load, ω (t), as the output and applied voltage, Vapp, as the input. The ultimate goal of this example is to control the angular rate by varying the applied voltage. Fig. 20 shows a simple model of the DC motor driving an inertial load J.

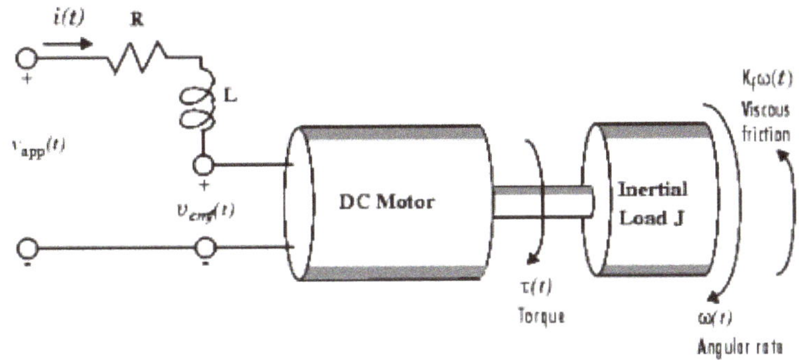


**Fig. 20. DC motor driving inertial load**

In this model, the dynamics of the motor itself are idealized; for instance, the magnetic field is assumed to be constant. The resistance of the circuit is denoted by R and the self- inductance of the armature by L. The important thing here is that with this simple model and basic laws of physics, it is possible to develop differential equations that describe the behavior of this electromechanical system. In this example, the relationships between electric potential and mechanical force are Faraday's law of induction and Ampere's law for the force on a conductor moving through a magnetic field.

A set of two differential equations describes the behavior of the motor. The first for the induced current, and the second for the angular rate,

$$\frac{di}{dt} = -\frac{R}{L}\cdot i(t) - \frac{K_b}{L}\cdot \omega(t) + \frac{1}{L}\cdot V_{app}$$
$$\frac{d\omega}{dt} = -\frac{K_F}{J}\omega(t) + \frac{K_m}{J}\cdot i(t) \tag{47}$$

The objective is to control angular velocity ω by manipulating applied voltage, Vapp. The nominal operating conditions are shown in Table 5.

**Table 5. Nominal dc motor operating conditions**

| | |
|---|---|
| $K_b$ = 0.015 (emf constant) | $K_m$ = 0.015 (torque constant) |
| $K_f$ = 0.2Nms | J = 0.2Kgm$^2$ / sec$^2$ |
| R = 2Ω | L = 0.5H |

The corresponding model under certain assumptions is converted into transfer function form as,

$$\frac{y(s)}{u(s)} = \frac{1.5}{(60.31 * s + s^2)} \tag{48}$$

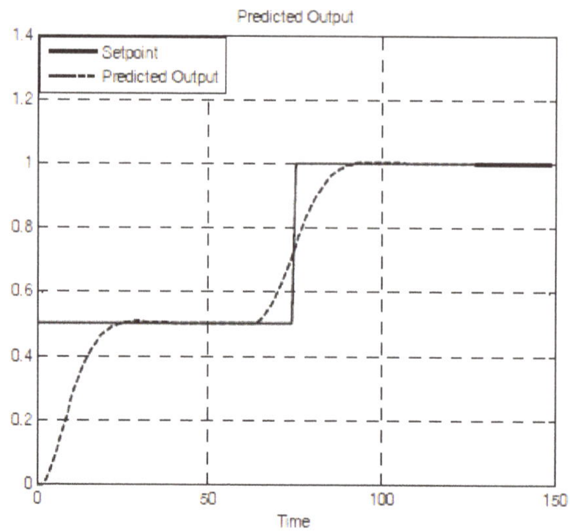


**Fig. 21. System output using NGPC**

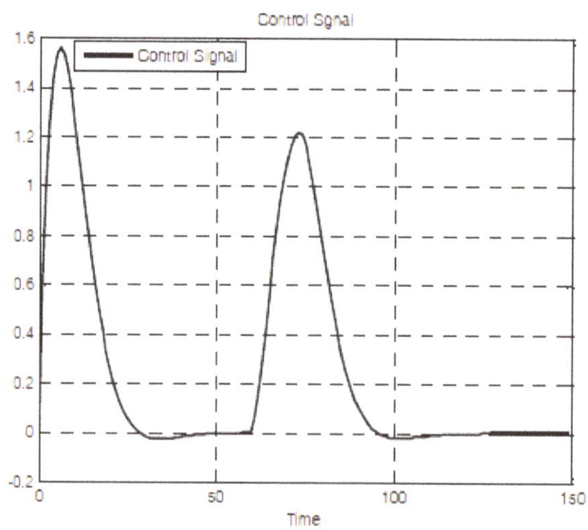


**Fig. 22. Control signal for system**

Fig. 21 shows the plant output for NGPC and Fig. 22 shows the control efforts taken by controller. Performance evaluation of the controller is carried out using ISE and IAE criteria. Table 6 gives ISE and IAE values for NGPC implementation for linear systems given by equation (48).

**Table 6. ISE and IAE Performance Comparison of NGPC for dc motor**

| Systems | Setpoint | NGPC | |
|---|---|---|---|
| | | ISE | IAE |
| System I | 0.5 | 1.505 | 212.5 |
| | 1 | 1.249 | 202.7 |

## 8. APPLICATION OF QUASI-NEWTON ALGORITHM WITH LEVENBERG-MARQUARDT ALGORITHM FOR NONLINEAR SYSTEM

In order to assess the efficiency of two algorithms, namely Newton Raphson and Levenberg Marquardt, they were applied and their outcomes were compared. The specifics regarding this implementation are provided. The conclusion provides a clear explanation of the usefulness of each algorithm. The Levenberg-Marquardt method achieves convergence in much less iterations compared to previous algorithms. The primary expense of the Newton Raphson algorithm lies in the computation of the Hessian matrix. However, due to this additional computational burden, the Levenberg Marquardt approach outperforms other techniques in terms of speed, particularly when dealing with low iteration numbers. As a result, it is a practical and effective solution for real-time control applications. A comparison is made between the simulation results obtained by the Newton-Raphson and Levenberg-Marquardt algorithms. The Levenberg-Marquardt method exhibits convergence towards an optimal solution. The performance of these two methods is also provided in terms of Integrated Squared Error (ISE) and Integrated Absolute Error (IAE).

### 8.1 Simulation Results

A multitude of physical plants demonstrate nonlinear activity. Linear models can provide an approximation of these relationships, but there are cases where a nonlinear model is preferred. This section demonstrates the process of training a neural network to accurately represent a complex plant with nonlinear characteristics. The trained model is subsequently utilized for Nonlinear Generalized Predictive Control (NGPC) purposes. Equation (43) represents the Duffing's equation, which is an extensively researched nonlinear system. The Newton-Raphson algorithm and Levenberg-Marquardt algorithm have been utilized to solve the system of equations represented by equation (43), and the outcomes have been compared. Figure 23 depicts the implementation of the Newton Raphson method, while Figure 24 illustrates the implementation of the LM algorithm. Figure 25. Illustrates the control measures implemented by the controller.

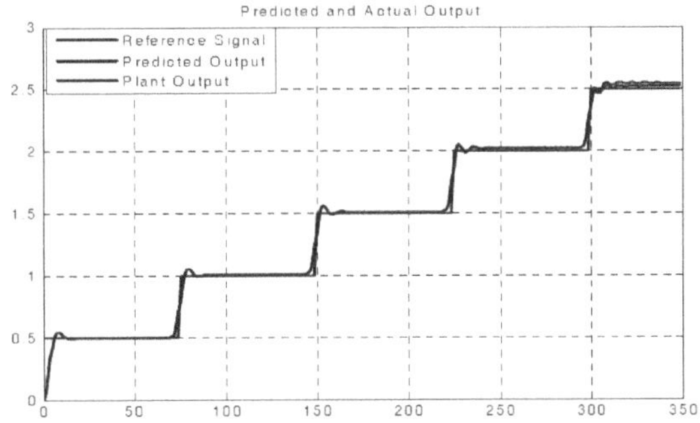


**Fig. 23. Predicted Output and Actual Plant Output for Newton Raphson implementation**

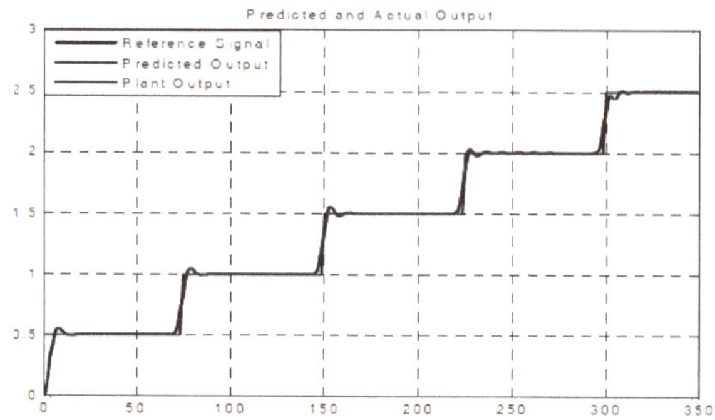


**Fig. 24. Predicted Output and Actual Plant Output for Levenberg Marquardt implementation**

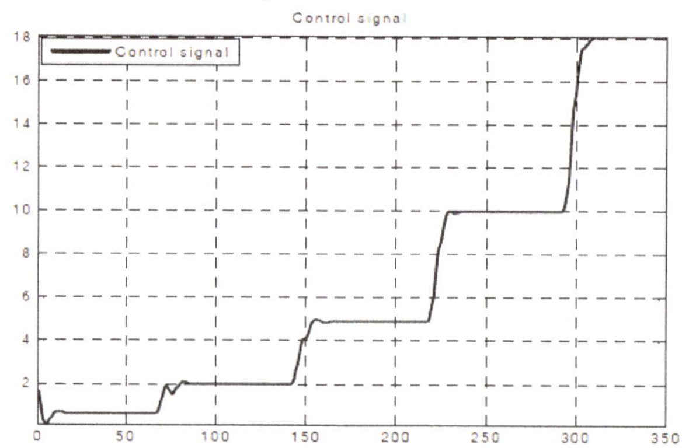


**Fig. 25. Control signal for system**

**Table 7. ISE and IAE Performance Comparison of Newton Raphson and Levenberg Marquardt algorithm**

| Set point | Newton Raphson | | Levenberg Marquardt | |
|---|---|---|---|---|
| | ISE | IAE | ISE | IAE |
| 0.5 | 0.8135 | 2.7950 | 0.8084 | 2.5537 |
| 1 | 0.1363 | 1.2961 | 0.1283 | 1.0315 |
| 1.5 | 0.1292 | 1.2635 | 0.1193 | 0.9931 |
| 2 | 0.1448 | 2.1251 | 0.1026 | 0.8791 |
| 2.5 | 0.0986 | 2.0367 | 0.0382 | 0.5326 |

## 9. CONCISE SUMMARY

This chapter presents the integration of predictive control and neural network to create a control approach called Neural Generalized Predictive Control (NGPC). The NGPC algorithm functions in two distinct modes: prediction and control. It produces a series of future control signals during each sample period to optimize the control effort of the controlled systems. The GPC (Generalized Predictive Control) and NGPC (Nonlinear Generalized Predictive Control) are utilized across many systems to assess their capacity. The output of the trained neural network serves as the anticipated output of the plant. The forecasted result is utilized in the algorithm for minimizing the cost function. The GPC criterion is minimized using two distinct methods: a Quasi Newton algorithm and a Levenberg Marquardt algorithm. The performance of these combinations has been evaluated based on the Integral Square Error (ISE) and Integral Absolute Error (IAE). The simulation result demonstrates that the Levenberg Marquardt algorithm outperforms the Newton Raphson optimization algorithm in terms of control performance.The performance of the NGPC is also evaluated using both highly nonlinear and linear industrial processes, such as the continuous stirred tank reactor (CSTR) and the DC motor. The simulation results demonstrate the superiority of NGPC over GPC in controlling both linear and nonlinear plants.

## REFERENCES


1. Chen, Z. ;Yuan, Z. & Yan, Z.(2002). "A survey on neural network based nonlinear predictive control", Engineering of Control China, Vol. 9, No. 4, pp. 7-11.
2. Clarke, D. W.; Mohtadi, C.& Tuffs, P. C.(1987). Generalized predictive control-part-I and Part-II the basic algorithms, Automatica, Vol. 23, pp. 137-163.
3. Dorf, R. C. and Bishop, R. H. (1998). Modern control systems, Addison-Wesley, Menlo Park, CA, USA.
4. Hashimoto, S. ; Goka, S.; Kondo, T. & Nakajima, K., (2008). Model predictive control of precision stages with nonlinear friction, Advanced Intelligent Mechatronics,.International Conference IEEE/ASME., Gunma Univ., Kiryu.
5. Kim J. H.; Ansung,D. & Gyonggi, D.(1998).Fuzzy model based predictive control, IEEE Transactions on Fuzzy Systems, pp. 405-409.
6. Mathworks USA,(2007) ,MATLAB 7.0.1 Software, Mathworks Natic, USA.
7. Nahas, E. P.; Henson, M. A. & Seborg, D. E. (1992). Nonlinear internal model control strategy for neural network models, Computers Chemical Engineering, Vol. 16, pp. 1039-1057.
8. Norgaard, M. (2000). Neural network based control system design toolkit Ver. 2.0" Technical Report, 00-E -892, Department of Automation, Technical University of Denmark.
9. Nørgaard,M.(2004),Neural Network Based System Identification Toolbox," Tech Report. 00-E- 891, Department of Automation, Technical University of Denmark.
10. Piche, S.; Sayyar-Rodsari,B.; Johnson,D.; Gerules, M.; Technol, P. & Austin, T. X(2000). Nonlinear model predictive control using neural network , IEEE Control Systems Magazine, Vol. 20, No. 3, pp. 53-62.

11. Qin, S. J. and. Badgwell, T.(2003).A Survey of Industrial model predictive control technology", Control Engineering Practice, Vol. 11, No. 7, pp. 733-764.

12. Raff, T. ; Sinz, D. & Allgower, F. (2008). Model predictive control of uncertain continuous- time systems with piecewise constant control input: A convex approach, American Control conference: Inst. for Syst. Theor. & Autom. Control, Univ. of Stuttgart, Stuttgart;

13. Rao, D. N.;Murthy, . M. R. K,; Rao, . S. R. M. & Harshal D. N. ,(2006). Neural generalized predictive control for real time application, International Journal of Information and Systems Sciences, Vol. 3, No. 1, pp. 1-15.

14. Rao, D. N. ; Murthy, M. R. K. ; Rao, S. R. M. &. Harshal, D. N (2007). Computational comparisons of GPC and NGPC schemes", Engineering Letters, Vol. 14.

15. Rossiter, A.( 2006 ).Recent Developments in predictive control, UKACC Control, Mini symposia, pp. 3-26.

16. Sorensen, P. H.; Norgaard, O. Ravn, & N. K. Poulsen,(1999). Implementation of neural network based nonlinear predictive control, Neurocomputing, Vol. 28, 1999, pp. 37- 51.

17. Soloway, D. & Haley, P. J. (1997). Neural generalized predictive control- A Newton-Raphson implementation", NASA Technical Report (110244).

18. Sun, X.; Chang, R.; He, P. & Fan,Y.(2002). Predictive control based on neural network for nonlinear system with time delays, IEEE Transactions on Neural Network, pp. 319- 322.

# 13 UTILIZING NEURAL NETWORKS AND SELF-ORGANIZING MAPS FOR FORECASTING, DIAGNOSIS, AND DECISION MAKING

## 1. INTRODUCTION: AN OVERVIEW

Advanced methods like back-propagation neural networks (BPNN), self-organizing maps (SOM), decision trees, and Bayesian networks have been thoroughly studied and efforts have been made to utilize them in identification, prediction, and control tasks. These techniques have been investigated by researchers such as Rumelhart et al. (1986), Kohonen (1995), Quinlan (1993), and Jensen (2001). Several applications of these methods in identification, prediction, and control have been explored by Bishop (1995), Kil & Shin (1996), and Pham & Liu (1995). This chapter covers three subjects: (1) the utilization of BPNN and selective learning for forecasting, (2) the application of SOM and ensemble learning for diagnostics, and (3) the implementation of SOM and Analytic Hierarchy Process (AHP) for decision making (Saaty, 1980).

The initial section outlines the utilization of Backpropagation Neural Network (BPNN) and selective learning approaches to enhance the predictive capability of BPNN in forecasting significant fluctuations in the stock market. The text describes three approaches: the selective-presentation approach, where training data with significant changes in the prediction-target time series are presented more frequently; the selective-learning-rate approach, where the learning rate for training data with minor changes is decreased; and the combination of both approaches. Forecasting daily stock values is employed as a challenging real-world task. The findings from multiple studies on stock-price prediction indicated that the performances of these two methodologies were comparable and superior to the conventional presenting strategy. Moreover, combining these methodologies resulted in additional enhancement of performance.

The second section presents an empirical investigation on medical diagnosis utilizing Self-Organizing Maps (SOM) and ensemble learning. We utilize Self-

Organizing Maps (SOM) for medical diagnosis, namely in cases of breast cancer, heart illness, and hypothyroid diagnosis. We compare the performance of SOM with that of decision trees and Bayesian networks. We employed the UCI data sets for the purpose of medical diagnostics. The diagnostic accuracy for breast cancer and hypothyroidism was quite good, but the accuracy for heart disease diagnosis was rather poor. Thus, we utilize ensemble learning techniques such as bagging and boosting of Self-Organizing Maps (SOM) for the purpose of diagnosing cardiac disease. The accuracy of SOM was significantly enhanced with the use of ensemble learning.

The third section provides an analysis of purchase decision making using the methods of SOM (Self-Organizing Maps) and AHP (Analytic Hierarchy Process). We have presented a way for making purchase decisions utilizing Self-Organizing Maps (SOM) and Analytic Hierarchy Process (AHP). Initially, we employ the Self-Organizing Map (SOM) algorithm to partition numerous products into multiple clusters. Furthermore, we choose other possibilities by utilizing the product maps. Ultimately, we select a final option from the available possibilities utilizing the Analytic Hierarchy Process (AHP). To illustrate the practical use of our method, we apply it to a specific scenario involving the purchase of a personal computer (PC).

We evaluated a total of 120 types of notebook computers. We assessed the efficacy of our approach through trials involving 117 participants, and subsequently verified its effectiveness.

## 2. UTILIZING NEURAL NETWORKS AND SELECTIVE LEARNING FOR FORECASTING

The use of back-propagation neural networks for prediction has been extensively studied, with research conducted by Weigend et al. (1990), Vemuri & Rogers (1994), and Mandic & Chambers (2001). Neural networks have also been applied to financial market prediction by Azoff (1994), Refenes & Azema-Barac (1994), White (1988), Baba & Kozaki (1992), Freisleben (1992), Tang et al. (1991), and Kohara (2002). Additionally, they have been used for electricity load forecasting by Park et al. (1991) and in other domains. In the conventional method, a neural network is trained by presenting all the training data equally in each cycle, regardless of the magnitude of changes in the prediction-target time series. Additionally, the learning rates for all the training data are the same. Furthermore, the process of network learning often concludes when the network's outputs align with the expected outputs, reaching the point of least mean squared error.

In our previous work(Kohara, 1995), we emphasized that the capacity to forecast significant alterations holds greater significance than the capacity to forecast minor alterations. When the training data is uniformly supplied with a similar learning rate, the Backpropagation Neural Network (BPNN) will effectively learn both tiny and large changes to the same extent, hence it does not have a greater ability to learn huge changes. We have conducted an investigation into selective learning strategies aimed at enhancing the predictive capabilities of neural networks in anticipating significant changes. In our prior work, we introduced the selective-presentation (Kohara, 1995) and selective-learning-rate (Kohara, 1996) methods and utilized them for stock market forecasting. The selective-presentation strategy involves presenting the training data that corresponds to significant changes in the prediction-target time series more frequently. The selective learning rate strategy involves reducing the learning rate for training data that corresponds to modest changes. In a prior study (Kohara, 1995), an alternative stopping criterion for financial predictions was also examined. Network learning is halted when it reaches the point where the profit is maximized during experimental stock-trading.

In our prior work, we suggested merging the selective-presentation and selective-learning-rate methods (Kohara, 2008). By integrating these two methodologies, we may effortlessly accomplish refined and incremental targeted learning of neural networks based on the extent of modification. Forecasting daily stock values was characterized as a complex and unpredictable real-world challenge.

## 2.1 The Selective-Presentation and Selective-Learning-Rate Approaches

In order to enhance the learning capability of neural networks about significant variations in prediction-target time series, we partition the training data into two categories: large-change data (L-data) and small-change data (S-data). The L-data (S-data) exhibit next-day changes that exceed (fall below) a predetermined threshold. In the selective-presentation strategy, the frequency of presenting L-data to neural networks is higher compared to S-data. As an illustration, the entire training data is shown every fifth learning cycle, whereas the L-data is shown every cycle. In the selective-learning-rate technique, all training data is provided in each cycle. However, the learning rate of the back-propagation training algorithm is reduced for S-data compared to L-data. The following text is an outline of these two techniques.

*Selective-Presentation* Approach

- Separate the training data into L-data and S-data.
- Train back-propagation networks with more presentations of L-data than of S-data.
- Stop network learning at the point satisfying a certain stopping criterion (e.g., stop at the point having the maximum profit).

*Selective-Learning-Rate* Approach

- Separate the training data into L-data and S-data.
- Train back-propagation networks with a lower learning rate for the S-data than for the L-data.
- Stop network learning at the point satisfying a certain stopping criterion (e.g., stop at the point having the maximum profit).

We combined these two approaches to achieve fine-tuned and step-by-step learning of neural networks according to the degree of change. The outline is as follows.

Combining *Selective-Presentation* and *Selective-Learning-Rate* Approaches

- Separate the training data into L-data and S-data.
- Separate L-data into two subsets: L1-data and L2-data, where changes in L2- data are larger than those in L1-data.
- Separate S-data into two subsets: S1-data and S2-data, where changes in S2-data are larger than those in S1-data.
- Train back-propagation networks with more presentations of L1- and L2-data than of S1- and S2-data, and with a lower learning rate for L1- and S1-data than for L2 and S2-data.
- Stop network learning at the point satisfying a certain stopping criterion (e.g., stop at the point having the maximum profit).

## 2.2 Evaluation through Experimental Stock-Price Prediction

We considered the following types of knowledge for predicting Tokyo stock prices. These types of knowledge involve numerical economic indicators (Kohara, 1995).

- If interest rates decrease, stock prices tend to increase, and vice versa.
- If the dollar-to-yen exchange rate decreases, stock prices tend to decrease, and vice versa.
- If the price of crude oil increases, stock prices tend to decrease, and vice versa. We used the following five indicators as inputs to the neural network.
    - TOPIX: the chief Tokyo stock exchange price index
    - EXCHANGE: the dollar-to-yen exchange rate (yen/dollar)
    - INTEREST: an interest rate (3-month CD, new issue, offered rates) (%)
    - OIL: the price of crude oil (dollars/barrel)
    - NY: New York Dow-Jones average of the closing prices of 30 industrial stocks (dollars)

TOPIX was the prediction target. EXCHANGE, INTEREST and OIL were chosen based on the knowledge of numerical economic indicators. The Dow-Jones average was used because Tokyo stock market prices are often influenced by New York exchange prices. We assume that tomorrow's change in TOPIX is determined by today's changes in the five indicators according to the knowledge. Therefore, the daily changes in these five indicators (e.g. Δ TOPIX(t) = TOPIX(t) - TOPIX(t-1)) were input into neural networks, and the next-day's change in TOPIX was presented to the neural network as the desired output (Fig. 1). The back-propagation algorithm was used to train the network. All the data of the daily changes were scaled to the interval [0.1, 0.9]. A 5-5-1 multi-layered neural network was used (five neurons in the input layer, five in the hidden layer, and one in the output layer).

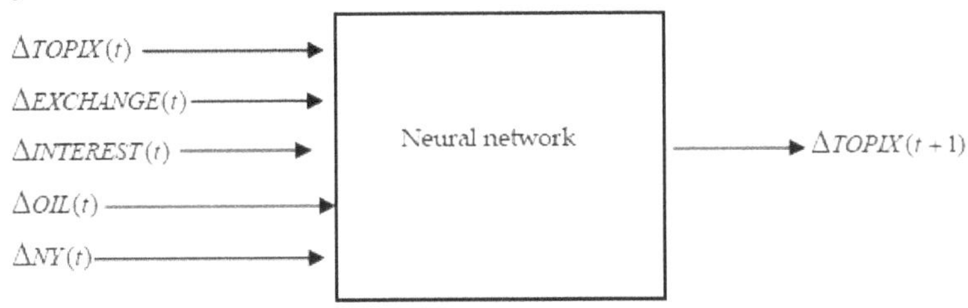


**Fig. 1. Neural prediction model**

## 2.3 Evaluation Experiments

We used data from a total of 409 days (from August 1, 1989 to March 31, 1991): 300 days for training, 30 days for validation (making decisions on stopping the network learning), and 79 days for making predictions. In Experiment 1, all training data were presented in each cycle with an equal learning rate ( ε = 0.7). In Experiment 2, L-data were presented five times as often as S-data. Here, the large-change threshold was 14.78 points (about US$ 1.40), which was the median of absolute value of TOPIX daily changes in the training data. In Experiment 3, the learning rate for the S-data was reduced up to 20% (i.e., ε = 0.7 for the L- data and ε = 0.14 for the S-data). In each experiment, network learning was stopped at the point having the maximum profit (the learning was stopped at the point having the maximum profit for the validation data during 8000 learning cycles). The prediction error and profit were monitored after every hundred learning cycles.

When a large change in TOPIX was predicted, we tried to calculate "Profit" as follows: when the predicted direction was the same as the actual direction, the daily change in TOPIX was earned, and when it was different, the daily change in TOPIX was lost. This calculation of profit corresponds to the following experimental TOPIX trading system. A buy (sell) order is issued when the predicted next-day's up (down) in TOPIX is larger than a preset value which corresponds to a large change. When a buy (sell) order is issued, the system buys (sells) TOPIX shares at the current price and subsequently sells (buys) them back at the next-day price.

Transaction costs on the trades were ignored in calculating the profit. The more accurately a large change is predicted, the larger the profit is.

In each experiment, the momentum parameter α was 0.7. All the weights and biases in the neural network were initialized randomly between -0.3 and 0.3. In each experiment the neural network was run four times for the same training data with different initial weights and the average was taken.

The experimental results are shown in Table 1. Multiple regression analysis (MR) was also used in the experiments. The "prediction error on large-change test data" is the mean absolute value of the prediction error for the test L-data. Applying our *selective-presentation* approach (Experiment 2) reduced the prediction error for test L-data and improved profits: the prediction-error on L-data was reduced by 7% (1- (21.3/22.9)) and the network's ability to make profits through experimental TOPIX-trading was improved by 30% (550/422) compared with the results obtained with the usual presentation approach (Experiment 1). The prediction error and profits in Experiment 3 (*selective-learning-rate* approach) were comparable to those in Experiment 2 (*selective-presentation* approach).

Combining *selective- presentation* with *selective-learning-rate* approaches further reduced the prediction error for test L-data and improved profits: the prediction-error was reduced by 10% and the network's ability to make profits was improved by 38% (Kohara, 2008).

**Table 1. Experimental results on daily stock price prediction**

| | MR | Experiment 1 | Experiment 2 | Experiment 3 |
|---|---|---|---|---|
| Presentation method | equal | equal | *selective* | equal |
| Learning rate | | equal | equal | *selective* |
| Prediction error for | 24.3 | 22.9 | 21.3 | 21.3 |
| large-change data | (1.06) | (1) | (0.93) | (0.93) |
| (relative value) | | | | |
| Profit on test data | 265 | 422 | 550 | 563 |
| (relative value) | (0.62) | (1) | (1.30) | (1.33) |

## 2.4 Summary of Section 2

We described selective learning techniques for forecasting. In the first approach, training data corresponding to large changes in the prediction-target time series are presented more often, in the second approach, the learning rate for training data corresponding to small changes is reduced, and in the third approach, these two techniques are combined. The results of several experiments on stock-price prediction showed that the performances of these two approaches were similar and both better than the usual presentation approach, and combining them further improved the performance. Next, we will apply these techniques today's stock market and other real-world forecasting problems. We also plan to develop a forecasting method that integrates statistical analysis with neural networks.

## 3. DIAGNOSIS WITH SELF-ORGANIZING MAPS AND ENSEMBLE LEARNING

We applied decision trees, Bayesian networks and SOM to medical diagnosis such as breast cancer, heart disease and hypothyroid diagnosis. The UCI data sets (Merz et al., 1997) were used as medical diagnosis problem. The accuracy of breast cancer and hypothyroid diagnosis was comparatively high and the accuracy of heart disease diagnosis was comparatively low. Therefore, we applied ensemble learning such as *bagging* (Breiman, 1994) and *boosting* (Schapire, 1990) of SOM to heart disease diagnosis. The accuracy with ensemble learning of SOM was much improved.

### 3.1 Medical Diagnosis with Decision Trees, Self-Organizing Maps, and Bayesian Networks

*Viscovery SOMine 4.0* was used as SOM software and *See5 release 1.19* was used as decision tree software with default parameter values. We used *BayoNet 3.0.7* as Bayesian network software and constructed Bayesian networks semi-automatically. All problems were evaluated with 10-fold cross-validation. Experimental results on accuracy for medical diagnosis are shown in Table 2. Accuracy for breast cancer diagnosis (breast-w) was comparatively high and accuracy for heart disease diagnosis (heart-c) was comparatively low. In these problems, the accuracy with SOM was comparatively high and the accuracy with decision trees was comparatively low.

**Table 2. Experimental results on accuracy for medical diagnosis (1)**

| | Decision trees | SOM | Bayesian networks |
|---|---|---|---|
| Software | See5 release 1.19 | Viscovery SOMine 4.0 | BayoNet 3.0.7 |
| Breast-w | 95.3% | <u>99.1%</u> | 96.0% |
| Heart-c | 72.7% | <u>89.4%</u> | 80.6% |

Therefore, we apply SOM and decision trees to hypothyroid data. All problems were evaluated with 10-fold cross-validation. Experimental results on accuracy are shown in Table 3. The accuracy for hypothyroid diagnosis was comparatively high. In this problem, the accuracy with decision trees was better than that with SOM.

**Table 3. Experimental results on accuracy for medical diagnosis (2)**

| | Decision trees | SOM |
|---|---|---|
| Software | See5 release 1.19 | Viscovery SOMine 4.0 |
| Hypothyroid | <u>99.5%</u> | 96.8% |

## 3.2 Related Work

Tsoumakas et al. dealt with the combination of classification models that have been derived from running different (heterogeneous) learning algorithm on the same data set (Tsoumakas et al., 2004). They used WEKA implementations of the 10 base-level classification algorithms: decision tree (C4.5), support vector machine (SVM), naïve Bayes, k nearest neighbor, radial basis function and so on. Experimental results on accuracy for the same medical data as we used are shown in Table 4. Both worst and best results stated in the reference are shown.

Zhang & Su extended decision trees to represent a joint distribution and conditional independence, called conditional independence trees (CITrees) and reported that the CITree algorithm outperforms C4.5 and naïve Bayes significantly in classification accuracy (Zhang & Su, 2004). Their experimental results on accuracy for the same medical data as we used are also shown in Table 4.

Garcia-Pedrajas et al. presented a new approach to crossover operator in genetic evolution of neural networks and reported that their approach was compared to a classical crossover with an excellent performance (Garcia-Pedrajas et al., 2006). Their experimental results on accuracy for the same medical data as we used are also shown in Table 4.

Radivojac et al. investigated the problem of supervised feature selection within the filtering framework (Radivojac et al., 2004). In their approach, applicable to the two-class problems, the feature strength is inversely proportional to the p-value of the null hypothesis. Their experimental results for accuracy of heart disease data performed using naïve Bayes and SVM are shown in Table 4.

Robnik-Sikonja improved random forests (Breiman, 2001) and the experimental results on accuracy for breast cancer data are shown in Table 4 (Robnik-Sikonja, 2004). As mentioned above, the accuracy for breast cancer was high up to 99.5% (Garcia-Pedrajas et al., 2006) and the accuracy for hypothyroid was also high up to 99.7% (Tsoumakas et al., 2004). On the other hand, the accuracy for heart disease was comparatively low. The accuracy was 89.4% at most which was attained by SOM.

**Table 4. Experimental results on accuracy for medical diagnosis (3)**

| Reference | Tsoumakas et al., 2004 | Zhang & Su, 2004 | Garcia-Pedrajas et al., 2006 |
|---|---|---|---|
| Classifier | Voting | DT, NB | NN, GA |
| Breast-w | 97.0% to 98.0% | 94.3% to 97.1% | 93.8% to <u>99.5%</u> |
| Heart-c | 81.8% to 88.4% | 78.1% to 84.4% | 86.7% to 89.1% |
| Hypothyroid | 97.2% to <u>99.7%</u> | 93.1% to 93.2% | 94.4% to 94.9% |

| Reference | Radivojac et al., 2004 | Robnik-Sikonja, 2004 | This chapter |
|---|---|---|---|
| Classifier | NB, SVM | RF | SOM |
| Breast-w | Not available | 96.6% to 96.7% | 99.1% |
| Heart-c | 83.5% to 83.9% | Not available | 89.4% |
| Hypothyroid | Not available | Not available | 96.8% |
| Voting: voting of heterogeneous classifiers including DT, NB, SVM etc., DT: decision trees, NB: naïve Bayes, SVM: support vector machine, NN: neural networks, GA: genetic algorithm, RF: random forest. | | | |

Quinlan investigated ensemble learning such as bagging and boosting of C4.5 (Quinlan, 1996). Experimental results on accuracy for the same medical data as we used are shown in Table 5. Bagging and boosting improved average of accuracy by about 1%. Therefore, we have decided to apply ensemble learning of SOM to heart disease data.

**Table 5. Experimental results on accuracy for medical diagnosis (4)**

| | C4.5 | Bagged C4.5 | Boosted C4.5 |
|---|---|---|---|
| Breast-w | 94.7% | 95.8% | 95.9% |
| Heart-c | 77.0% | 78.5% | 78.6% |
| Hypothyroid | 99.5% | 99.6% | 99.6% |
| Average | 90.4% | 91.3% | 91.4% |

## 3.3 Ensemble Learning of Self-Organizing Maps

The steps of bagging and boosting for 10-fold cross-validation we used are as follows.

**Bagging**

**Step 1:** Divide all data into ten sets.

**Step 2:** Select one set for test data and use the remains for training set.

**Step 3:** Resample training subset from the training set mentioned in Step 2.

**Step 4:** Train SOM using the training subset resampled in Step 3.

**Step 5:** Repeat Step 3 to Step 4 nine times more and train the other nine SOM.

**Step 6:** Input test data to ten SOM and obtain results with mean summed outputs by ten SOM.

**Step 7:** Repeat Step 2 to Step 6 nine times more and obtain average accuracy for test data.

**Boosting**

**Step 1:** Divide all data into ten sets.

**Step 2:** Select one set for test data and use the remains for training set.

**Step 3:** Train SOM using training set, input test data to the SOM and obtain results with outputs by the SOM.

**Step 4:** Duplicate the misclassified data in Step 3 to the training set used in Step 3 and obtain modified training set.

**Step 5:** Repeat Step 3 to Step 4 nine times more to the other nine SOM.

**Step 6:** Input test data to ten SOM and obtain results with mean summed outputs by ten SOM.

**Step 7:** Repeat Step 2 to Step 6 nine times more and obtain average accuracy for test data.

We resampled 60% data of the training set in the Step 3 of bagging. When there were not misclassified data in the Step 5 of boosting, we didn't repeat Step 3 to Step 4 and the number of SOM in the Step 6 of boosting was not always ten.

Experimental results on accuracy for heart disease data with ensemble learning of SOM are shown in Table 6. Bagging and boosting improved accuracy by 1.7% and 5.0%, respectively. Especially, boosted SOM was very effective in heart disease data and its accuracy reached 94.4% which was better than that described in references mentioned above by about 5 to 10%.

**Table 6. Experimental results on accuracy for medical diagnosis (5)**

| | SOM | Bagged SOM | Boosted SOM |
|---|---|---|---|
| Heart-c | 89.4% | 91.1% | 94.4% |

### 3.4 Summary of Section 3

We applied SOM to medical diagnosis such as breast cancer, heart disease and hypothyroid diagnosis, comparing with decision trees and Bayesian networks. We found that their accuracy of breast cancer and hypothyroid diagnosis was comparatively high and their accuracy of heart disease diagnosis was comparatively low. Then, we applied ensemble learning such as bagging and boosting of SOM to heart disease diagnosis. We found that their accuracy was much improved. Next, we will apply SOM and ensemble learning to the other medical diagnosis of the UCI data sets such as lung cancer and diabetes. We also plan to develop a diagnosis method that integrates decision trees and Bayesian networks with SOM.

## 4. UTILIZING SELF-ORGANIZING MAPS WITH ANALYTIC HIERARCHY PROCESS FOR DECISION MAKING

Kotler (2002) states that marketing researchers have created a model of the purchase decision process that consists of different stages. The consumer progresses through five distinct stages: problem awareness, information retrieval, assessment of options, buy determination, and post-purchase behavior. Consumer choice making involves five consecutive sets. The initial set comprises the complete range of brands accessible to the buyer. The individual customer will only become acquainted with a limited number of these brands (awareness set). Certain brands will satisfy the initial purchasing criteria (consideration set). As the individual acquires further information, only a select number will persist as formidable candidates inside the decision set. All the brands in the choice set may be deemed acceptable. The individual selects a definitive option from this collection. Multiple sophisticated decision support systems (DSS) have been suggested to assist in various issues (refer to Table 7). In 2006, Kohara and Isomae introduced a system for supporting purchase decisions that combines Self-Organizing Maps (SOM) with Analytic Hierarchy Process (AHP). Initially, we partition a multitude of items (collective set) into multiple clusters utilizing Self-Organizing Maps (SOM). Next, we choose a set of options (known as a choice set) by utilizing the product maps. Ultimately, we select a final option from the available possibilities via the Analytic Hierarchy Process (AHP). To illustrate the practical use of our method, we apply it to a scenario involving the purchase of a personal computer (PC). We evaluated a total of 120 varieties of notebook computers. We assessed the efficacy of our approach by conducting trials with 117 individuals, and subsequently verified its effectiveness.

**Table 7. Examples of intelligent decision support systems**

| Application | Approach | Reference |
|---|---|---|
| Bankruptcy prediction | CBR and AHP | Park & Han, 2002 |
| Weather prediction | Fuzzy reasoning | Riordan & Hansen, 2002 |
| Economic prediction | Neural networks | Kohara, 2002 |
| Electricity business planning | Knowledge-based system | Ha et al., 2003 |
| Economic market model | Multi-agent system | Walle & Moldovan, 2003 |
| Clinical decision support | AHP | Suka et al., 2003 |
| Purchase decision support | SOM and AHP | Kohara & Isomae, 2006 |

## 4.1 Self-Organizing Maps of Personal Computers

The SOM algorithm is based on unsupervised, competitive learning. It provides a topology preserving mapping from the high dimensional space to map units. Map units, or neurons, usually form a two-dimensional lattice and thus the mapping is a mapping from high dimensional space onto a plane. The property of topology preserving means that the mapping preserves the relative distance between the points. Points that are near each other in the input space are mapped to nearby map units in the SOM. The SOM can thus serve as a cluster analyzing tool of high-dimensional data. We considered one hundred and twenty kinds of notebook PCs which were sold in Japan on June 2004. We clustered these PCs using the following features: CPU speed (GHz), main memory capacity (MB), HDD storage capacity (GB), weight (kg), price (yen), battery life (hours), and so on.

We used the above features in two ways: continuous and classified data input. For classified data of CPU speed, we divide into three classes: under 1, 1 to 2, and over 2 GHz. For classified data of main memory capacity, we divide into two classes: 256 and 512 MB. For classified data of HDD storage capacity, we divide into three classes: under 40, 40 to 60, and over 60 GB. For classified data of weight, we divide into five classes: under 1, 1 to 2, 2 to 3, 3 to 4, and over 4 Kg. For classified data of price, we divide into six classes: under 100, 100 to 150, 150 to 200, 200 to 250, 250 to 300, and over 300 thousand yen. For classified data of battery life, we divide into six classes: under 1, 1 to 2, 2 to 3, 3 to 4, 4 to 5, and over 5 hours. *Viscovery SOMine 4.0* was used as SOM software. Fig. 2 and Fig. 3 show self-organizing map and an example of component map of PCs with classified data inputs, respectively. Fig. 4 and Fig. 5 show self-organizing map and an example of component map of PCs with continuous data inputs, respectively. There were five clusters in Fig. 2. When inspecting component maps, the feature of each cluster is clear. For example, when inspecting "under 1 GHz (1-GHz)" component map (see Fig. 3), we understand that one of the features of Cluster 5 is that CPU speed is under 1 GHz. In Fig. 3, originally red color (here, black) neurons correspond to under 1 GHz class and originally blue color (here, dark grey) neurons correspond to the other class.

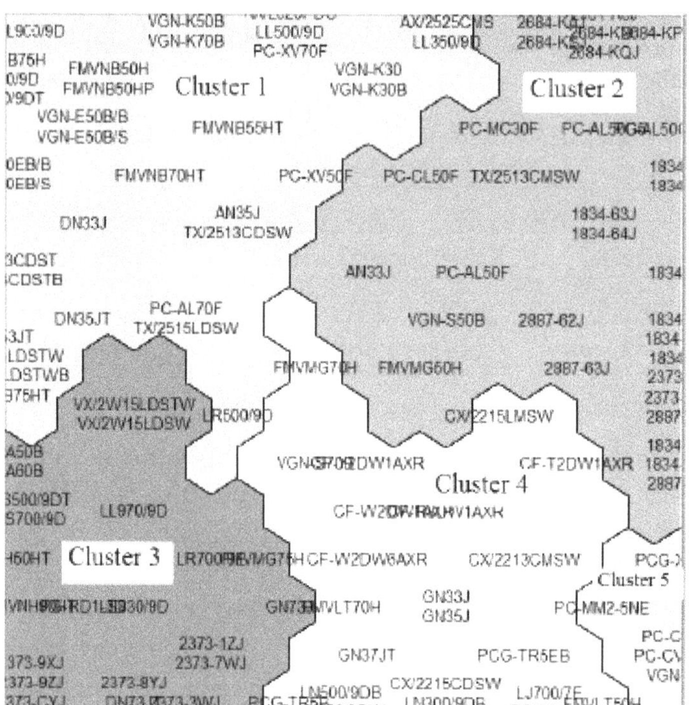


**Fig. 2. Self-organizing map of PCs with classified data inputs**

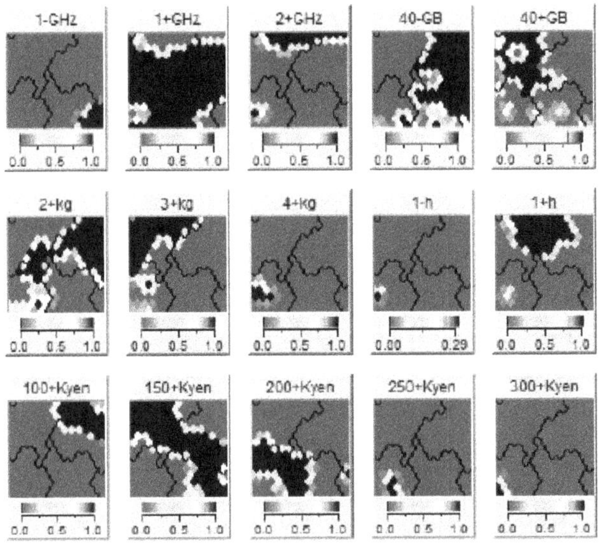


**Fig. 3. Component maps of PCs with classified data inputs**

There were four clusters in Fig. 4. In CPU (GHz) component map of Fig. 5, originally red (here, black) neurons correspond to 2.6 and more GHz and originally blue (here, dark grey) neurons correspond to 0.9 GHz CPU speed. Originally green and yellow (here, light grey) neurons correspond to intermediate values of CPU speed. When inspecting CPU component map of Fig. 5, the feature of each cluster is not clear. So, classified data input is better than continuous data input for clustering PCs. From now, we used classified data input only. We inspected every component map and understand that features of Cluster 1 to Cluster 5 are as in Table 8.

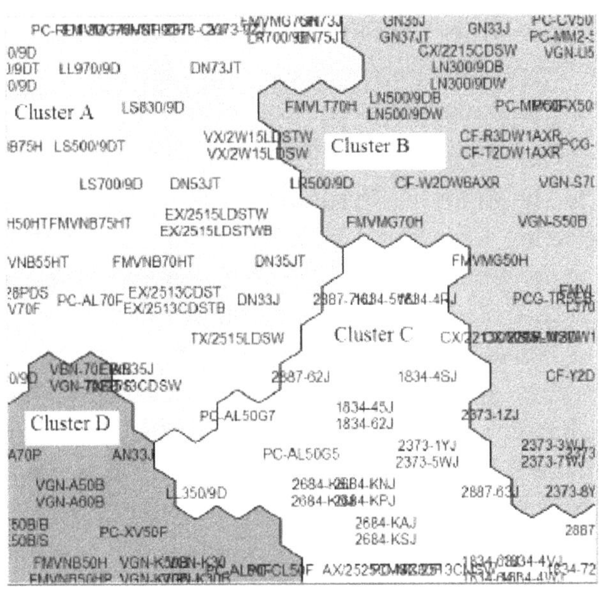


**Fig. 4. Self-organizing map of PCs with continuous data inputs**

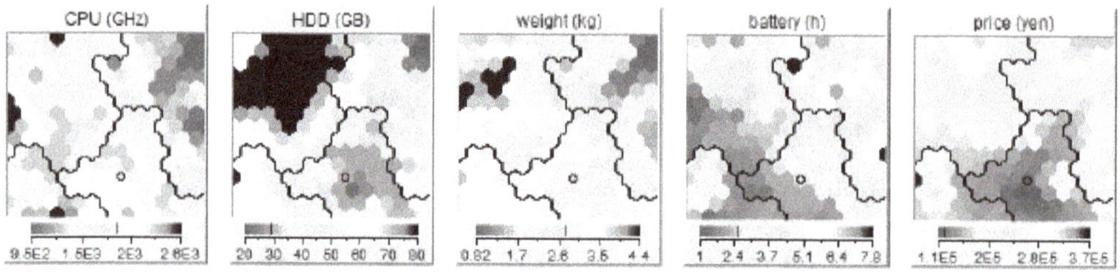


**Fig. 5. Component maps of PCs with continuous data inputs**

**Table 8. Features of clusters with classified data inputs**

| | Features | Main feature |
|---|---|---|
| Cluster 1 | 1 to 2 GHz (CPU), 40 to 60 GB (HDD), 3 to 4 Kg (weight), 150 to 200 thousand yen (price) | High performance |
| Cluster 2 | under 40 GB (HDD), 256 MB (main memory), 100 to 150 thousand yen (price) | Low performance and low price |
| Cluster 3 | over 60 GB (HDD), 512 MB (main memory), over 200 thousand yen (price) | Highest performance and high price |
| Cluster 4 | 1 to 2 Kg (weight), over 4 hours (battery life) | High mobility |
| Cluster 5 | under 1 GHz (CPU), under 1 Kg (weight), 150 to 200 thousand yen (price) | Small size |

## 4.2 Analytic Hierarchy Process for Decision Making

The Analytic Hierarchy Process (AHP) is a decision-making approach that use hierarchical structures to model a problem and establish priorities for possible solutions based on user input. Saaty (1980) demonstrated that the process of assigning weights to activities in multi-criteria decision-making may be efficiently managed through hierarchical structuring and pairwise comparisons. Pairwise comparisons include making judgments between two specific elements instead than trying to rank an entire list of elements.

Figure 6 depicts the Analytic Hierarchy Process (AHP) model used to address the issue of purchasing a personal computer. On the first level, the objective was to purchase a personal computer. On the second level, four criteria were established, and on the third level, five possibilities were identified. In this study, we employed four specific criteria: affordability, portability, efficiency, and aesthetic preference. In this context, great mobility refers to having a low weight and a long-lasting battery. High performance refers to a combination of fast CPU speed, ample main memory capacity, and extensive HDD storage capacity. We can utilize the PC maps to choose different options through numerous methods. We utilized the neighborhood view function of Viscovery SOMine software to choose five alternatives. This function shows all nodes that are topologically comparable to a reference node. Additionally, we performed AHP analysis, which can be found in Tables 9 to 14. In this instance, the primary factor of consideration is the price, as indicated in Table 9. Consequently, PC 2 has been designated as the ultimate selection, as indicated in Table 14.

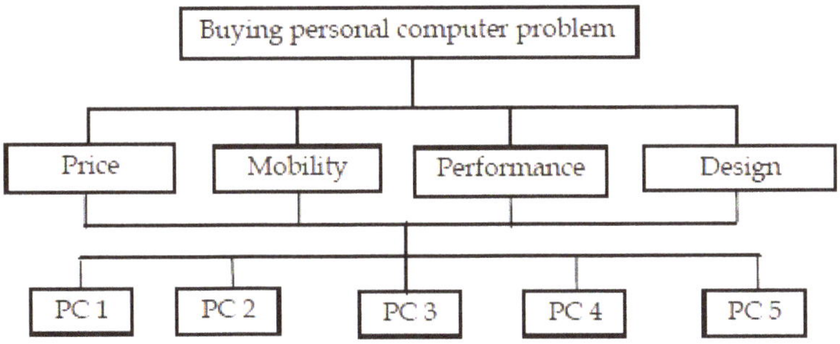


**Fig. 6. AHP model for a buying PC problem**

**Table 9. Pairwise comparison matrix among four criteria**

| | Price | Mobility | Performance | Design | Weights |
|---|---|---|---|---|---|
| Price | 1 | 7 | 3 | 5 | 0.565 |
| Mobility | 1/7 | 1 | 1/5 | 1/3 | 0.055 |
| Performance | 1/3 | 5 | 1 | 3 | 0.262 |
| Design | 1/5 | 3 | 1/3 | 1 | 0.118 |
| Consistency index = 0.039 | | | | | |

**Table 10. Pairwise comparison for price**

| | PC 1 | PC 2 | PC 3 | PC 4 | PC 5 | Weights |
|---|---|---|---|---|---|---|
| PC 1 | 1 | 1/5 | 1/4 | 1/4 | 1 | 0.063 |
| PC 2 | 5 | 1 | 3 | 3 | 5 | 0.456 |
| PC 3 | 4 | 1/3 | 1 | 1 | 4 | 0.209 |
| PC 4 | 4 | 1/3 | 1 | 1 | 4 | 0.209 |
| PC 5 | 1 | 1/5 | 1/4 | 1/4 | 1 | 0.063 |
| Consistency index =0.031 | | | | | | |

**Table 11. Pairwise comparison for mobility**

| | PC 1 | PC 2 | PC 3 | PC 4 | PC 5 | Weights |
|---|---|---|---|---|---|---|
| PC 1 | 1 | 5 | 5 | 1 | 1/3 | 0.221 |
| PC 2 | 1/5 | 1 | 1 | 1/5 | 1/5 | 0.056 |
| PC 3 | 1/5 | 1 | 1 | 1/5 | 1/5 | 0.056 |
| PC 4 | 1 | 5 | 5 | 1 | 1/3 | 0.221 |
| PC 5 | 3 | 5 | 5 | 3 | 1 | 0.448 |
| Consistency index =0.049 | | | | | | |

**Table 12. Pairwise comparison for performance**

| | PC 1 | PC 2 | PC 3 | PC 4 | PC 5 | Weights |
|---|---|---|---|---|---|---|
| PC 1 | 1 | 2 | 2 | 1/2 | 1/2 | 0.186 |
| PC 2 | 1/2 | 1 | 1 | 1/2 | 1/2 | 0.121 |
| PC 3 | 1/2 | 1 | 1 | 1/2 | 1/2 | 0.121 |
| PC 4 | 2 | 2 | 2 | 1 | 1/2 | 0.246 |
| PC 5 | 2 | 2 | 2 | 2 | 1 | 0.326 |
| Consistency index = 0.034 | | | | | | |

**Table 13. Pairwise comparison for design**

| | PC 1 | PC 2 | PC 3 | PC 4 | PC 5 | Weights |
|---|---|---|---|---|---|---|
| PC 1 | 1 | 3 | 3 | 1/3 | 5 | 0.247 |
| PC 2 | 1/3 | 1 | 1 | 1/5 | 3 | 0.104 |
| PC 3 | 1/3 | 1 | 1 | 1/5 | 3 | 0.104 |
| PC 4 | 3 | 5 | 5 | 1 | 6 | 0.497 |
| PC 5 | 1/5 | 1/3 | 1/3 | 1/6 | 1 | 0.048 |
| Consistency index =0.040 | | | | | | |

**Table 14. Final results of AHP for a buying PC problem**

| Alternatives | Results (rank) |
|---|---|
| PC 1 | 0.126 (5) |
| PC 2 | 0.305 (1) |
| PC 3 | 0.165 (3) |
| PC 4 | 0.253 (2) |
| PC 5 | 0.151 (4) |

## 4.3 Evaluation Experiments

We assessed the effectiveness of our decision support technique through experimental purchase decision making using Self-Organizing Maps (SOM) and Analytic Hierarchy Process (AHP) as described by Kohara and Isomae (2006). There are 117 individuals engaged in conducting evaluation experiments. The individuals in question were undergraduates studying in the Department of Electrical, Electronics and Computer Engineering at Chiba Institute of Technology. The aforementioned issue with the purchase of a personal computer was utilized. Following the completion of the studies, we proceeded to inquire the subjects about certain matters. 89.7% of the participants indicated that the self-organizing map of personal computers (PCs) is highly beneficial or beneficial for choosing certain options. 85.4% of the participants indicated that AHP is highly valuable or valuable for making a final decision among the options, and 91.5% of the participants agreed with the outcome.

### 4.4 Summary of Section 4

We described a purchase decision support method using SOM and AHP. First, we divide many products into several clusters using SOM. Secondly, we select some alternatives using the product maps. Finally, we make a final choice from the alternatives using AHP. As an example of real-world applications, we apply our method to a buying personal computer problem. We evaluated our method through experiments conducted by 117 people and confirmed its effectiveness. We will apply our method into another buying product problem.

## 5. CONCISE SUMMARY

This chapter has discussed the use of Backpropagation Neural Networks (BPNN) and selective learning for forecasting, Self-Organizing Maps (SOM) and ensemble learning for diagnostics, and SOM and Analytic Hierarchy Process (AHP) for decision making. Initially, we outlined the utilization of Backpropagation Neural Network (BPNN) and selective learning approaches to enhance the predictive capabilities of BPNN in forecasting significant fluctuations in the stock market. Additionally, we conducted an empirical investigation on medical diagnosis using Self-Organizing Maps (SOM) and ensemble learning techniques. Furthermore, we elucidated the process of purchase decision making using SOM (Self-Organizing Maps) and AHP (Analytic Hierarchy Process). These strategies can be utilized for different forecasting, diagnostic, and decision-making challenges.

## REFERENCES

1. Azoff, E. (1994). Neural Network Time Series Forecasting of Financial Markets, John Wiley and Sons, West Sussex
2. Baba, N. & Kozaki, M. (1992). An intelligent forecasting system of stock price using neural networks, Proceedings of International Conference on Neural Networks, pp. I-371- I-377, Singapore
3. Bishop, C. (1995). Neural Networks for Pattern Recognition, Oxford University Press Breiman L. (1994). Bagging predictors. Technical Report 421, Statistics Department, University of California, Berkley
4. Breiman, L. (2001). Random forests. Machine Learning, 45, (2001) 5-23
5. Freisleben, B. (1992). Stock market prediction with backpropagation networks. Lecture Notes in Computer Science, Vol. 604, 451-460, Springer-Verlag, Heidelberg
6. Garcia-Pedrajas, N.; Ortiz-Boyer, D. & Hervas-Martinez, C. (2006). An alternative approach for neural network evolution with a genetic algorithm: Crossover by combinatorial optimization. Neural Networks, Vol. 19, No. 4, (2006) 514-528

7. Ha, L.; Forgionne, G. & Wang, F. (2003). Facilitating Electronic Business Planning with Decision Making Support Systems, Lecture Notes in Artificial Intelligence, Vol. 2774, 45-51, Springer-Verlag, Heidelberg
8. Jensen, F. (2001). Bayesian Networks and Decision Graphs, Springer
9. Kil, D. & Shin, F. (1996). Pattern Recognition and Prediction with Applications to Signal Characterization, American Institute of Physics Press
10. Kohara, K. (1995). Selective Presentation Learning for Forecasting by Neural Networks, Proceedings of International Workshop on Applications of Neural Networks to Telecommunications, pp. 316-323, Stockholm
11. Kohara, K. (1996). Selectively Intensive Learning to Improve Large-Change Prediction by Neural Networks, Proceedings of International Conference on Engineering Applications of Neural Networks, pp. 463-466, London
12. Kohara, K. (2002). Neural networks for economic forecasting problems. In: Expert Systems - The Technology of Knowledge Management and Decision Making for the 21st Century-, Leondes, C. T. (Ed.) Academic Press. San Diego, CA
13. Kohara, K. & Isomae, M. (2006). Purchase Decision Support with Self-Organizing Maps and Analytic Hierarchy Process, Proceedings of International Conference on Knowledge Engineering and Decision Support, pp. 151-157, Lisbon
14. Kohara, K. (2008). Combining Selective-Presentation and Selective-Learning-Rate Approaches for Neural Network Forecasting of Stock Markets, Proceedings of International Workshop on Artificial Neural Networks and Intelligent Information Processing, pp. 3-9, Madeira Kohonen, T. (1995). Self-Organizing Maps, Springer
15. Kotler, P. (2002). Marketing Management, eleventh edition, Prentice Hall
16. Mandic, D. & Chambers, J. (2001). Recurrent Neural Networks for Prediction, John Wiley & Sons
17. Merz, C.; Murphy, P. & Aha, D. (1997). UCI repository of machine learning databases. Dept of ICS, University of California, Irvine
18. Park, C-S. & Han, I. (2002). A CBR with feature weights derived by analytic hierarchy process for bankruptcy prediction. Expert Systems with Applications, Vol. 14, No. 3, (2002) 255-264
19. Park, D.; El-Sharkawi, M.; Marks II, R.; Atlas, L. & Damborg, M. (1991). Electric load forecasting using an artificial neural network. IEEE Transactions on Power Systems,Vol. 6, No. 2, (1991) 442-449
20. Pham, D. & Liu, X. (1995). Neural Networks for Identification, Prediction and Control, Springer

21. Quinlan, J. (1993). C4.5: Programs for Machine Learning, Morgan Kaufmann
22. Quinlan, J. (1996). Bagging, boosting and C4.5, Proceedings of the 13th National Conference on Artificial Intelligence and the 8th Innovative Applications of Artificial Intelligence Conference, pp. 725-730
23. Refenes, A. & Azema-Barac, M. (1994). Neural network applications in financial asset management. Neural Computing & Applications, Vol. 2, No. 1, (1994) 13-39
24. Riordan, D. & Hansen, B. K. (2002). A fuzzy case-based system for weather prediction. Engineering Intelligent Systems, Vol. 3, (2002) 139-145
25. Rumelhart, D.; Hinton, G. & Williams, R. (1986). Learning internal representations by error propagation. In: Parallel Distributed Processing 1, Rumelhart, D.; McClelland, J. & the PDP Research Group (Ed.), MIT Press, Cambridge, MA Saaty, T. (1980). The Analytic Hierarchy Process, McGraw-Hill
26. Schapire, R. (1990). The strength of weak learnability. Machine Learning, 5, (1990) 197-227
27. Suka, M.; Ichimura, T. & Yoshida, K. (2003). Clinical Decision Support System Applied the Analytic Hierarchy Process. Lecture Notes in Artificial Intelligence, Vol.2774, 417- 423, Springer-Verlag, Heidelberg
28. Tang, Z., Almeida, C. & Fishwick, P. (1991). Time series forecasting using neural networks vs. Box-Jenkins methodology. Simulation, Vol. 57, No. 5, (1991) 303-31
29. Tsoumakas, G.; Katakis, I. & Vlahavas, I. (2004). Effective Voting of Heterogeneous Classifiers. Lecture Notes in Artificial Intelligence, Vol. 3201, 465-476, Springer-Verlag, Heidelberg
30. Radivojac, P.; Obradovic, Z.; Dunker, A. & Vucetic, S. (2004). Feature Selection Filters Based on the Permutation Test. Lecture Notes in Artificial Intelligence, Vol. 3201, 334-346, Springer-Verlag, Heidelberg
31. Robnik-Sikonja, M. (2004). Improving Random Forests. Lecture Notes in Artificial Intelligence, Vol. 3201, 359-370, Springer-Verlag, Heidelberg
32. Vemuri, V. & Rogers, R. (Ed.) (1994). Artificial Neural Networks: Forecasting Time Series, IEEE Press, Los Alamitos, CA
33. Walle, B. V. & Moldovan, M. (2003). An Information Market for Multi-agent Decision Making: Observation from a Human Experiment. Lecture Notes in Artificial Intelligence, Vol. 2774, 66-72, Springer-Verlag, Heidelberg
34. Weigend, A., Huberman, B. & Rumelhart, D. (1990). Predicting the future: a connectionist approach. International Journal of Neural Systems, Vol. 1, No. 3, (1990) 193-209

35. White, H. (1988). Economic prediction using neural networks: the case of IBM daily stock return, Proceedings of International Conference on Neural Networks, pp. II-451-II- 458, San Diego, CA

36. Zhang, H. & Su, J. (2004). Conditional Independence Trees. In: Boulicaut J-F, Esposito F, Giannotti F, Pedreschi D (eds). Lecture Notes in Artificial Intelligence, Vol. 3201, 513-524, Springer-Verlag, Heidelberg

# 14 DIFFICULTIES POSED BY MIDDLEWARE IN THE CONTEXT OF THE INTERNET OF THINGS

## 1. INTRODUCTION: AN OVERVIEW

Advancements in networking, sensor, and RFID technologies now make it possible to connect different physical items to the IT infrastructure. This has the potential to bring about the reality of the Internet of Things and Ubiquitous Computing concepts. Furthermore, this presents opportunities for expanding the scope of industrial automation, such as implementing automated monitoring, control, and maintenance planning for industrial resources and processes. A significantly larger quantity of resources, including equipment, infrastructural elements, materials, and products, can be linked to IT systems, enabling monitoring and potential control. This evolution will also inevitably generate a need for extensive integration with diverse external resources, including data storages, information services, and algorithms, which may be available in different units within the same business, in other organizations, or on the Internet.

The integration of computing and physical systems has the potential to create a challenging situation known as "the nightmare of ubiquitous computing" (Kephart & Chess, 2003). In this scenario, human operators will struggle to handle the intricate interactions within the system, and even architects will find it difficult to predict and design for such complexity. The concept of autonomic computing, as proposed by IBM (Kephart & Chess, 2003), emphasizes the necessity for computing systems to operate autonomously with little human intervention. This involves the establishment of higher-level policies rather than direct administration. Consequently, the computing systems will possess the ability to manage themselves, encompassing self-configuration, self-optimization, self-protection, and self-healing, as envisioned by IBM. The IBM vision highlights that for a complex system to be able to manage itself during operation, its components need to possess a certain level of autonomy. Consequently, we anticipate that software agent technologies will have a significant role in constructing these intricate systems. The agent-based method in software engineering is known for its ability to simplify the design of intricate systems (Jennings, 2001). In the topic of multi-agent systems, there is a strong focus on developing decentralized systems that can handle spontaneous configuration, tolerate partial failures, and provide adaptive rearrangement of the entire system (Mamei & Zambonelli, 2006).

An inherent challenge in ubiquitous computing systems is the presence of heterogeneity, which refers to the diversity in components, standards, data formats, protocols, and other aspects. This diversity poses considerable barriers to achieving interoperability across the components. Today, semantic technologies are considered a crucial solution for addressing the challenges of interoperability and integration in a diverse and linked world of objects and systems. According to Lassila (2005), semantic technologies are considered to be a more effective method for achieving interoperability compared to current standards-based approaches. The Internet of Things should actually evolve into the Semantic Web of Things, as suggested by Brock and Schuster in 2006. We adhere to this perspective. In addition, we utilize semantic technologies to not only simplify the identification of diverse components and data integration, but also to govern and synchronize the behavior of these components (i.e., specifying the intended behavior in a prescriptive manner, using declarative semantic programming).

It is well acknowledged that attaining interoperability in ubiquitous environments cannot be accomplished by imposing rigorous standards and expecting everyone to conform. Interoperability necessitates the presence of middleware to serve as the adhesive that connects diverse components. Having a reliable collection of middleware that provides application programming interfaces, communication, and other services to applications would make it easier to create applications and transition from static programming methods to a flexible and dynamic composition capability (Buckley, 2006).

This chapter outlines our conceptualization of a middleware for the Internet of Things, which serves as the foundation for our research project UBIWARE. The project aims to develop a next-generation middleware platform that enables the creation of self-managed complex systems, specifically in the industrial sector. These systems will be composed of distributed, heterogeneous, shared, and reusable components of various types, such as smart machines and devices, sensors, RFIDs, web services, software applications, humans with their interfaces, and others. This middleware will facilitate the automatic discovery of different components and the configuration of a system with intricate functionality, which is based on the individual functionalities of the components.

## 2. THE UTILIZATION OF SEMANTIC WEB TECHNOLOGY IN THE CONTEXT OF THE INTERNET OF THINGS

### 2.1 Semantic Web technology for ubiquitous computing

As stated by Buckley (2006), the true potential of the Internet of Things lies in the interconnectedness of devices. Interoperability necessitates that service clients possess prior knowledge of the functionalities provided by service providers, and semantic modeling should enable service requestors to comprehend the offerings of service providers. This is a crucial matter for transitioning to an open-world approach, where new or updated devices and services can emerge unpredictably, and towards device networks that can adjust dynamically to changes in context as required by application scenarios. Furthermore, this also has ramifications for the prerequisites of middleware, as it is necessary to establish a connection between the devices, which might be perceived as services, and the applications. Devices within the Internet of Things may require the capability to establish communication with other devices located in any part of the globe. This necessitates the implementation of a naming and addressing system, as well as mechanisms for searching and finding. The correlation between devices and identity, established by naming and addressing, presents many privacy and security obstacles. A cohesive collection of middleware, providing application programming interfaces, communication capabilities, and other services to applications, will streamline the process of developing services and applications. We must transition from static programming methodologies to a flexible and dynamic composition capability.

In (Lassila & Adler, 2003), the ubiquitous computing is presented as an emerging paradigm qualitatively different from current personal computing scenarios by involving dozens and hundreds of devices (sensors, external input and output devices, remotely controlled appliances, etc). A vision was presented for a class of devices, so called "Semantic Gadgets", that will be able to combine functions of several portable devices users have today. Semantic Gadgets will be able to automatically configure themselves in new environments and to combine information and functionality from local and remote sources. Semantic Gadgets should be capable of semantic discovery and device coalition formation: the goal should be to accomplish discovery and configuration of new devices without "a human in the loop." Authors pointed out that a critical to the success of this idea is the existence or emergence of certain infrastructures, such as the World Wide Web as a ubiquitous source of information and services and the Semantic Web as a more machine- and automation-friendly form of the Web.

In subsequent works, Lassila (2005a) and Lassila (2005b) explored the potential use of Semantic Web technologies in the context of mobile and ubiquitous computing. They argued that ubiquitous computing poses significant challenges in terms of interoperability. This effort is driven by the necessity to enhance the automation of user actions by enhancing the compatibility and exchange of data between systems, apps, and information sources. One crucial aspect of achieving the Semantic Web is the concept of "serendipitous interoperability". This refers to the capability of software systems to identify and employ services that were previously unknown and not taken into account during the design phase. In order to achieve this, it is vital to have more powerful methods of describing the meaning of services. These methods should allow for entirely automated discovery and activation, and eliminate any needless interaction with human users. Enhancing interoperability by refraining from making predetermined assumptions about how devices should communicate with each other will make it more feasible to create dynamic ubiquitous computing scenarios without the need for choreographing. The use of Semantic Web technology provides a more robust and effective method for achieving interoperability compared to current standards-based approaches.

## 2.2 Semantic Web technology for coordination

The agent-based approach has been recommended as a suitable method for constructing sophisticated, distributed software-based systems (Jennings, 2001). From an implementation perspective, agents represent a progressive advancement in software engineering methodologies and programming languages. This advancement follows the trend of enhancing localization and encapsulation in the fundamental components of programming models (Jennings, 2000). Following the example of structures in programming languages like C, which localize data, and objects in languages like C++ and Java, which localize both data and code, agents further localize their purpose, thread of control, and action choices. An agent is typically described as a self-contained computer system that operates inside a certain environment and has the ability to independently and adaptively take actions in order to achieve its intended goals (Wooldridge, 1997).

The challenge of transitioning from the problem domain to the solution domain is widely acknowledged as a significant issue in the field of software and systems engineering. Thus, in software design, the most effective abstractions are the ones that reduce the semantic gap between the units of analysis that are naturally used to conceptualize the problem and the structures found in the solution paradigm (Jennings, 2000). Having the same notion, specifically an agent, as the core focus in both problem analysis and solution design and execution can greatly facilitate the process of designing an effective solution and managing its complexity. On the other hand, the object-oriented approach is based on the underlying machine architecture and relies on implementation-level ontological primitives like object, method, invocation, and so on. Since the initial phases of software development rely on purposeful concepts like stakeholders, goals, and plans, there is an inevitable vacuum that must be filled. (Bresciani et al., 2004) asserted that the agent-oriented programming paradigm is uniquely capable of smoothly and flawlessly incorporating intentional models from early development stages into implementation and run-time phases. The agent-oriented method delays the transfer from domain notions to machine concepts until the design and implementation stage of individual agents, assuming that they would be implemented in an object-oriented programming language.

While the flexibility of agent interactions offers numerous benefits in creating complex systems, it also introduces unpredictability in the runtime system. This is because agents operate autonomously, resulting in unknown patterns and effects of their interactions (Jennings, 2000). This highlights the necessity for an efficient system of organizing, collaborating, and bargaining. (Although these concepts are technically separate, the term "coordination" is frequently employed as a broad term embracing all three. Therefore, for the purpose of conciseness, we shall also use it in this manner.) In his work, Jennings (2000) explores the practice of overcoming these challenges in certain systems and applications. This is achieved by implementing interaction protocols that can be formally analyzed, adopting inflexible and predetermined organizational structures, and/or restricting the nature and extent of agent interactions. Nevertheless, Jennings argued that these limitations also curtail the efficacy of the agent-based approach. Consequently, to fully unlock its capabilities, it is necessary to implement more enduring solutions.

The available literature sketches two major directions of search for such a longer term solution:

- D1: Social level characterization of agent-based systems. E.g. (Jennings, 2000) stressed the need for a better understanding of the impact of sociality and organizational context on an individual's behavior and of the symbiotic link between the behavior of the individual agents and that of the overall system.
- D2: Ontological approaches to coordination. E.g. (Tamma et al., 2005) asserted a need for common vocabulary for coordination, with a precise semantics, to enable agents to communicate their intentions with respect to future activities and resource utilization and get them to reason about coordination at run time. Also (Jennings et al., 1998) put as an issue to resolve the question about how to enable individual agents to represent and reason about the actions, plans, and knowledge of other agents to coordinate with them.

In our work, we attempt to provide a solution advancing into both D1 and D2 and somewhat integrating both. Some basic thinking, leading our work, follows.

## 3. THE VISION AND APPROACH

The ultimate objective is to achieve the vision of the Global Understanding Environment (GUN) (Terziyan, 2003; Terziyan, 2005; Kaykova et al., 2005a). We initiated the initial phase of the SmartResource initiative from 2004 to 2006. Figure 1 illustrates our research plan.

The objective of the Global Understanding Environment (GUN) is to enable the accessibility, proactivity, and cooperation of diverse resources, including physical objects, digital content, and human beings, through the internet. The three core principles of this platform are Interoperability, Automation, and Integration. Interoperability in GUN necessitates the use of Semantic Web standards, metadata based on RDF, ontologies, and semantic adapters for the resources. The implementation of automation in GUN necessitates the proactive allocation of resources through the utilization of agent technologies. The process of integrating in GUN involves the use of ontology-based business process modeling and integration, as well as multi-agent technologies to coordinate business operations across resources.

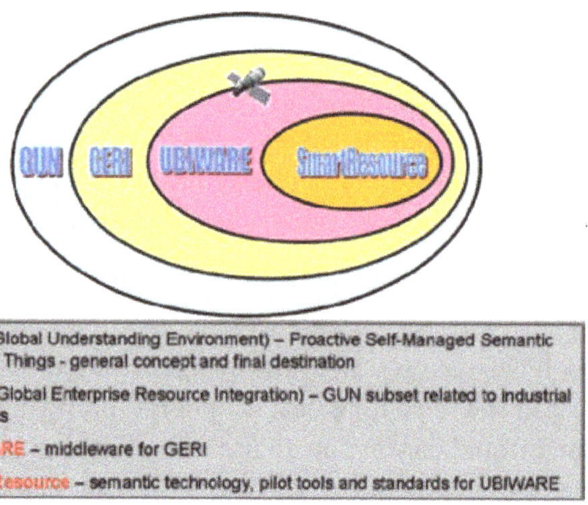


**Fig. 1. The research roadmap towards GUN.**

When implementing Semantic Web in the field of ubiquitous computing, it is essential that Semantic Web is capable of describing resources not only as passive functional or non-functional entities, but also their behavior, including proactivity, communication, and coordination. The term "global" in GUN has a dual connotation. Firstly, it suggests that resources possess the capability to interact and collaborate on a global scale, including the entire company and extending beyond its boundaries. Furthermore, it suggests a comprehensive comprehension on a worldwide scale. This implies that resource A possesses comprehensive knowledge of (1) the characteristics and condition of resource B, (2) the potential and realized actions of B, and (3) the operational procedures in which A and B, along with perhaps additional resources, are collaboratively engaged.

The primary layers of GUN are depicted in Figure 2. Adapters, or interfaces, can connect different resources to the Semantic Web environment. These resources may include sensors with digital output, data structuring using XML, and semantic adapter components that convert XML to Semantic Web format. Each resource will be allocated software agents that may monitor data from the adapter to determine the resource's state, make decisions on behalf of the resource, and seek external assistance if necessary. The use of agent technologies in GUN enables the movement of service components across different platforms, decentralized service discovery, the employment of FIPA communication protocols, and the integration/composition of services through multiple agents.

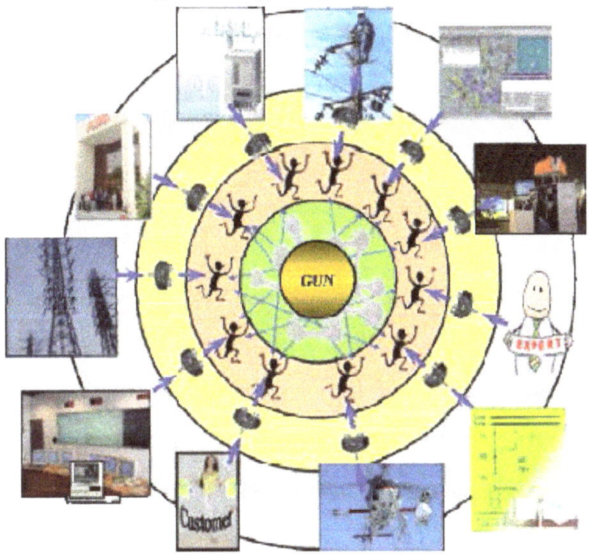


**Fig. 2. Main layers of GUN.**

When implementing the GUN concept, every conventional system component transforms into an agent-driven "intelligent resource", meaning it becomes proactive and capable of self-management. This can also be implemented using recursion. As an illustration, the interface of a system component has the potential to function as an intelligent resource in its own right. This means that it can possess its own responsible agent, sensors and actuators that are semantically adaptable, a history of its actions, commitments with other resources, and the ability to self-monitor, self-diagnose, and engage in self-maintenance activities. This would ensure a great degree of dynamism and flexibility in the interface. This technique certainly offers distinct benefits in comparison to other software technologies that are essential components of it, such as OOSE, SOA, Component-based SE, Agent-based SE, and Semantic SE. This approach can easily be used to different conceptual domain models. Domain ontology can be seen as an intelligent asset that facilitates the inclusion of numerous ontologies in a system design. This enables interoperability, on-the-fly mapping, and maintenance through communication between relevant agents.

Our goal is to extend the concepts of automatic discovery, selection, composition, orchestration, integration, invocation, execution monitoring, coordination, communication, negotiation, and context awareness, which have primarily been applied to the Semantic Web-Services domain, to a broader "Semantic Web of Things" domain. In addition, we aim to enhance this list by incorporating automated self-management capabilities such as self-organization, diagnostics, forecasting, control, configuration, adaptability, tuning, maintenance, and learning.

According to a more global view to the Ubiquitous Computing technology:

- UBIWARE will classify and register various ubiquitous devices and link them with web resources, services, software and humans as business processes' components;
- UBIWARE will consider sensors, sensor networks, embedded systems, alarm detectors, actuators, communication infrastructure, etc. as "smart objects" and will provide similar care to them as to other resources.
- Utilization of the Semantic Web technology should allow:
- Reusable configuration patterns for ubiquitous resource adapters;
- Reusable semantic history blogs for all ubiquitous components;
- Reusable semantic behavior patterns for agents and processes descriptions;
- Reusable coordination, design, integration, composition and configuration patterns;
- Reusable decision-making patterns;
- Reusable interface patterns;

Reusable security and privacy policies. Utilization of the Distributed AI technology should allow:

- Proactivity and autonomic behavior;
- Communication, coordination, negotiation, contracting;
- Self-configuration and self-management;
- Learning based on live blog histories;
- Distributed data mining and knowledge discovery;
- Dynamic integration;
- Automated diagnostics and prediction;
- Model exchange and sharing.

## 4. THE CORE OF THE MIDDLEWARE

### 4.1 Challenges

The primary goals of the Ubiware core (UbiCore) are as follows. In order to enable every resource to operate intelligently, a software agent must be connected to it. This will allow the resource to actively perceive, monitor, and manage its own condition. Additionally, it will be able to communicate with other components, combine its own knowledge and external information, and utilize this knowledge for self-diagnosis and self-maintenance. The system must facilitate the automatic discovery of resources and enable the configuration of a complicated feature by leveraging the atomic functionalities of the resources. The primary objective is to guarantee a consistent and organized functioning of the components and the entire system. This is achieved by ensuring that the smart resources adhere to their designated organizational responsibilities and by maintaining a comprehensive grasp of the system's overall structure among the resources. The latter implies that resource A possesses comprehensive knowledge of (1) the attributes and condition of resource B, (2) the potential and actual actions of B, and (3) the operational procedures in which A and B, along with potentially additional resources, are collectively engaged.

Researchers in the field of software agents have developed multiple Agent Programming Languages (APLs) that focus on internal designs and implementation methodologies. Some examples of these languages include AGENT-0 (Shoham, 1993), AgentSpeak(L) (Rao, 1996), 3APL (Dastani et al., 2004), and ALPHA (Collier et al., 2005). All of those languages are declarative and rule-based, and they are founded on the first-order logic of n-ary predicates. Each of these is also influenced by the Beliefs-Desires-Intentions architecture (Rao & Georgeff, 1991). For instance, in the ALPHA agent program, there are declarations that specify the beliefs and goals of the agent. Additionally, there are declarations of a set of rules, which include belief rules (that generate new beliefs based on existing ones), reactive rules (that immediately invoke certain actions), and commitment rules (that establish a commitment to invoke an action). Sensors, which perceive the environment and generate new beliefs, and actuators, which implement the actions to be invoked, are external code components written in Java.

Based on the review of the languages mentioned above, we list the following important features of them, which should also be realized in UBIWARE's language for specification of behavior models:

- Ability to specify beliefs (something that the agent believes to be true) and goals (something that the agent does not believe to be true but wants to eventually become true).

- Ability to describe behavior rules, i.e. actions taken when a certain condition is met (can be either presence or absence of certain beliefs or presence of certain goals). Ability to have as the result of firing rule all of following:
- Adding/removing beliefs
- Engaging sensors and actuators
- Creating commitments (actions to be executed later)
- Ability to describe plans, i.e. predefined sequences of actions.
- Ability to describe commitments that are:
- Executed when certain condition is met
- Dropped when certain condition is met (or is not met anymore)

The agent-oriented method delays the conversion from domain concepts to machine concepts until the design and implementation of individual agents. An advantage of utilizing an APL is that the transition is further delayed, until the specific sensors and actuators are implemented. However, it appears that this benefit is the only one being taken into consideration. We have not found any ways in literature that expand the role of APL code beyond the development stage. APL code is often authored by the agent's developer and can be either compiled into an executable program or interpreted at runtime. It is an inherent and unchanging property of the agent. The APL code is strictly confined to the agent and is not expected to originate from external sources during runtime. Furthermore, it is not shared with any other agents in any manner.

Exporting and sharing APL code would likely be logical in this context. The methodologies for designing agent-based systems, such as OMNI (Vazquez-Salceda, 2005), define an organizational position using a set of rules. Additionally, an APL is a language that operates based on rules. Utilizing an APL (Access Control Policy Language) to define a position appears to be a logical and intuitive approach. The distinction lies in the fact that APL code associated with a position should inherently be owned and regulated by the organization, and can be accessed by the agents performing the function, possibly even during runtime. Having access during the execution of a program would also allow the organization to modify the role code if necessary.

The second inherent concept is that the agents have the ability to access the APL code of a role, not only to perform that task, but also to collaborate with the agents assuming that role. One such approach is for an agent to transmit a portion of its APL code to another agent in order to convey its intentions regarding future actions, eliminating the requirement for a separate content language. Alternatively, if the code of a job is publicly available inside the organization, agents can refer to it to get insight into how to engage with or anticipate the actions of an agent assuming that position.

However, when thinking about using the existing APLs in such a manner, there are at least two issues:

- The code in an APL is, roughly speaking, a text. However in complex systems, a description of a role may need to include a huge number of rules and also a great number of beliefs representing the knowledge needed for playing the role. Also, in a case of access of the code by agents that are not going to enact this role, it is likely that they may wish to receive only a relevant part of it, not the whole thing. Therefore, a more efficient, e.g. a database-centric, solution is probably required.
- When APL code is provided by an organization to an agent, or shared between agents, mutual understanding of the meaning of the code is obviously required. While using first-order logic as the basis for an APL assures understanding of the semantics of the rules, the meaning of predicates used in those rules still needs to be consistently understood by all the parties involved. On the other hand, we are unaware of tools allowing unambiguous description of the precise semantics of n-ary predicates.

To address these two problems, we propose developing an APL using the W3C's Resource Description Framework (RDF). RDF exclusively utilizes binary predicates, meaning that triples are used. However, it is still possible to encode n-ary predicates using various methods. RDF offers methods for efficient database storage and querying, as well as for explicitly describing semantics, such as utilizing OWL. The Semantic Agent Programming Language (S-APL) is our proposal for an RDF-based APL.

## 4.2 Proposed solution

The answer to the requirements mentioned above is our Semantic Agent Programming Language (S-APL). S-APL is built on top of these axioms:

- Everything is a belief. All other mental attitudes such as desires, goals, commitments, behavioral rules are just complex beliefs.
- Every belief is either a semantic statement (subject-predicate-object triple) or a linked set of such statements.
- Every belief has the context that restricts the scope of validity of that belief. Beliefs have any meaning only inside their respective contexts.
- Statements can be made about contexts, i.e. contexts may appear as subjects or/and objects of triples. Such statements give meaning to contexts. This also leads to a hierarchy of contexts (not necessarily a tree structure though).

- There is the general context G, which is the root of the hierarchy. G is the context for global beliefs (as opposed to local ones). Nevertheless, every local belief, through the hierarchical chain of its contexts, is linked to G.

The notation that is selected for use in S-APL is a subset of Notation3 (Berners-Lee, 2006). Notation3 was proposed by Berners-Lee as an alternative to the dominant notation for RDF which is RDF/XML. Notation 3 is a language which is more compact and probably better readable than RDF/XML, and is also extended to allow greater expressiveness.

UbiCore uses a basic 3-layer agent structure that is common to the APL approach (see e.g. Collier et al., 2005). The architecture can be seen in Figure 3. There is a behavior engine implemented in Java, a declarative middle-layer, and a set of so-called Reusable Atomic Behaviors (RABs). A RAB is a Java component and its main purpose is to act as sensor or actuator. However, we do not restrict RABs to be only sensors or actuators, i.e. components concerned with the agent's environment. A RAB can also be a reasoner (data-processor) if some of the logic needed is impossible or is not efficient to realize with S-APL, or if one wants to enable an agent to do some other kind of reasoning beyond the rule-based one.

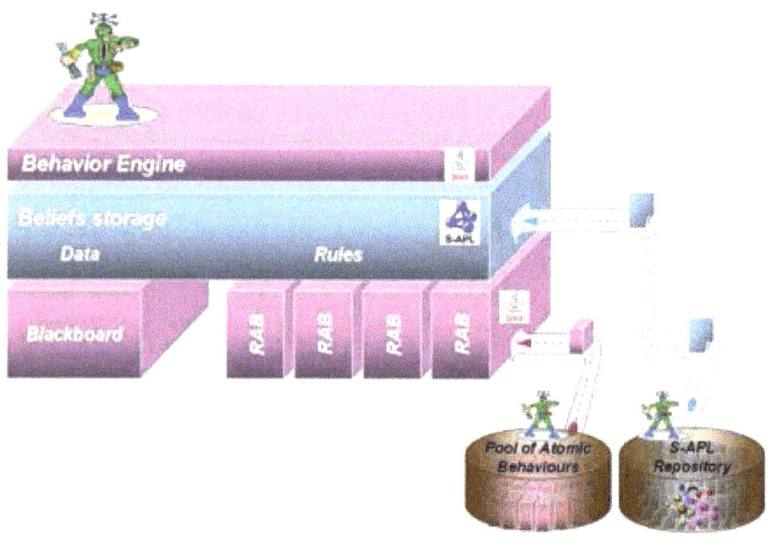


**Fig. 3. The agent architecture.**

The middle layer of the architecture acts as beliefs storage. What differentiates S-APL from traditional APLs is that S-APL is RDF-based. This provides the advantages of the semantic data model and reasoning. Another advantage of S-APL is that the data and the program code are treated in a similar way. Data and code use the same storage instead of two separate ones. This also means that: a rule upon its execution can add or remove another rule, the existence or absence of a rule can be used as a premise of another rule, and so on.

In traditional APL approaches this cannot be performed, because the rules are separated from the data. S-APL is very symmetric with respect to this - anything that can be done to a simple statement can also be done to any belief structure of any complexity. An S-APL agent can obtain data and rules also by querying S-APL repositories. For example a repository can contain a list of ready-made scripts that correspond to organizational roles. In our implementation, such querying is performed as inter-agent action with FIPA ACL messaging but does not involve any query or content languages beyond S-APL itself. As can be seen from Figure 3, agents also can load RABs remotely. This is done as an exception mechanism triggered when a rule prescribes engaging a RAB while the agent does not have it available. Thus, organizations are able to provide not only the rules to follow but also the tools needed for that.

An Ubiware-based application consists of a set of S-APL scripts (behavior models and data) and a set of Reusable Atomic Behaviors. In order to increase the reusability of the platform, we provide some ready-made scripts and behaviors. Therefore, our platform as such can be seen as consisting of the following three elements: (1) the behavior engine, (2) a set of "standard" SAPL models, (3) a set of "standard" RABs. The set of currently provided standard S-APL models includes for example communication models. The set of currently provided standard RABs includes RABs for such operations like downloading a web document, sending an email, loading and transforming into an S-APL presentation a text table document, loading an XML document, etc (see details in Katasonov, 2008).

Technically, our implementation is built on the top of the Java Agent Development Framework (JADE) (Bellifemine at al., 2007), which is a Java implementation of IEEE FIPA specifications. JADE provides communication infrastructure, agent lifecycle management, agent directory-based discovery and other standard services.

## 5. INTELLIGENT RESOURCE VISUALIZATION

### 5.1 Motivation

One of the reasons for intelligent resource visualization springs from the necessity of resource search and browsing processes enhancement. In (Marcos et al., 2005), several features of Classical Model of information search are criticized. First of all, this model does not adequately distinguish between the needs of a user and what a user must specify to get it. Very often, users may not know how to specify a good search query, even in natural language terms. Analyzing what is retrieved from the first attempt is usually not used to select useful results, but to find out what is there to be searched over. A second important criticism of the Classical Model is that any knowledge generated during the process of formulation a query is not used later in the sequence of search process steps, to influence the filtering step and presenting step of the search results, or to select the results. Finally, Classical Model provides an essentially context-free process. There is no proper way in which knowledge of the task context and situation, and user profile can have an influence on the information search process.

Thus, when we come to the vision of GUN (where all the resources of the virtual and the real world are connected and interoperate with each other) and the amount of information becomes huge, we have to elaborate new visualization techniques that simplify the search and browsing processes through reducing amount of queries via context-dependent resource visualization. Following this approach, we have a need to visualize the resource properties (in different from "directed arc between objects" representation way), the various relations between resources and the inter-resource communication process. And even more, we have a need to make this visualization more context-dependent, to be able to represent information in a handy way and adequate to a certain case (context), to reach the plasticity of UIs (Thevenin & Coutaz, 1999). Thus, the main focus in GUI development will be concentrated on the resource visualization aspects and we have a challenging task of semantically enhanced context-dependent multidimensional resource visualization.

## 5.2 4i: A new form of resource visualization

This part will study dynamic context-aware A2H (Agent-to-Human) interaction in UBIWARE, and will elaborate on a technology which we refer to as 4i (FOR EYE) technology. 4i enables creation of a smart human interface through flexible collaboration of an Intelligent GUI Shell, various visualization modules, which we refer to as MetaProvider- services, and the resources of interest.

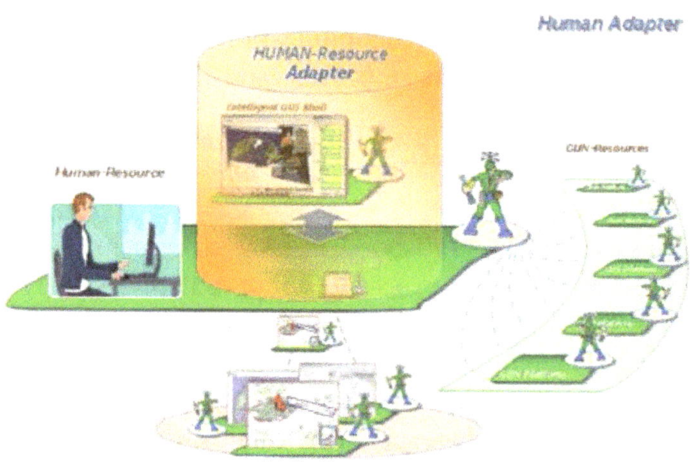


**Fig. 4. Human-Resource Adapter based on 4i (FOR EYE) technology.**

In UBIWARE, humans are important resources, which can play several distinct roles: a resource under care, a service provider, a user, and an administrator.

Obviously, the humans need some graphical interfaces to interact with the rest of the system. The same person can play several roles, switch between them depending on the context, and, in result, require different interfaces at different times. In addition, a UBIWARE-based system presents a large integration environment with potentially huge amounts of heterogeneous data. Therefore, there is a need for tools facilitating information access and manipulation by humans. A semantic context-aware multimodal visualization approach would provide an opportunity for creating smart visual interfaces able of presenting relevant information in a more suitable and more personalized form.

From the GUN point of view, a human interface is a special case of a resource adapter (Human-Resource AdaPter - HRAP). We believe, however, that it is unreasonable to embed all the data acquisition, filtering and visualization logic into such an adapter. Instead, external services and application should be effectively utilized. Therefore, the intelligence of a smart interface (HRAP) will be a result of collaboration of multiple agents: the human's agent, the agents representing resources of interest (those to be monitored or/and controlled), and the agents of various visualization services – MetaProviders via an agent of HRAP (see Figure 4). This approach makes human interfaces different from other resource adapters and indicates a need for devoted research.

Based on 4i technology, an infrastructure will be embedded into UBIWARE enabling effective realization of the following system functions:

- Visualization of data provided by a service in response to a request;
- Search, retrieving and visualization of data required by a human expert,
- Providing access to contextual information, and visualization of it;
- Visualization of resource registration, configuration, and security policy establishment processes;
- Resource discovery via MetaProviders (because they act as thematic portals).

## 5.3 4i Infrastructure

As previously stated, a key requirement for visual interfaces is the ability to display information based on a selected contextual property of a resource. The interface should allow users to easily choose a context for data representation and should also support cases where multiple contextual properties are selected for complex views and filtering. MetaProviders are unique portals that offer resources with a specialized visualization view, which can fulfill such criteria. The name "MetaProvider" reflects its function of granting access to and presenting several resources, each of which serves as a provider of its own information or data. Every

GUN resource possesses its own distinct physical and digital place. However, this does not imply that there should be only a single method for obtaining access to them. MetaProvider is an interface mediator that allows users to mark or pick a resource (object) on its interface and provides a link to the original location of the resource. Put simply, it permits the registration of resources to enable future access to their data. Simultaneously, each resource has the ability to be registered on many MetaProviders in various perspectives. The primary characteristic of MetaProviders is that every person responsible for managing a GUN-Resource must register the resource itself. It leads to rapid accumulation of data that may be accessed by MetaProvider. Each user or resource has immediate access to the relevant information of others. However, this interoperability introduces a new demand for both the MetaProviders and users. In order to achieve semantic interoperability, it is necessary for them to have a shared ontology. In addition to semantic interoperability, GUN-Resources possess proactive and goal-driven characteristics, and are equipped with a Resource Agent for communication between resources (R2R) and agents (A2A).

4i (FOR EYE) is a combination of the GUN Resource Platform Intelligent GUI Shell and MetaProviders. It is a smart middleware that allows users to access and combine different MetaProviders based on their specific needs. The MetaProviders are visualization modules/platforms that display resource data in a filtered and context-dependent manner. 4i also provides integration on two levels: it integrates the information and data of the resources to be visualized, and it integrates the views of resource representation with a convenient resource browsing feature. Please refer to Figure 5 for a visual representation of this integration. The interface's context-awareness and intelligence introduce a new capability that allows the user to obtain not only raw data, but also integrated information that is relevant to a chosen situation. The GUI Shell enables the user to dynamically switch between MetaProviders to better depict information based on the context or nature of the resource. The primary functions of MetaProvider include the following four roles:

- Context-aware resource visualization module that presents information regarding to specified context in more suitable and personalized for user form;
- Interface for integrated information visualization with intelligent context-aware filtering mechanism to present only relevant information, avoiding a glut of unnecessary information;
- Visual Resource Platform that allows resource registration, search, access and modification of needed information/data in a space of registered resources;
- Mediator that facilitates resource to resource (R2R) communication.

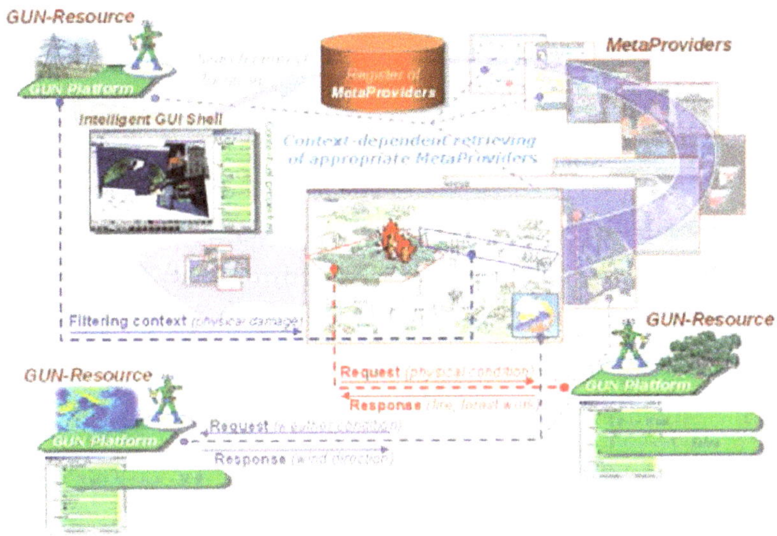


**Fig. 5. Intelligent Interface for Integrated Information (4i technology).**

## 6.3 Concise Summary

4i (FOR EYE) technology is a fundamental component of UBIWARE. Technology serves as a foundation for adapting human resources and will be further developed in accordance with the ideas and goals of UBIWARE. The intelligence of this intelligent interface is achieved through the collaboration of multiple agents: the human agent, the agents representing the resources of interest (those to be monitored, controlled, or requiring human assistance), and the agents of various visualization services - MetaProviders, facilitated by the Human-Resource Adapter agent. On the other hand, based on the 4i framework, dedicated interfaces (MetaProviders) will be created to enable a human (Platform administrator) to customize the functionality of a UBIWARE-based system.

# 6. RESOURCE HISTORIES

## 6.1 Motivation

In UBIWARE, every resource is represented by a software agent. Among major responsibilities of such an agent is monitoring the condition of the resource and the resource's interactions with other components of the system and humans. The beliefs storage of the agent will, therefore, naturally include the history of the resource, in a sense "blogged" by the agent. Obviously, the value of such a resource history is not limited to that particular resource. A resource may benefit from the information collected with respect to other resources of the same (or similar) type, e.g. in a situation which it faces for the first time while others may have faced that situation before. Also, mining the data collected and integrated from many resources may result in discovery of some knowledge important at the level of the whole ubiquitous computing system. A scalable solution requires mechanisms for inter-agent information sharing and data mining on integrated information which would allow keeping the resource histories distributed without need to copy those histories to a central repository.

The work in this area is aimed at answering the following research question:

- How to semantically markup the history of a resource in a system, in order to make it reusable for other resources and at the system-level?
- What mechanisms are needed for effective and efficient sharing of information between the agents representing different resources?

## 6.2 General approach

The chosen strategy for this assignment is to refrain from creating unique protocols and languages for transferring information between agents. Instead, the aim is to maximize the utilization of existing capabilities established within the UbiCore framework outlined in Section 4.

Evidently, there is a discernible resemblance in the following. An agent must consistently examine its own belief system to assess the conditions specified in its behavior rules and determine which rules should be implemented. Conversely, when one agent requests information from another agent, it essentially interrogates the knowledge repository of that other agent.

Our strategy is to build the external querying process to closely resemble the agent's own querying of its belief base to verify the prerequisites for executing a rule. In addition, we intend to utilize the Semantic Agent Programming Language (S-APL) not only for defining the actions of the agents, but also as the language for communication between the agents, replacing FIPA-SL or similar languages. The benefits of this are self-evident, as the UBIWARE platform will achieve maximum symmetry and expressive capability. The agents will have the capability to exchange queries with one other, not just to obtain factual information (both current and historical) about the external world (the domain), but also, for instance:

- Query if the other agent knows a plan for achieving a certain goal,
- Query if the other agent knows a rule that should be applied in a particular situation.

## 6.3 Ontonuts

This section presents the Ontonuts concept, which aims to simplify the organization of modular scripts and plans in the UBIWARE platform. The concept bears some resemblance to the concept of Enterprise Java Beans, as it is inherently semantic and possesses its own distinctive architectural characteristics. Ontonut concepts, in general, embody a specific capacity that is characterized by a defined input and an anticipated result. Next, we modify Ontonuts to address the issue of distributed querying. In UBIWARE, there are two primary perspectives on distributed querying: adapter-based and service-based. The former addresses the integration of external data sources such as databases, files, and online services into the platform, while the later focuses on the querying between agents. However, both aim to achieve the same objective: to ensure that distributed querying is seamless for an agent (see to Figure 6).

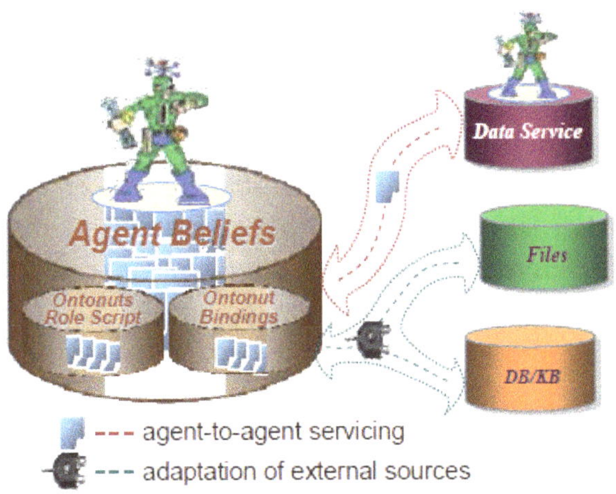


***Data* Fig. 6. Distributed querying in UBIWARE.**

The process of agent-to-agent querying adheres to the servicing paradigm, specifically focusing on data servicing as stated in the work of Quilitz and Leser (2008). The process of adapting external sources, as detailed in Langegger et al, 2007, addresses the issue of connecting and interacting with resources that exist outside of the platform. This involves talking with these resources in their original language. From the perspective of the business logic developer, it is important for every remote resource to be transparent. This ensures that the business logic script remains modular and straightforward. Ontonuts serve as a multifunctional component, acting as a wrapper, adaptor, and connection all in one.

For instance, a database may store data in its original format regarding the events of a particular machine. Let us examine the scenario in which an agent must promptly analyze a succession of events and, as a result, routinely execute queries. An agent can achieve a flat solution by incorporating all database calls and transformations into the same S-APL code used for analysis. Alternatively, it can declare an Ontonut and describe a pattern that precisely defines the permissible query types. Subsequently, the code responsible for conducting the analysis will be completely segregated.

The Ontonuts technology is composed of a Semantic Agent Programming Language (S-APL) script and Reusable Atomic Behaviors (RABs). This allows for dynamic addition, removal, or configuration of the technology. Ontonuts, when operating on the platform, can be conceptualized as functioning in the capacity of an engine or agent. Ontonuts enable the componentization of S-APL code from a programming perspective by applying a semantic annotation to it. These portions of code that have been annotated are referred to as capabilities, which serve as the equivalent of functions in procedural programming. The capabilities can be flexibly integrated into plans and executed by the Ontonuts engine.

The Ontonut capabilities consist of S-APL descriptions that have clearly specified preconditions and effects. The semantic annotation of Ontonut, through the identification of preconditions and effects, enables the automatic planning and composition of an agent's operations in order to accomplish a specific objective. The script has an S-APL function that generates the desired outcome depending on the precondition. The UBIWARE platform's data architecture is based on triples, meaning that goals, preconditions, and effects are defined as sets of triples in S-APL. Let's consider a given initial data collection {A1 A2 A3}, a goal G1 defined as {C1 C2 C3}, and two ontologies O1 and O2, which are specified as:

```
O1 :type :Ontonut
O1 :precondition {A1 A2 A3}
O1 :effect {B1 B2 B3}
O1 :script {{A1 A2 A3}=>... =>{B1 B2 B3}}

O2 :type :Ontonut
O2 :precondition {B1 B2 B3}
O2 :effect {C1 C2 C3}
O2 :script {{B1 B2 B3}=>...=>{C1 C2 C3}}
```

The appearance of the goal G1 will activate Ontonuts engine that will match G1 against available effects and apply planning, which will result in an execution plan: O1=>O2=>G1.

## 6.4 Conclusion

The objective of this endeavor is to organize the S-APL code of the UBIWARE agent with the aim of streamlining agent programming and facilitating automated goal-oriented planning. Despite its limited scope, the work addresses fundamental issues related to agent planning and semantic programming. The primary subject is the automatic planning of distributed inquiries, particularly in the context of distributed RDF queries and virtual graphs. In this scenario, where the queried graph lacks underlying content, RDF queries are employed to construct SQL queries for the relevant database.

This strategy does not include the mapping of models. Instead, it utilizes patterns to specify the desired outcome and applies queries or other code to populate the patterns as required by the agent.

# 7. MIDDLEWARE FOR PEER-TO-PEER DISCOVERY

## 7.1 Motivation

As multi-agent systems become more widespread, the number of multi-agent platforms being used will continue to rise. We recognize the need for the existence of systems that facilitate the establishment of links between various platforms. We propose enhancing UBIWARE by incorporating a feature that enables search agents to get not only local results but also global results from the full network of interconnected platforms. The objective is to enhance the capability of UBIWARE to enable it to manage inter-platform communication and discovery in a peer-to-peer (P2P) manner. This expansion will enable the establishment of a connection between distinct UBIWARE systems, facilitating the interchange of information and collaboration among agents located on different platforms.

The objective is to develop methods that will expand the scope of semantic resource discovery in UBIWARE through peer-to-peer discovery. These mechanisms must allow an agent to:

- To discover agents playing a certain organizational role,
- To discover an agent (or agents) possessing certain needed information,
- To discover resources (through its agents) of certain type or possessing certain properties (e.g. a device in state X, or a web-resource providing some information searched for).
- In all cases, existence of a central Directory Facilitator is not assumed, so the discovery is to be performed in peer-to-peer manner. This has at least two goals:
- Improving the survivability of UBIWARE-based systems. P2P is a complementary mechanism which can be utilized in an exception situation where the central Directory Facilitator became for some reason unavailable.
- Discovery across UBIWARE-based systems. There could be several UBIWARE-based agent platforms (applications) started up independently, each having own Directory Facilitator. While JADE (underlying agent framework of UBIWARE) supports communication among agents residing on different platforms, it does not provide for the cross-platform discovery.

## 7.2 Inter-platform and agent discovery in multi-platform environment

Let us examine the scenario in which multiple autonomous UBIWARE-based agent platforms coexist in the network, each with a group of agents operating and registered with DF.

Inter-agent discovery and communication are seamlessly facilitated within each platform, as the platform's components handle these tasks. Is it feasible (and if so, how) to enable agents from different platforms to search remote DFs and communicate with agents located on foreign platforms? Implementing this method on the UBIWARE platform will improve its functionality. If an agent is unable to find a service or access necessary information within its own platform, it can search the DFs on other known remote systems to find the required data. Agents might derive advantages from this opportunity in two scenarios:

The service or information needed by an agent is not available on the platform. Either the agent who had access to this service/information is no longer on the platform or cannot be reached.

The parallels between peer-to-peer file sharing networks and multi agent systems are evident. P2P networks facilitate the exchange of messages between nodes to identify their neighboring nodes and the resources they collectively own. In multi-agent systems, the agents strive to identify and employ other agents and their services, hence improving the overall functionality of the platform. In this section, we examine the current methods employed in constructing file sharing peer-to-peer (P2P) networks that can be applied to solve the problem. These methods include the centralized approach of a general Directory Facilitator, as well as solutions derived from pure peer-to-peer file sharing networks that are connected to JADE's DF federations mechanism.

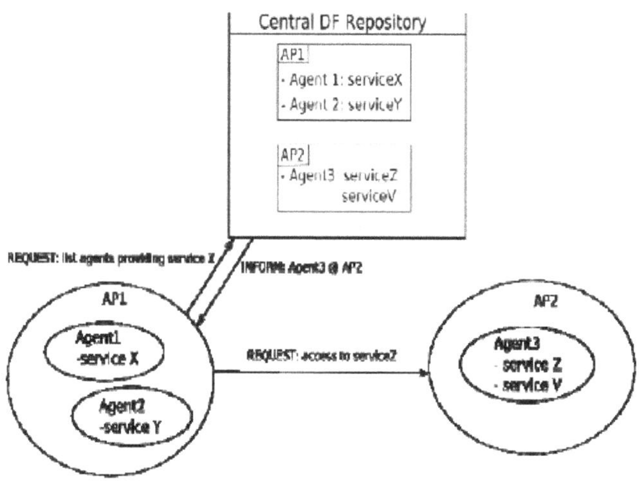


**Fig. 7. Multi-platform environment with centralized DF.**

The inter-platform agent discovery in a centralized approach is derived from the hybrid paradigm used in Napster platform for file sharing peer-to-peer networks. The hybrid architecture involves the central server maintaining a repository of the resources that each node is currently sharing. Transferring this concept to a multi-platform setting involves selecting one of the current platforms to function as a central server. Each agent that desires to broadcast its services contacts the central server and enrolls with the global Directory Facilitator. In order to identify an agent fulfilling a specific role, agents send a request to the central server, which will respond with a list of addresses (AIDs) of agents that offer the desired service. Figure 7 illustrates the concept of a multi-platform ecosystem with a central Directory facilitator.

Federated Directory Facilitators provide an alternative method for achieving inter-platform agent discovery. As per the FIPA specifications, Directory Facilitators should possess the capability to mutually register with one another, thereby establishing a federation within a single platform that is also distributed. When a DF agent establishes a registration with another DF agent, it becomes a subordinate and will henceforth receive all search requests from its superior DF agent, while also sending them to each of its subordinates. The dissemination of the query across the federation is regulated by a parameter called MaxDepth in the search restrictions. This parameter determines the number of times the question to be delivered. The MaxResults option restricts the quantity of agents' descriptions that will be provided to the agent making a query. Figure 8 illustrates an instance of an Agent Platform that includes many federated DFs, each of which has a set of agents registered with it.

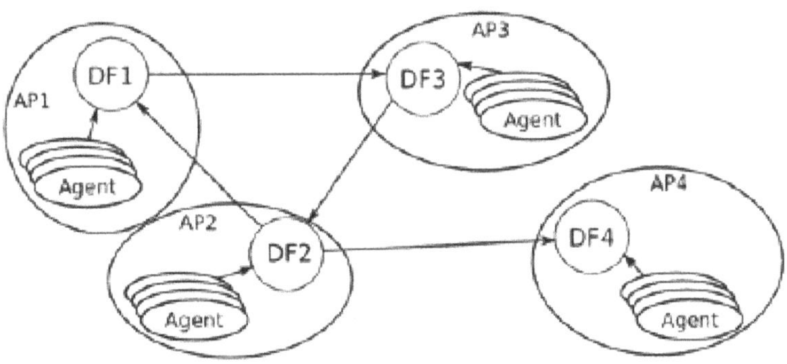


**Fig. 8. Federation of Directory Facilitators.**

One alternative method is to establish a dynamic peer-to-peer topology. We have adopted the concept of the Gnutella system, as proposed by Klingberg and Manfredi in 2002. Gnutella is a distributed peer-to-peer system, meaning it functions without the need for a central server. Each participant fulfills the roles of both a server and a client, leading to the introduction of the word "servent". All servants collaborate by sharing resources, enabling others to easily access and download them. All servants execute two types of operations by utilizing a message forwarding system. Firstly, by the exchange of messages, servants enhance or sustain their level of connectedness. Furthermore, the search queries are exchanged to discover necessary resources that may be shared by certain network-connected servents. The protocol mandates the presence of at least one continuously active node within the network, which furnishes a new participant with the addresses of the already functioning servents. Upon initialization, each servant acquires a set of addresses and establishes connections with them. To identify other participants, the system initiates the PING / PONG procedure, as seen in Figure 9.

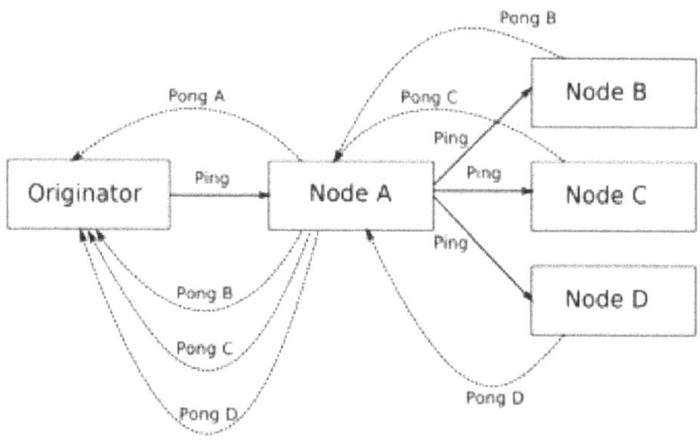


**Fig. 9. Propagation of PING and PONG messages in Gnutella discovery process.**

### 7.3 Conclusions

We have presented three different approaches of building distributed peer-to-peer infrastructure in multiplatform environments. By the means of inter-platform discovery we give agents the opportunity to communicate, share services and resources beyond the boundaries of their home platforms.

Future work will include incorporating one of the described methods into the UBIWARE prototype. We also plan to conduct further research upon improving the efficiency of created network of agent platforms.

## 8. SUMMARY AND FUTURE RESEARCH

This chapter outlines the obstacles that must be overcome in order to realize the vision of the Internet of Things and Ubiquitous Computing. The current advancements in networking, sensor, and RFID technologies enable the integration of different physical objects into the IT infrastructure. Nevertheless, the intricacy of such a system can potentially become overpowering and uncontrollable. Hence, there is a requirement for computing systems that possess the ability to operate autonomously with minimal human intervention. This intervention is primarily focused on establishing higher-level guidelines rather than engaging in direct administration. We propose that the incorporation of multi-agent systems principles can effectively address this complexity, as it enables the creation of intricate systems.

Another obstacle that must be confronted is the issue of resource heterogeneity. Today, semantic technologies are considered a crucial solution for addressing the challenges of interoperability and integration in a diverse and linked world of objects and systems. Semantic technologies are purported to offer a more robust method of achieving interoperability compared to current standards-based approaches. Due to this rationale, we are of the opinion that Semantic Web technologies will have a significant impact on the realization of the Internet of Things.

We are of the opinion that implementing inflexible criteria is not the appropriate approach to attain interoperability. Instead, we recommend utilizing middleware as a cohesive agent that connects diverse components.

Our concept of a middleware for the Internet of Things, which has served as the foundation for our research project Ubiware, is shaped by these principles. Ubiware is a crucial component in the realization of our broader objective known as the Global Understanding Environment (GUN). The objective of the Global Understanding Environment (GUN) is to enable the accessibility, proactivity, and cooperation of diverse resources, including physical objects, digital content, and human beings, through the internet. The three core principles of this platform are Interoperability, Automation, and Integration.

The core is the most crucial component of the middleware. Within the Ubiware project, we commonly use the term "UbiCore" to refer to it. The objective of UbiCore is to enable all resources to possess intelligence by connecting them to a software agent. This intelligence would allow the resources to actively perceive, monitor, and manage their own condition. Additionally, they would be able to communicate with other components, integrate their own experiences and external functionality, and utilize these for self-diagnosis and self-maintenance.

To articulate our intentions, it was necessary to possess a language. There are multiple agent programming languages (APLs) available, such as AGENT-0, AgentSpeak(L), 3APL, and ALPHA. All of these languages are declarative and rule-based, and they are founded on the first-order logic of n-ary predicates. Each of these is additionally influenced by the Beliefs-Desires-Intentions framework. However, none of them take into account the potential for sharing the APL code with other agents or leaving the agent in the runtime.

Exporting and sharing APL code would be logical due to two primary factors. Firstly, this technique can be utilized to define organizational responsibilities, as organizational roles are defined using a set of rules and APL is a rule-based language. Additionally, the agents have the ability to retrieve a role's APL code not only for the purpose of performing that duty, but also for the purpose of collaborating with the agents who are assuming that role. By employing this method, an agent is able to convey its intentions on forthcoming actions.

When considering the utilization of the current APLs as described above, there are at least two challenges that arise. First and foremost, the code in an APL is essentially a textual representation. However, in intricate systems, a depiction of a function may necessitate incorporating a vast array of regulations as well as a substantial quantity of beliefs that indicate the requisite information for fulfilling the role. Thus, it is likely that a more efficient approach, such as one that is focused on databases, is needed. Furthermore, in cases where an organization provides APL code to an agent or shares it across agents, it is essential to have a shared understanding of the code's meaning.

To address these two problems, we propose developing an APL using the W3C's Resource Description Framework (RDF). RDF exclusively use binary predicates, which are represented as triples. The Semantic Agent Programming Language (S-APL) is our proposal for an RDF-based APL. The decision to utilize Notation3 as the foundation of this language was made due to its compactness and superior readability compared to RDF/XML.

Our APL technique utilizes a standard 3-layer agent structure. The system consists of a Java-based behavior engine, a declarative middle-layer, and a collection of Java components that serve as sensors and actuators. We designate the latter as Reusable Atomic Behaviors (RABs). A RAB, or Reasoning Agent Component, is a component that is involved with the agent's surroundings and reasoning processes. The intermediate layer functions as a repository for beliefs. The distinguishing factor between S-APL and traditional APLs is that S-APL is based on RDF (Resource Description Framework). This offers the benefits of the semantic data model and the process of logical inference.

Our platform's architecture dictates that an application using it will comprise a collection of S-APL documents, which include data and behavior models, as well as a set of atomic behaviors specific to that application. There exists a collection of standardized RABs and a collection of standardized S-APL scripts. They form the foundation of the Ubiware core. In addition, the user has the ability to provide their own S-APL scripts and/or RABs.

We assert that the concept of Internet of Things necessitates a fresh methodology when it comes to representing resources visually. The conventional model of information retrieval has certain drawbacks. Initially, the user faces challenges in converting their search concept into the appropriate search query. Oftentimes, the initial search is employed solely to ascertain the content that is available for exploration. Furthermore, the classical approach implements a process that is independent of any specific environment.

To address the aforementioned drawbacks of the classical model, we propose the introduction of the For Eye (4i) idea. 4i is conducting research on a dynamic context-aware Agent-to-Human (A2H) interface in Ubiware. 4i facilitates the development of an intelligent human interface by allowing for flexible collaboration between an Intelligent GUI Shell, many visualization modules (referred to as MetaProvider-services), and the desired resources.

MetaProviders are modules that visually represent resource data in a context-dependent and filtered manner. They integrate the resources to be visualized and give a convenient way to browse across resource representations. The GUI Shell is utilized to connect MetaProviders together.

The presence of an agent representing each resource implies that the agent have information about the current status of that resource. The data regarding this state could prove advantageous for other agents. Other agents can utilize this information in a novel setting, but some may have encountered that situation previously. Furthermore, the process of extracting valuable insights from the aggregated data acquired from various sources can lead to the identification of crucial knowledge that is significant for the entire ubiquitous computing system.

We contend that the establishment of a central repository is an inappropriate strategy. Instead, we suggest the concept of distributed resource histories, which rely on a transparent process of inter-agent information sharing and data mining. To accomplish this objective, we propose the introduction of the Ontonut idea.

The Ontonuts technology is built using a combination of a Semantic Agent Programming Language (S-APL) script and Reusable Atomic Behaviors (RABs). This allows for dynamic addition, removal, or configuration of the technology. Each Ontonut symbolizes a certain ability to access certain information. An Ontonut is characterized by its precondition, effect, and script property annotations. A precondition is a specific state that must be met in order to execute the desired functionality of the ontonut. The effect of performing this Ontonut refers to the ensuing data that can be acquired. The script attribute specifies the method for acquiring the data. One component of the Ontonuts technology is a planner that autonomously generates a querying strategy based on the available ontonuts and the intended aim supplied by the agent.

Multiple Ubiware-based platforms may potentially exist in the future. Our objective is to develop methods that will expand the scope of semantic resource discovery in Ubiware through peer-to-peer discovery. We conducted an analysis of three different methods: Centralized Directory Facilitator, Federated Directory Facilitators, and the establishment of a dynamic peer-to-peer topology. We assert that this form of revelation should not rely on a centralized Directory Facilitator. This will enhance the system's ability to survive.

Our future focus will be on the core extension. Presently, we are engaged in the development of an extension for an agent observable environment. This presents novel opportunities for the organization and automatic adjustment of systems. We intend to do additional research in the field of peer-to-peer inter-platform discovery with the goal of enhancing the efficiency of the established network of agent platforms. One other subject of our research is the field of self-configuration and autonomous application composition.

## REFERENCES


1. Bellifemine, F. L., Caire G., Greenwood, D. (2007). *Developing Multi-Agent Systems with JADE*, Wiley, ISBN 978-0470057476
2. Berners-Lee, T., (2006). *Notation* 3, online (May 2009): http://www.w3.org/DesignIssues/Notation3.html
3. Bresciani, P., Perini, A., Giorgini, P., Giunchiglia, F., and Mylopoulos, J. (2004) Tropos: An agent-oriented software development methodology. *Autonomous Agents and Multi- Agent Systems* 8(3): 203-236
4. Brock, D.L., Schuster, E. W., Allen, S.J., and Kar, Pinaki (2005) An Introduction to Semantic Modeling for Logistical Systems, *Journal of Business Logistics*, Vol.26, No.2, pp. 97- 117 ( available in: http://mitdatacenter.org/BrockSchusterAllenKar.pdf ).
5. Buckley, J. (2006) From RFID to the Internet of Things: Pervasive Networked Systems, *Final Report on the Conference organized by DG Information Society and Media*, Networks and Communication Technologies Directorate, CCAB, Brussels (online :http://www.rfidconsultation.eu/docs/ficheiros/WS_1_Final_report_27_Mar.pdf ).
6. Collier, R., Ross, R., O'Hare, G. (2005). Realising reusable agent behaviours with ALPHA. *In: Eymann, T., Klugl, F., Lamersdorf, W., Klusch, M., Huhns, M.N. (eds.) MATES 2005*. LNCS (LNAI), vol. 3550, pp. 210-215. Springer, Heidelberg
7. Dastani, M., van Riemsdijk, B., Dignum, F., Meyer, J.J. (2004). A programming language for cognitive agents: Goal directed 3APL. *In: Dastani, M., Dix, J., El Fallah-Seghrouchni, A. (eds.) PROMAS 2003*. LNCS (LNAI), vol. 3067, pp. 111-130. Springer, Heidelberg
8. Jennings, N.R., Sycara K. P., and Wooldridge, M. (1998). A roadmap of agent research and development. *Autonomous Agents and Multi-Agent Systems 1(1)*: 7-38.
9. Jennings, N.R. (2000) On agent-based software engineering. *Artificial Intelligence 117(2)*: 277- 296.
10. Jennings, N.R. (2001) An agent-based approach for building complex software systems. *Communications of the ACM 44(4)*: 35-41.
11. Katasonov, A. (2008). *UBIWARE Platform and Semantic Agent Programming Language (S-APL). Developer's guide*, Online: http://users.jyu.fi/~akataso/SAPLguide.pdf.
12. Kaykova O., Khriyenko O., Kovtun D., Naumenko A., Terziyan V., and Zharko A. (2005a) General Adaption Framework: Enabling Interoperability for Industrial Web Resources, *In: International Journal on Semantic Web and Information Systems, Idea Group*, Vol. 1, No. 3, pp.31-63.

13. Kephart J. O. and Chess D. M. (2003). The vision of autonomic computing, *IEEE Computer*, Vol. 36, No. 1, pp. 41-50
14. Klingberg, T., Manfredi, R. (2002) *Gnutella 0.6*, online: http://rfc-gnutella. sourceforge. net/src/rfc-0_6-draft.html
15. Langegger, A., Blochl, M., Woss, W., (2007). Sharing Data on the Grid using Ontologies and distributed SPARQL Queries, *Proceedings of 18th International Conference on Database and Expert Systems Applications*, pp.450-454, Regensburg, Germany, 3-7 Sept. 2007
16. Lassila, O. (2005a) Applying Semantic Web in Mobile and Ubiquitous Computing: Will Policy-Awareness Help?, in Lalana Kagal, Tim Finin, and James Hendler (eds.): *Proceedings of the Semantic Web Policy Workshop, 4th International Semantic Web Conference*, Galway, Ireland, pp. 6-11.
17. Lassila, O. (2005b) Using the Semantic Web in Mobile and Ubiquitous Computing, in: Max Bramer and Vagan Terziyan (eds.): *Proceedings of the 1st IFIP WG12.5 Working Conference on Industrial Applications of Semantic Web*, Springer IFIP, pp. 19-25.
18. Lassila, O., and Adler, M. (2003) Semantic Gadgets: Ubiquitous Computing Meets the Semantic Web, In: D. Fensel et al. (eds.), Spinning the Semantic Web, MIT Press, pp. 363-376.
19. Mamei, M, Zambonelli F. (2006). *Field-Based Coordination for Pervasive Multiagent Systems*, Soringer, ISBN 9783540279686, Berlin
20. Quilitz, B., Leser, U. (2008) Querying Distributed RDF Data Sources with SPARQL, The Semantic Web: Research and Applications, *5th European Semantic Web Conference, ESWC 2008*, Tenerife, Canary Islands, Spain, June 1-5, 2008, pp.524-538.
21. Rao, A.S. and Georgeff, M.P.(1991) Modeling rational agents within a BDI architecture. *Proc. 2nd International Conference on Principles of Knowledge Representation and Reasoning (KR'91)*, pp. 473-484.
22. Rao, A.S. (1996) AgentSpeak(L): BDI agents speak out in a logical computable language. *Proc. 7th European Workshop on Modelling Autonomous Agents in a Multi-Agent World,* LNCS vol.1038, pp. 42-55
23. Tamma, V.A.M., Aart, C., Moyaux, T., Paurobally, S., Lithgow-Smith, B., and Wooldridge, M. (2005) An ontological framework for dynamic coordination. *Proc. 4th International Semantic Web Conference'05*, LNCS vol. 3729, pp. 638-652.
24. Terziyan V. (2003) Semantic Web Services for Smart Devices in a "Global Understanding Environment", *In: R. Meersman and Z. Tari (eds.), On the Move to*

*Meaningful Internet Systems 2003: OTM 2003 Workshops*, Lecture Notes in Computer Science, Vol. 2889, Springer-Verlag, pp.279-291.

25. Terziyan V. (2005) Semantic Web Services for Smart Devices Based on Mobile Agents, *In: International Journal of Intelligent Information Technologies*, Vol. 1, No. 2, Idea Group, pp. 43-55.
26. Thevenin, D. and Coutaz, J., (1999). Plasticity of User Interfaces: Framework and Research Agenda. *In Proceedings of Interact'99*, vol. 1, Edinburgh: IFIP, IOS Press, 1999, pp. 110-117.
27. Vázquez-Salceda, J., Dignum, V., and Dignum, F. (2005) Organizing multiagent systems. *Autonomous Agents and Multi-Agent Systems 11(3)*: 307-360.
28. Wooldridge, M. (1997) Agent-based software engineering. *IEE Proceedings of Software Engineering 144(1)*: 26-37.

# 15 METHODS OF ARTIFICIAL INTELLIGENCE IN FAULT-TOLERANT CONTROL

## 1. AN OVERVIEW OF FAULT TOLERANT CONTROL

The growing need for high-quality goods, reliable systems, and uninterrupted plant operations has led engineers and scientists to focus increasingly on developing methods and systems capable of effectively managing specific types of failures. Furthermore, the worldwide crisis intensifies competition among industries, making plant shutdowns impractical due to resulting production losses and subsequent market absence. Essential services like power grids, water supplies, transportation systems, communication networks, and commodity production cannot be disrupted without jeopardizing human health and social stability.

However, contemporary technologies and demanding working conditions heighten the likelihood of system malfunctions, which can result in the loss of human life and equipment. Additionally, hazardous settings in locations like nuclear or chemical facilities impose stringent limitations on human labor. Automation and intelligent systems are crucial in all of these contexts to minimize the impact of errors.

An essential advantage of the Fault Tolerant Control (FTC) approach is its ability to ensure uninterrupted operation of the plant, even in the presence of faults or performance degradation in the process. This method mitigates the risk of a fault escalating into a more severe failure. To summarize, the primary benefits of installing a Fault-Tolerant Control (FTC) system are as follows, as stated by Blanke et al. (1997):

- Plant availability and system reliability in spite of the presence of a fault.
- Prevention to develop a single fault in to a system failure.
- The use of information redundancy to detect faults instead of adding more hardware.
- The use of reconfiguration in the system components to accommodate a fault.
- FTC admits degraded performance due to a fault but maintain the system availability.

- Is cheap because most of the time no new hardware will be needed.

Some areas where FTC is being used more often are: aerospace systems, flight control, automotive engine systems and industrial processes. All of these systems have a complex structure and require a close supervision; FTC utilizes plant redundancy to create an intelligent system that can supervise the behavior of the plant components making these kinds of systems more reliable.

Since a few years ago, emerging FTC techniques have been proposing new controller designs capable to tolerate system malfunctions and maintain stability and desirable performance properties. In order to achieve its objectives, two main tasks have to be considered on an active FTC system: fault detection and diagnosis and controller reconfiguration. The main purpose of fault detection and diagnosis is to detect, isolate and identify the fault, determining which faults affect the availability and safety of the plant. The controller reconfiguration task accommodates the parameters in order to reduce the fault effects.

Although several schemes of FTCS have been proposed, most of them are closely related to a general architecture. (Blanke et al., 1997) introduce an approach for the design of an FTC system, shown in figure 1, which included three operational levels: single sensor validation, fault detection and isolation using analytical redundancy, and an autonomous supervision and reconfiguration system. The single sensor validation level involves the control loop with actuators, sensors, the controller and the signal conditioning and filtering. The second level (FDI) is composed of detectors and effectors that will perform the remedial actions. And finally, the supervision level deals with state-event logic in order to describe the logical state of controlled objects.

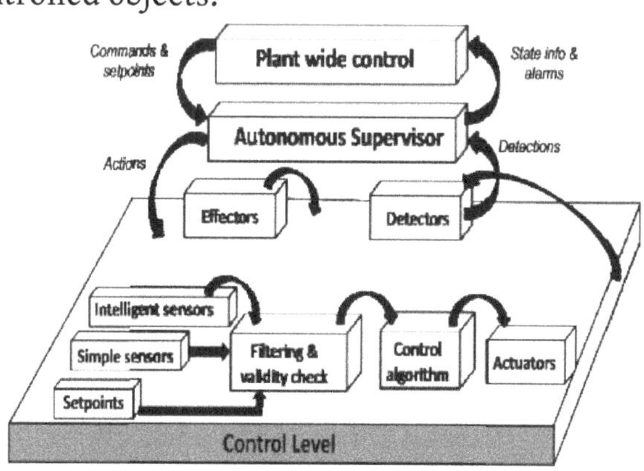


**Fig. 1. Architecture for Fault Tolerant Autonomous Control Systems proposed by (Blanke, 1997).**

A slightly different architecture is presented in (Kar ai et al, 2003). They introduce a scheme of Fault-Adaptive Control Technology (FACT), centered on model-based approaches for fault detection, fault isolation and estimation, and controller selection and reconfiguration for hybrid systems (see figure 2). Hybrid models derived from hybrid bond graphs are used to model the continuous and discrete system dynamics. The supervisory controller, modeled as a generalized finite state automaton, generates the discrete events that cause reconfigurations in the continuous energy-based bond graph models of the plant. Fault detection involves a comparison between expected behaviors of the system, generated from the hybrid models, with actual system behavior.

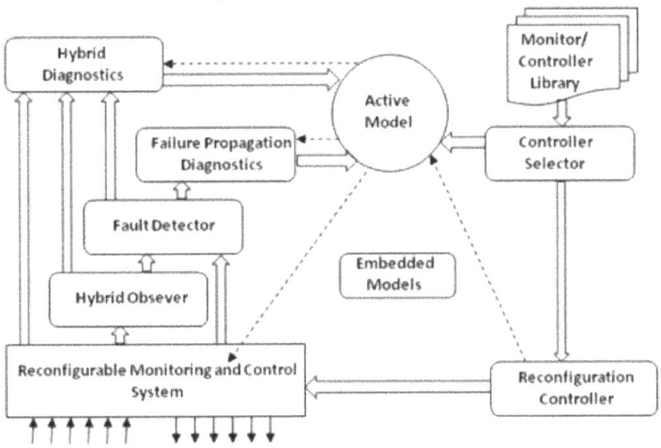


**Fig. 2. Architecture for Fault-Adaptive Tolerant Control Technology (FACT) proposed by (Karsai et al, 2003).**

## 2. CATEGORIZATION OF FAULT TOLERANT CONTROL METHODS

Some authors have proposed different classifications for the FTC methods (Blanke et al., 2003; Eterno et al., 1985; Farrel et al., 1993; Lunze & Richter, 2006; Patton, 1997; Stengel, 1991). The classification shown in figure 3 includes all the methods explained by these authors. We can also find a recent and very complete survey of FTC methods and applications in (Zhang & Jiang, 2008).

Regarding the design methods, fault tolerant control can be classified into two main approaches: active or passive. In Active Fault Tolerant Control (AFTC), if a fault occurs, the control system will be reconfigured using some properties of the original system in order to maintain an acceptable performance, stability and robustness. In some cases degraded system operations have to be accepted (Blanke et al., 2001; Patton, 1997; Mahmoud et al., 2003). In Passive Fault Tolerant Control (PFTC) the system has a specific fixed controller to counteract the effect and to be robust against certain faults (Eterno et al., 1985).

To implement the AFTC approach two tasks are needed: fault detection and isolation and controller reconfiguration or accommodation. FDI means early detection, diagnosis, isolation, identification, classification and explanation of single and multiple faults; and can be accomplished by using the following three methodologies (Venkatasubramanian et al., 2003a, 2003b, 2003c):

*Quantitative Model-Based*: requires knowledge of the process model and dynamics in a mathematical structural form. Also, the process parameters, which are unknown, are calculated applying parameter estimation methods to measured inputs and outputs signals of the process. This approach uses analytical redundancy that can be obtained by implementing Kalman filters, observers and parity space.

*Qualitative Model-Based*: Are based on the essential comprehension of the process physics and chemical properties. The model understanding is represented with quality functions placed in different parts of the process. This methodology can be divided in abstraction hierarchies and causal models. Abstraction hierarchies are based on decomposition and the model can establish inferences of the overall system behavior from the subsystems law behavior. This can be done using functional or structural approaches. Causal models take the causal system structure to represent the process relationships and are classified in diagraphs, fault trees and qualitative physics.

*Process History-Based*: uses a considerable amount of the process historical data and transform this data into a priori knowledge in order to understand the system dynamics. This data transformation is done using qualitative or quantitative methods. The quantitative methods are divided in expert systems (solves problems using expertise domain) and trend modeling (represents only significant events to understand the process). Quantitative methods can be statistical (use PCA, DPCA, PLA, CA) and non statistical (neural networks) to recognize and classify the problem.

After the detection and isolation of the fault, a controller reconfiguration or accommodation is needed. In controller accommodation, when a fault appears, the variables that are measured and manipulated by the controller continue unaffected, but the dynamic structure and parameters of the controller change (Blanke et al., 2003). The fault will be accommodated only if the control objective with a control law that involves the parameters and structure of the faulty system has a solution (Blanke et al., 2001). In order to achieve fault accommodation, two approaches can be used: adaptive control and switched control. Adaptive control means to modify the controller control law to handle the situation where the system's parameters are changing over time. It does not need a priori information about the parameters limits. The goal is to minimize the error between the actual behavior of the system and the desirable behavior. In the other hand, switched control is determined by a bank of controllers designed for specifics purposes (normal operation or fault) that switch from one to another in order to control a specific situation (Lunze & Richter, 2006).

Meanwhile, controller reconfiguration is related with changing the structure of the controller, the manipulated and the measured variables when a fault occurs (Steffen, 2005). This is achieved by using the following techniques:

*Controller Redesign*. The controller changes when a fault occurs in order to continue achieving its objective (Blanke et al., 2003). This can be done by using several approaches: pseudo inverse methods (modified pseudo inverse method, admissible pseudo inverse method), model following (adaptive model following, perfect model following, eigen structure assignment) and optimization (linear quadratic design, model predictive control) (Caglayan *et al.*, 1988; Gao & Antsaklis, 1991; Jiang, 1994; Lunze & Richter, 2006; Staroswiecki, 2005).

*Fault Hiding Methods.* The controller continues unchanged when a fault is placed, because a reconfiguration system hides the fault from the controller. This method can be realized using virtual actuators or virtual sensors. (Lunze & Richter, 2006; Steffen, 2005).

*Projection Based Methods.* A controller is designed a priori for every specific fault situation and replaces the nominal controller if that specific fault occurs. This can be done by a bank of controllers and a bank of observers (Mahmoud et al., 2003).

*Learning Control.* This methodology uses artificial intelligence like neural networks, fuzzy logic, genetic algorithms, expert systems and hybrid systems which can learn to detect, identify and accommodate the fault (Polycarpou & Vemuri, 1995; Stengel, 1991; Karsai et al, 2003).

*Physical Redundancy.* This is an expensive approach because it uses hardware redundancy (multiple sensor or actuators) and decision logic to correct a fault because it switches the faulty component to a new one. An example of this is the voting scheme method (Isermann et al., 2002; Mahmoud et al., 2003).

On the other hand, passive FTC is based on robust control. In this technique, an established controller with constant parameters is designed to correct a specific fault to guarantee stability and performance (Lunze & Richter, 2006). There is no need for online fault information. The control objectives of robust control are: stability, tracking, disturbance rejection, sensor noise rejection, rejection of actuator saturation and robustness (Skogestad & Postlethwaite, 2005). Robust control involves the following methodologies:

*H∞ controller.* This type of controller deals with the minimization of the H-infinity- norm in order to optimize the worst case of performance specifications. In Fault Tolerant Control can be used as an index to represent the attenuation of the disturbances performances in a closed loop system (Yang & Ye, 2006) or can be used for the design of robust and stable dynamical compensators (Jaimoukha et al., 2006; Liang & Duan, 2004).

*Linear Matrix Inequalities (LMIs).* In this case, convex optimization problems are solved with precise matrices constraints. In Fault Tolerant Control is implemented to achieve robustness against actuator and sensor faults. (Zhang et al., 2007).

*Simultaneous Stabilization.* In this approach multiple plants must achieve stability using the same controller in the presence of faults. (Blondel, 1994).

*Youla-Jabr-Bongiorno-Kucera (YJBK) parameterization.* This methodology is implemented in Fault Tolerant Control to parameterize stabilizing controllers in order to guarantee system stability. YJBK in summary is a representation of the feedback controllers that stabilize a given system (Neimann & Stoustrup, 2005).

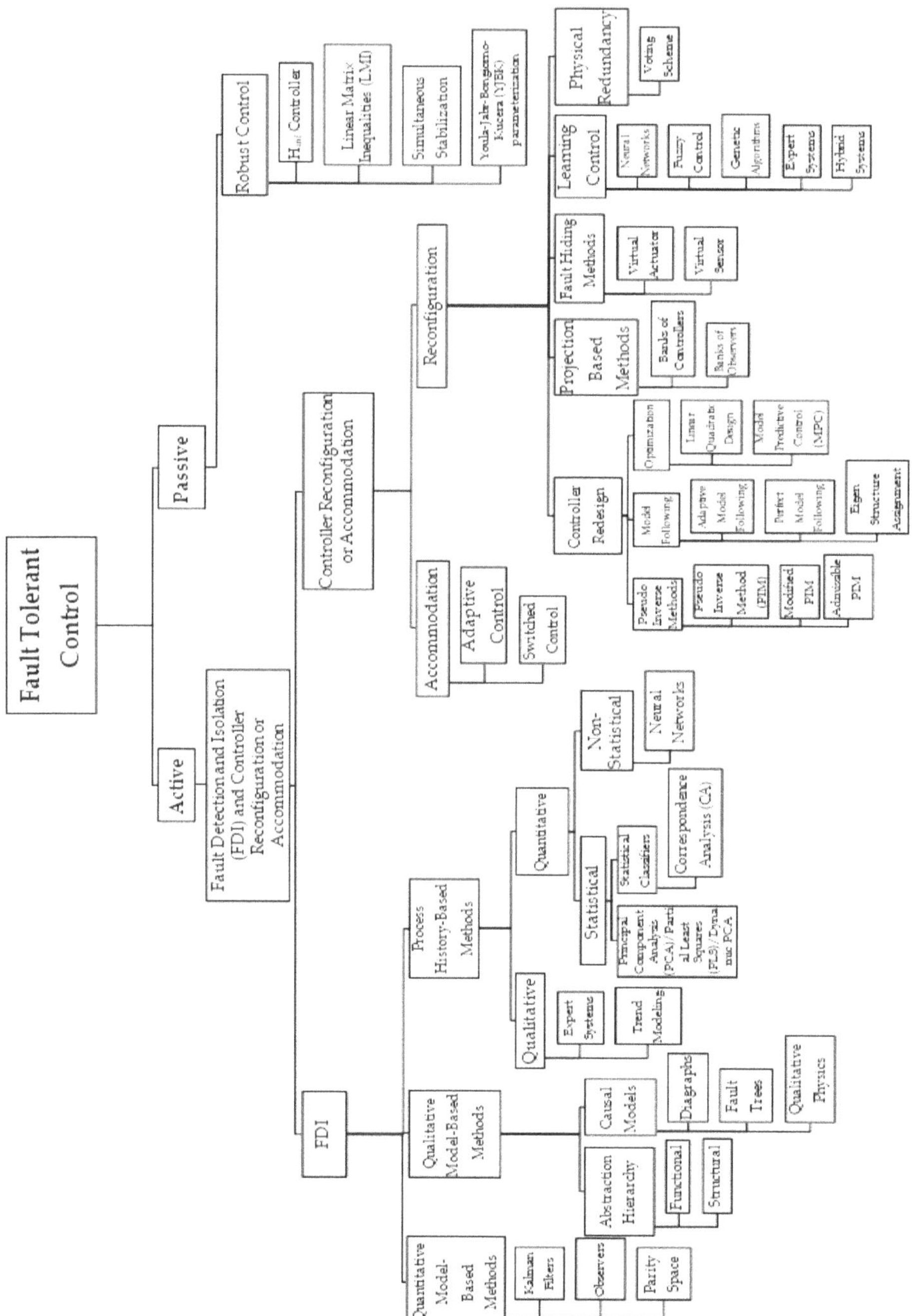

g. 3. FTC classification approaches

**Fig. 3. FTC classification approaches**

## 3. METHODS OF ARTIFICIAL INTELLIGENCE

The use of AI in fault tolerant control has been suggested in the past (Bastani & Chen, 1988). Methods such as Neural Networks (NNs), Fuzzy Logic and Neuro-Fuzzy Systems, offer an advantage over traditional methods (state observers, statistical analysis, parameter estimation, parity relations, residual generation, etc) because can reproduce the behavior of non linear dynamical systems with models extracted from data. This is a very important issue in FTC applications on automated processes, where information is easily available, or processes where accurate mathematical models are hard to obtain. In the other hand, AI optimization tools such as Genetic Algorithms (GAs) provide a powerful tool for multiobjective optimization problems frequently found on FTC.

### 3.1 Neural Networks

Artificial Neural Networks (ANNs) are mathematical models that try to mimic the biological nervous system. An artificial neuron have multiple input signals *x1*, *x2*, ...,*xn* entering the neuron using connection links with specific weights *w1*, *w2*, ..., *wn* or $\sum_{i=1}^{n} w_n x_i$ named the net input, and also have a firing threshold *b*, an activation function *f* and an output of the neuron that is represented by $y = f\left(\sum_{i=1}^{n} w_i x_i - b\right)$. The firing threshold *b* or bias can be represented as another weight by placing an extra input node $x_0$ that takes a value of 1 and has a $w_0$=-*b*. (Nguyen et al., 2002). This can be represented in the figure 4.

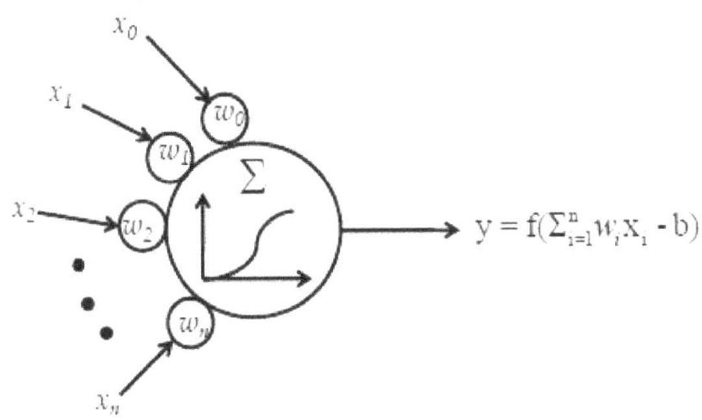


**Fig. 4. Artificial Neuron.**

A neural network with more than one input layer of neurons, a middle layer called the hidden layer and an output layer is named a multi-layer neural network.

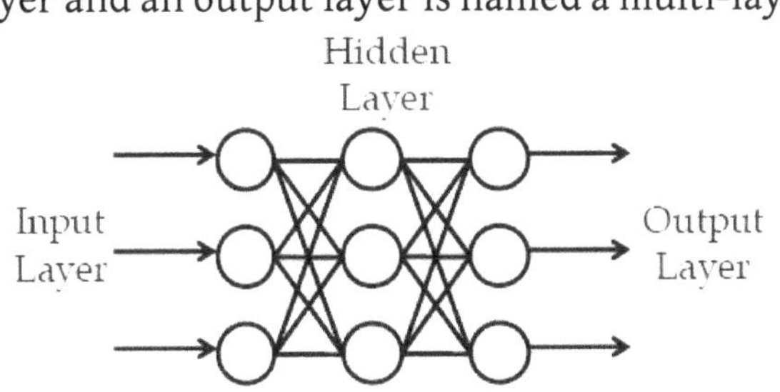


**Fig. 5. Multi-layer neural network.**

A neural network can have a feedback or a feed forward structure. In the feedback structure the information can move back and forward. In the feedforward structure, the information moves only forward from the input nodes through the outputs nodes with no cycles in the network (Ruan, 1997).

The neural networks need to be trained from examples, in a process called supervised learning. Once a successfully training is done, the neural network is ready if and only if the networks reproduce the desired outputs from the given inputs. The most common methodology for this kind of learning is the backpropagation algorithm, where the weights of the neural network are determined by using iteration until the output of the network is the same as the desired output (Rumelhart et al., 1986). In addition, unsupervised learning uses a mechanism for changing values of the weights according to the input values, this mechanism is named self-organization. An example of this algorithm is the Hebbian learning algorithm (Ruan, 1997).

### 3.1.1 Neural Networks in Fault Tolerant Control

Artificial neural networks have been applied in fault tolerant control because they are helpful to identify, detect and accommodate system faults. The application of ANNs to FTC can be divided in three groups. The first group includes neural networks used as fault detectors by estimating changes in process models dynamics *(Polycarpou* & Helmicki, *1995; Patton et al., 1999; Polycarpou, 2001; Gomaa, 2004)*. The second group includes neural networks used as controllers (*Wang & Wang, 1999; Pashilkar et al., 2006*), and the third group integrates neural networks which performs both functions: fault detection, and control (*Perhinschi et al., 2007; Yen & DeLima 2005*).

*(Polycarpou* & Helmicki, *1995)* proposed a construction of automated fault detection and accommodation architecture that uses on-line approximators and adaptive-learning schemes. The online approximator is a neural network model that monitors changes in the system dynamics due to a failure.

*(Patton et al., 1999)* use a scheme of neural network to detect and isolate a fault in two steps: residual generation and decision making. In the first step a residual vector characterizes the fault and then the second step process the residual vector information in order to locate the fault and the time of occurrence. Once the residual is trained, qualitative knowledge of the plant can be added. This combination of qualitative and quantitative approached is helpful to decrease the number of false alarms in the fault decision making step.

*(Polycarpou, 2001)* proposed a methodology for fault accommodation of a multivariable nonlinear dynamical system using a learning approach that monitors and approximates any abnormal behavior using neural networks and adaptive

nonlinear estimation. When a fault occurs the neural network is used to estimate the nonlinear fault function supplying a framework for fault identification and accommodation. The neural network at the beginning of the monitoring stage is capable of learning the modeling errors in order to improve the system robustness.

*(Gomaa, 2004)* recommended a fault tolerant control approach based on multi-ANN system faulty models. The nominal plant is nonlinear and is vulnerable to faults. A feedforward neural network is trained as the nominal model; two PID controllers are used, one for the nominal plant and the other for the neural network imitating the nominal plant (reference model). Both PIDs controllers were tuned using genetic algorithms. If there exist a difference between the nominal plant (*yp*) and the reference model (*yrm*) a nonzero residual is generated. Then, depending on the magnitude of the residual an ANN faulty model and its respective compensation path are selected to repair the fault and improve the system operating conditions. This can be observed in figure 6.

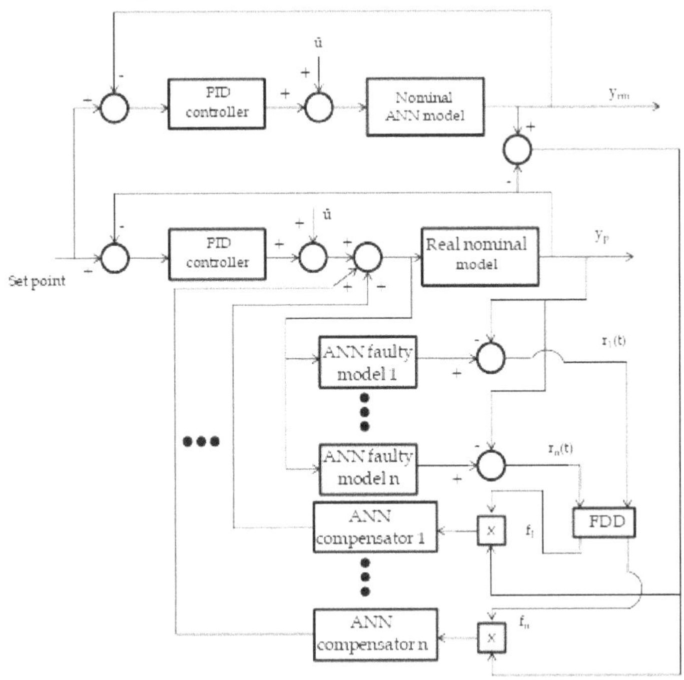


**Fig. 6. Multi-ANN faulty models FTC scheme (Gomaa, 2004).**

*(Wang & Wang, 1999)* proposed a Neural Network-FTC where the nominal system is controlled by a Pseudolinear Neural Network (PNN) based one-step-ahead controller that uses a modified gradient approach. When the system is with no fault, a PNN model connected in parallel to the plant model can be trained and used for the design of the control algorithm. The PNN model is helpful for the detection of a fault. In addition, when a fault is present, a residual signal is generated and an extra neural network based fault compensation loop is imported in order to

provide the closed loop stability. This last neural network is a two layers perceptron network and its weights are updated using the modified gradient approach. This FTC system is shown in figure 7.

*(Pashilkar et al. 2006)* proposed a neural controller that improves the fault tolerant potential of a fighter aircraft during landing. The faults are caused by severe winds or stuck control surfaces and can be divided in single faults (aileron or elevator stuck) or double fault (aileron and elevator stuck). This neural network controller employs a feedback error learning method with a dynamic radial basis function neural network. The neural network uses on-line training and not a-priori training. This kind of controller helped to improve the capability of handling large faults and also helps to achieve the desired requirements.

*(Perhinschi et al., 2007)* presented a methodology for detection, identification and accommodation of sensor and actuator failures inside fault tolerant control laws. The fault detection and identification uses neural estimators. The accommodating control laws design for the actuator fault is done using nonlinear dynamic inversion with neural network augmentation. Whereas the accommodation of sensor fault is accomplished by changing the failed sensor output for neural estimates calculated in the detection and identification process. This approach can handle sensor and actuator faults successfully. It uses membership functions to describe the mathematical model of process.

*(Yen & DeLima, 2005)* presented a neural network trained on-line with a global dual heuristic programming architecture. This approach has also a supervision structure made from decision logic. This supervision level is very efficient to identify the controller faults in early stages and can supply new values to improve the convergence utilizing dynamic model bank information.

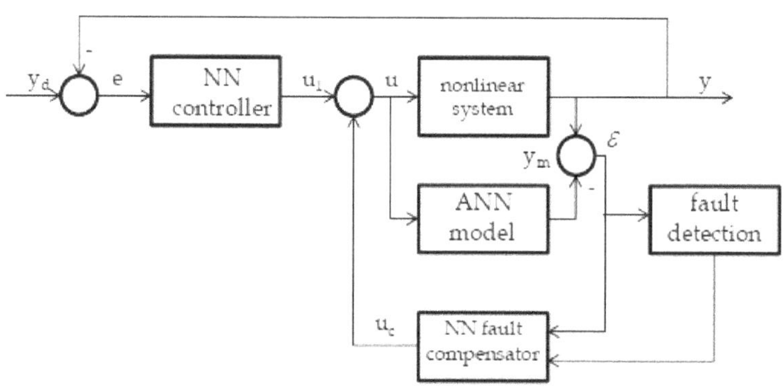


**Fig. 7. Neural Network FTC scheme proposed by (Wang & Wang, 1999).**

## 3.2 Fuzzy Logic

Fuzzy logic is a research field of fuzzy set theory; this methodology uses heuristic and mathematical tools to approximate reasoning in order to achieve knowledge processing when lack of information exists. In addition, this approach tries to mimic the control logic rules used by persons.

In Fuzzy systems, a mathematical model of the system is not needed because the desired input and output relationships are defined by "if" and "then" rules. Also, due to the lack of system information it is not possible to precise the exact value of a variable "x" but it is possible to know the membership function. The membership function is created to describe range of values that the variable can have, this is known as input fuzzification (Passino & Yurkovich, 1997; Ruan, 1997; Nguyen, 2002). This membership function can have different shapes that are defined according to a specific problem; the shapes forms are: triangular, Gaussian trapezoidal, sigmoidal S and sigmoidal Z functions.

Once the membership functions of variables are established, the inference rules "if (antecedent)-then (consequent)" are set. Then an inference method is used to quantify each premise, this is an application of a fuzzy operator in the "if" part of the rule and can be the math function "or" – "and", once this is done, the inference method fire rules, which is known as the implication method. Two inference methods are commonly used: Mamdani method, where the "if" and "then" parts are fuzzy expressions, and the Takagi-Sugeno method where the "if" part of the rule is a fuzzy expression, but the "then" part of the rule is a deterministic expression (Passino & Yurkovich, 1997; Nguyen, 2002). After applying the chosen inference method, the output fuzzy set is form. This can be shown in the next rules:

If *(0.4 or 0.2)* then *input (0.4)*

If *max (0.4, 0.2)* then *input (0.4)*

Finally*:* If *(0.4)* then *input (0.4)*

The above means that if the antecedent part of the rule is partially true which means that have a value minor than one, then the output fuzzy set will be truncated by the implication method, as showed in the next figure:

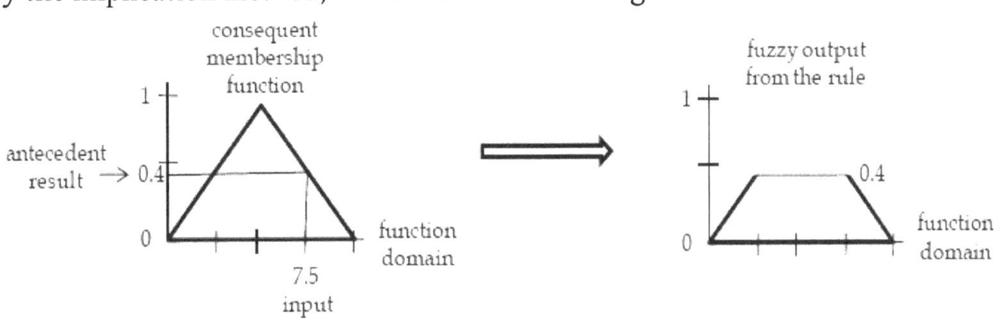


**Fig. 8. Implication method representation.**

The last step in the inference method is the aggregation of all the fuzzy outputs from the rules, as can be seen in the next figure:

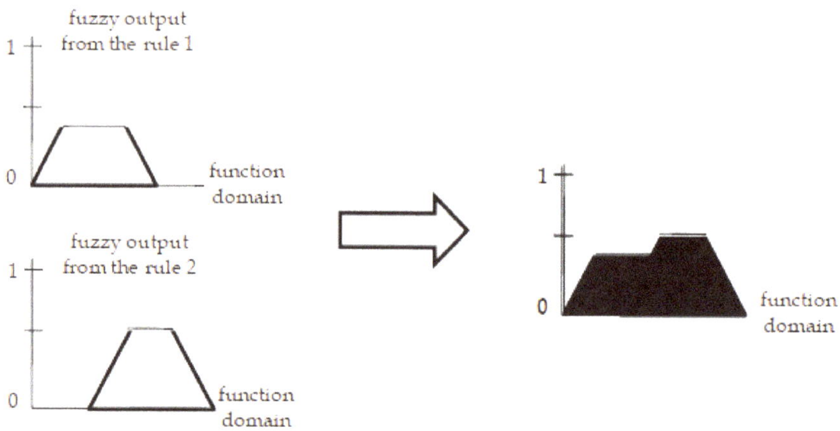


**Fig. 9. Fuzzy outputs aggregation representation.**

Finally, the fuzzy controller implements defuzzification, where the input of the defuzzification process is a fuzzy set and the output is just a number. This can be done using several methods as: center of area method also known as the center or gravity or centroid method, the height-center of area, the max criterion, the first of maxima and the middle of maxima method. The deffuzification result will be the output of the fuzzy system (Passino & Yurkovich, 1997; Nguyen, 2002). This is showed in the following figure:

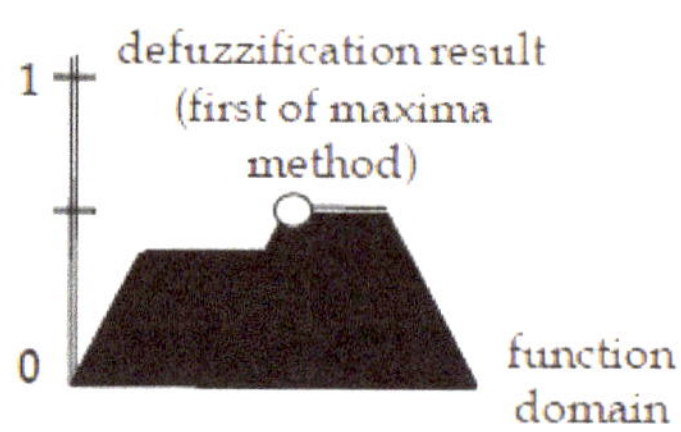


**Fig. 10. Defuzzification representation.**

In summary, the main advantages of Fuzzy Logic are: robustness, easy controller reconfiguration, works well with multi input multi output systems and adequate to model complex nonlinear systems.

### 3.2.1 Fuzzy Logic in Fault Tolerant Control

Fuzzy models are capable of handling nonlinear dynamic systems and the nonlinearities of the faults. In the following we describe several works reported with fault detectors and/or controllers based on fuzzy logic.

*(Kwong et al., 1995)* presented an expert supervision approach of Fuzzy learning system for fault tolerant control in aircrafts. This approach, shown in figure 11, uses a fuzzy model reference learning controller (FMRLC) to reconfigure the aircraft nominal controller in order to accommodate actuator failures without needing specific information about the failures. This approach also demonstrated that

the FMRLC can be improved using fault detection and identification information to acquire an adaptive system that can reconfigure its performance level when a failure occurs.

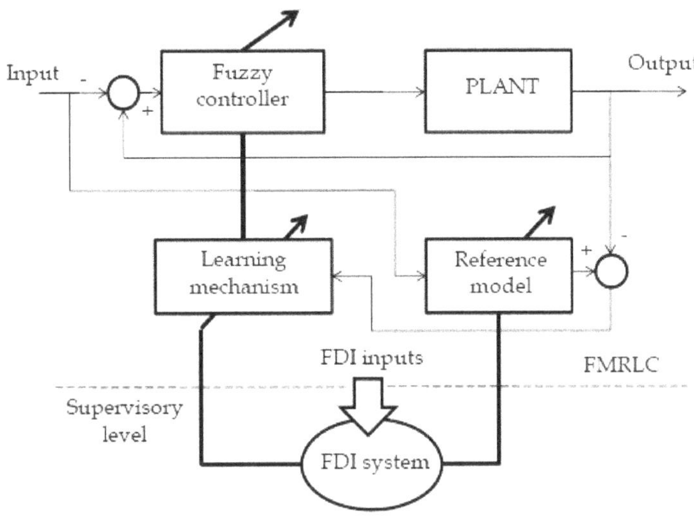


**Fig. 11. Supervisory FMRLC scheme proposed by (Kwong et al., 1995).**

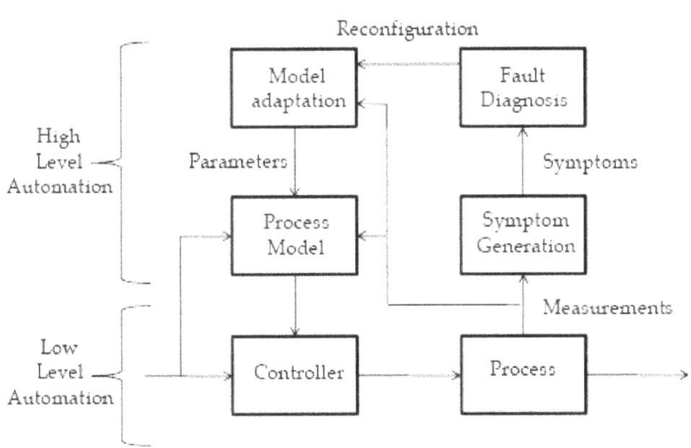


**Fig. 12. Model-based control, diagnosis and reconfiguration scheme proposed by (Ballé al., 1998).**

*(Ballé et al., 1998)* presented a model-based adaptive control and reconfiguration approach based on fault detection and diagnosis implemented to a heat exchanger. The fault detection and adaptive tasks are based on Takagi-Sugeno (T-S) fuzzy model of the process and the fault diagnosis task uses a self-organizing fuzzy structure. In this approach an on-line adaptation scheme is used for reconfiguration when the measurements are erroneous because of the sensor faults. If all the input signal and the disturbances can be measured the model is used for model-based predictive control. When a sensor fault occurs, the signals are not longer used for prediction. Instead of these measurements, a reduced model of the system process is required to adapt these erroneous signals as external disturbances. The T-S fuzzy model is

also used for residuals generation in order to detect and isolate sensor faults. Then the decision making is sustained by identifying particular patterns of the residual that are link to different faults. This is done by using a fuzzy classification tree that can learn sample data patterns. The next figure shows the explained approach:

*(Patton et al., 1999)* proposed the use of Takagi-Sugeno fuzzy observers for nonlinear dynamic systems. In this methodology the plant is represented by a T-S fuzzy model and a fuzzy observer estimate the system state vector. The fuzzy model is locally observable in order to design the fuzzy observer. A linear time invariant observer is created with every real of the fuzzy model. Then, the whole complete observer's dynamic is a weighted sum of the individual observers. A residual is generated from the observers and is compared against the fuzzy model output. If the residual is nonzero a fault is present in the system. *(Holmes & Ray, 2001)* proposed a damage mitigating control system in which the main objective is to accomplish optimal performance of the system with a high level of reliability, availability, component durability and maintainability. This scheme was implemented in a two-tier structure. A lineal robust sample data controller was proposed for the lower tier to track a reference trajectory vector. On the other hand, for the upper tier, a fuzzy logic based damage controller was implemented. This controller adjusts the system dynamic performance with the structural durability of critical components. This scheme demonstrated that a substantial gain in structural durability of critical components can be reached without significant performance degradation.

*(Diao & Passino, 2001)* implemented an intelligent Fault Tolerant Control scheme using a hierarchical learning structure for the design of the plant. This learning structure was a Takagi-Sugeno Fuzzy system. Then, the fault tolerant control system is based on stable adaptive fuzzy/neural control that has online learning properties which are used to acquire the unknown dynamics developed as consequence of system faults.

*(Diao & Passino, 2002)* presented a fault tolerant control scheme using adaptive estimation and control methods based on the learning properties of fuzzy systems and neural networks. An online approximation-based stable adaptive neural/fuzzy control was implemented in an input-output feedback time-varying nonlinear system. Also, the adaptive controller enhanced the fault tolerance capability by using the estimated information from the fault diagnosis unit, chosen by a multiple model interface with a supervisory expert approach.

*(Oudghiri et al., 2008)* proposed a Fault Tolerant scheme for lateral vehicle dynamics implementing static output feedback control with sliding mode observers in order to enhance the vehicle stability and handling in the presence of sensor fault. This approach uses three blocks, one for the fault and detection, a second for the static output feedback controller and a third for a switcher. The lateral vehicle

dynamics are represented in a nonlinear two degrees of freedom vehicle model implemented in a Takagi-Sugeno fuzzy model. The fault detection and isolation block uses a bank of observers' constructed using sliding mode in order to calculate the system state vector. Then, a set of diagnostic signal residuals are produced, associating the measured and the estimated outputs. After doing this, the sensor in which the fault happened is isolated.

## 3.3 Genetic Algorithms

Genetic Algorithms (GAs) are searching and optimizing algorithms motivated by natural selection evolution and natural genetics (Goldberg, 1989). The simplest GA follows the next steps: Generate a random initial population of chromosomes, calculate the fitness of every chromosome in the population, apply selection, crossover and mutation and replace the actual population with the new population until the required solution is achieved. The main advantages of GAs are: powerful computational effect, robustness, fault tolerance, fast convergence to a global optimal, capability of searching in complex landscape where the fitness function is discontinuous, can be combined with traditional optimization techniques (Tabu search) and have the ability to solve problem without needing human experts (Goldberg, 1989; Mitchell, 1996; Ruan, 1997).

### 3.3.1 Genetic Algorithms in Fault Tolerant Control

Recently genetic algorithms have been applied in fault tolerant control as a strategy to optimize and supervise the controlled system in order to accommodate system failures. Some applications of this technique are the following:

*(Schroder et al., 1998)* proposed a fault tolerant control technique for an active magnetic bearing. In this approach a nonlinear model of a turbo machine rotor from the rolls-royce lifted up by an active magnetic bearing was presented. This model is capable of modeling difference configuration of magnetic bearings. A multi-objective genetic algorithm was used to generate and adequate PID controller for the active magnetic bearing with different bearing configuration. Also the fault accommodation was done using a centralized fault compensation scheme.

*(Sugawara et al., 2003)* showed a fault tolerant control approach using multi-layer neural networks with a genetic algorithm. The proposed of this approach was to develop a self- recovery technique implemented for large scale neural networks programmed in a single ship to accommodate faults without the needing of a host computer. This FTC scheme uses hardware redundancy and weight retraining using a genetic algorithm in order to reconfigure the neural network to accommodate the fault. The objective of the genetic algorithm is to reduce the error between the actual output and the desired output.

## 4. THE PROPOSED METHODOLOGY, EXPERIMENTS AND RESULTS

We propose a new FTC schema, where a Model Reference Adaptive Control (MRAC) is used in combination with a neural network controller, in order to achieve a better performance when faults are present in the system. We use an experimental model of a heat exchanger where abrupt and gradual faults (also called *soft faults*) are induced in sensors and actuators. To compare our schema, we also have made experiments with a simple MRAC and MRAC-PID structures.

### 4.1 MRAC Controller

The Model Reference Adaptive Controller, shown in figure 13, implements a closed loop controller that involves the parameters that should be optimized, in order to modify the system response to achieve the desired final value. The adaptation mechanism adjusts the controller parameters to match the process output with the reference model output.

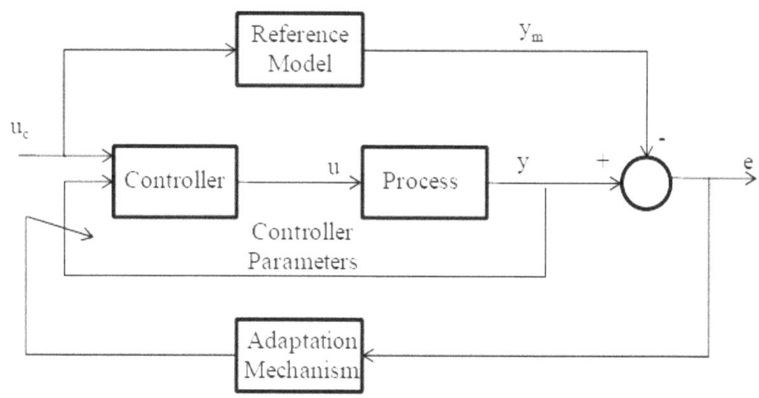


**Fig. 13. MRAC scheme (Nagrath, 2006).**

The controller error is calculated as follows:

$$e = y_{process} - y_{reference} \tag{1}$$

To reduce the error, a cost function was used, in the form of:

$$J(\theta) = \frac{1}{2}e^2(\theta) \tag{2}$$

The function above can be minimized if the parameters θ change in the negative direction of the gradient ***J***, this is called the gradient descent method and is represented by:

$$\frac{d\theta}{dt} = -\gamma \frac{\partial J}{\partial \theta} = -\gamma e \frac{\partial e}{\partial \theta} \tag{3}$$

where γ helps to adjust the speed of learning. The above equation is known as the MIT rule and determines how the parameter θ will be updated in order to reduce the error.

The implemented MRAC scheme in our process, shown in figure 14, has two adaptation parameters: adaptive feedfoward gain ($\theta_1$ )and adaptive feedback gain ($\theta_2$ ). These parameters will be updated to follow the reference model.

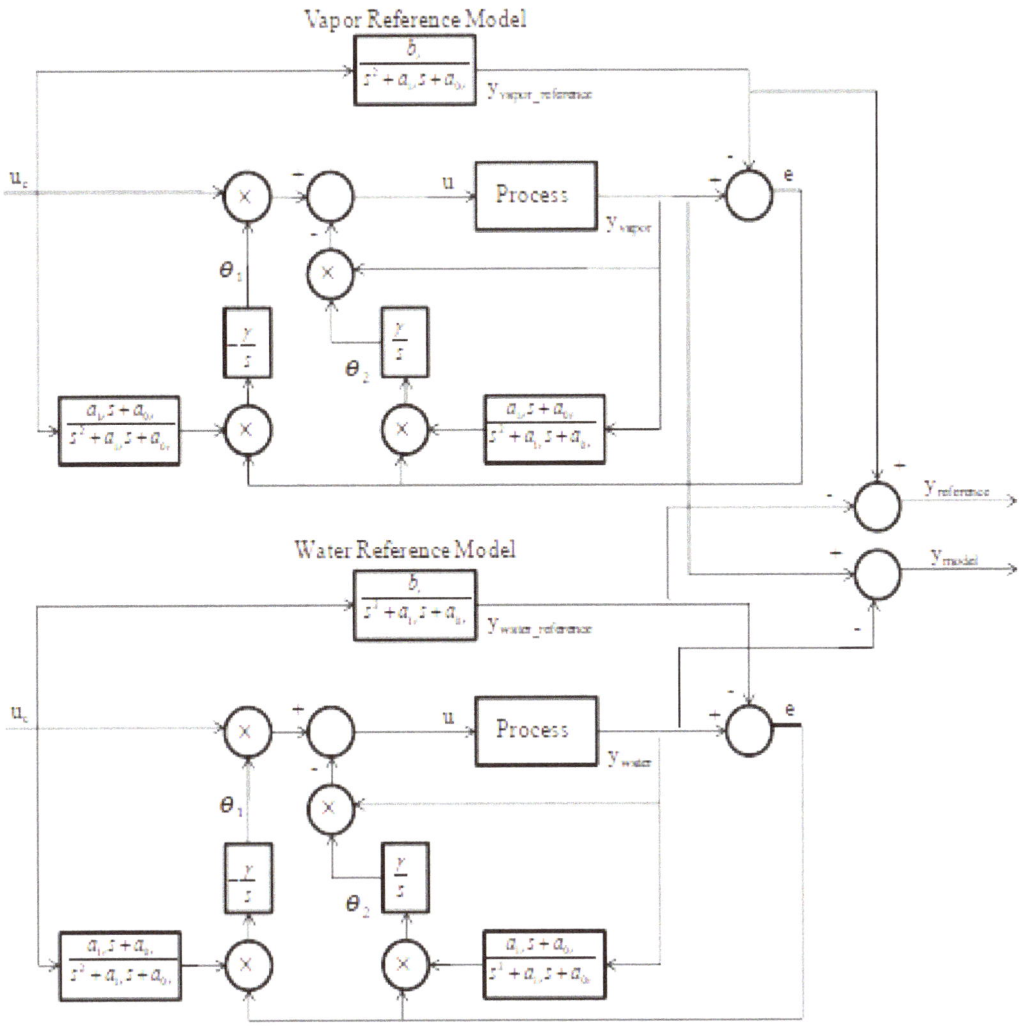


**Fig. 14. Fault tolerant MRAC scheme.**

## 4.2 PID Controller

To overcome the limitations of the simple MRAC structure, a classical PID controller, in the form of equation 4, was introduced in the feedforward part of simple MRAC scheme. The resulting structure is shown in figure 15. The desired response is introduced in terms of rise time, stabilization time and overshoot.

$$G_{PID\_controller} = K_p + K_i/_s + K_d s \tag{4}$$

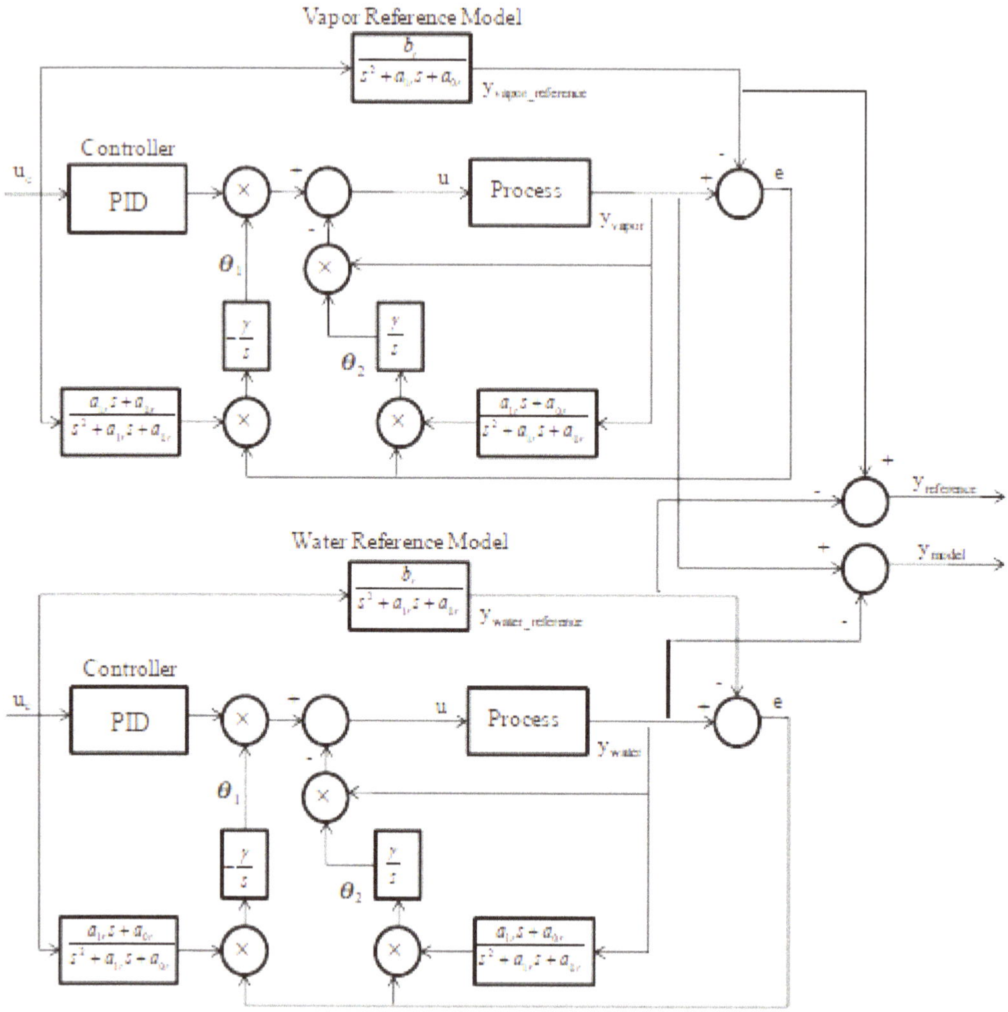


**Fig. 15. Fault tolerant MRAC-PID Controller structure.**

## 4.3 Neural Network Controller

In this scheme, we substitute the PID controller by a neural network (shown in figure 16). The neural network was trained with the original process inputs as well as the desired outputs.

## 4.4 Experiments

Two different faults were simulated: abrupt faults and gradual faults. In the abrupt faults case, the whole magnitude of the fault is developed in one moment of time and was simulated with a step function. On the other hand, gradual faults are developed during a period of time and are implemented with a ramp function.

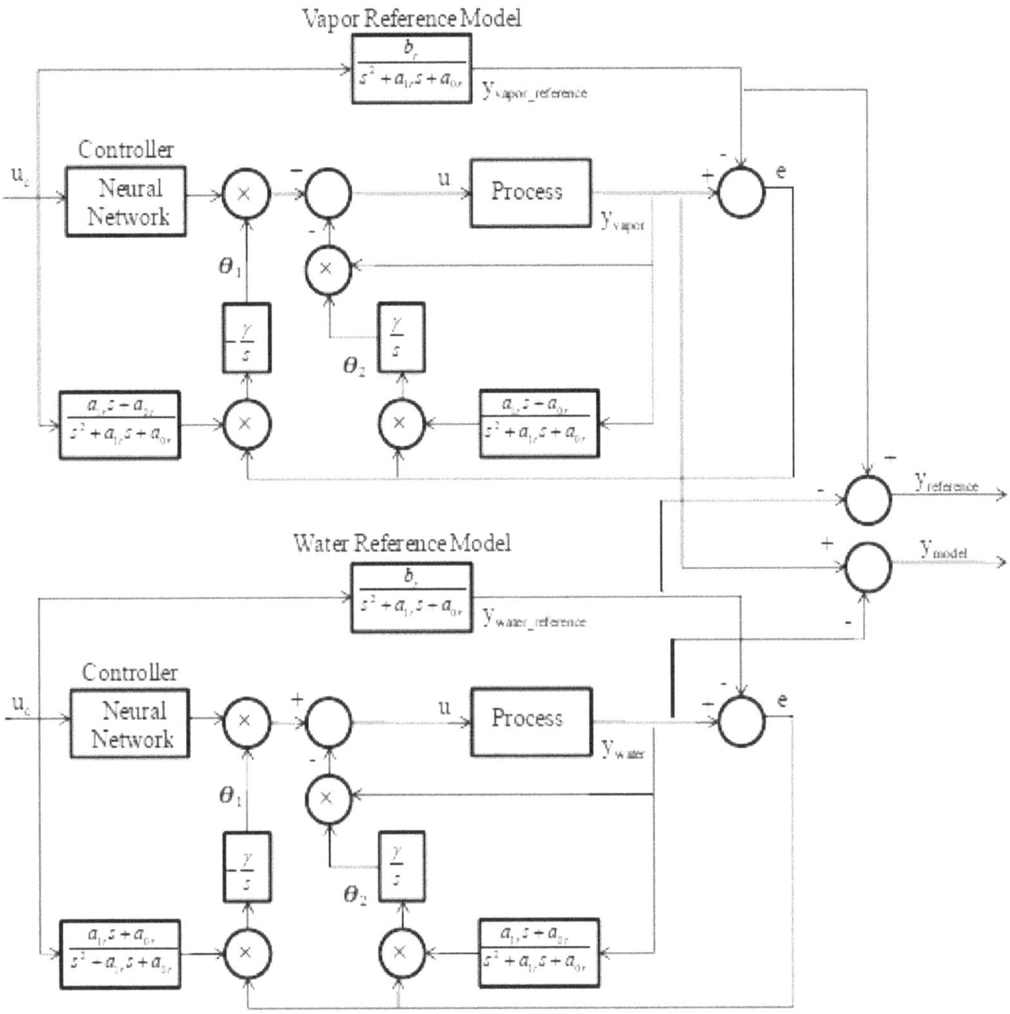


Both types of faults, abrupt and gradual, can be implemented in sensors (feedback), in which the properties of the process are not affected, but the sensor readings are mistaken. Also, can be implemented in actuators (process entry) in which the process properties are not affected either, but the process behavior can change or can be interrupted.

We use an industrial heat exchanger to test our approach (shown in figure 17). The process has two inputs: water and steam flows controlled by pneumatic valves, and one output, the water temperature measured by a termocouple. Variations in water and steam flows are determined by flow transmitters. To develop the continuous model of this process, an identification experiment was performed, where a Pseudo Random Binary Sequence (PRBS) was applied to water and steam valves, and variations in water temperature were recorded.

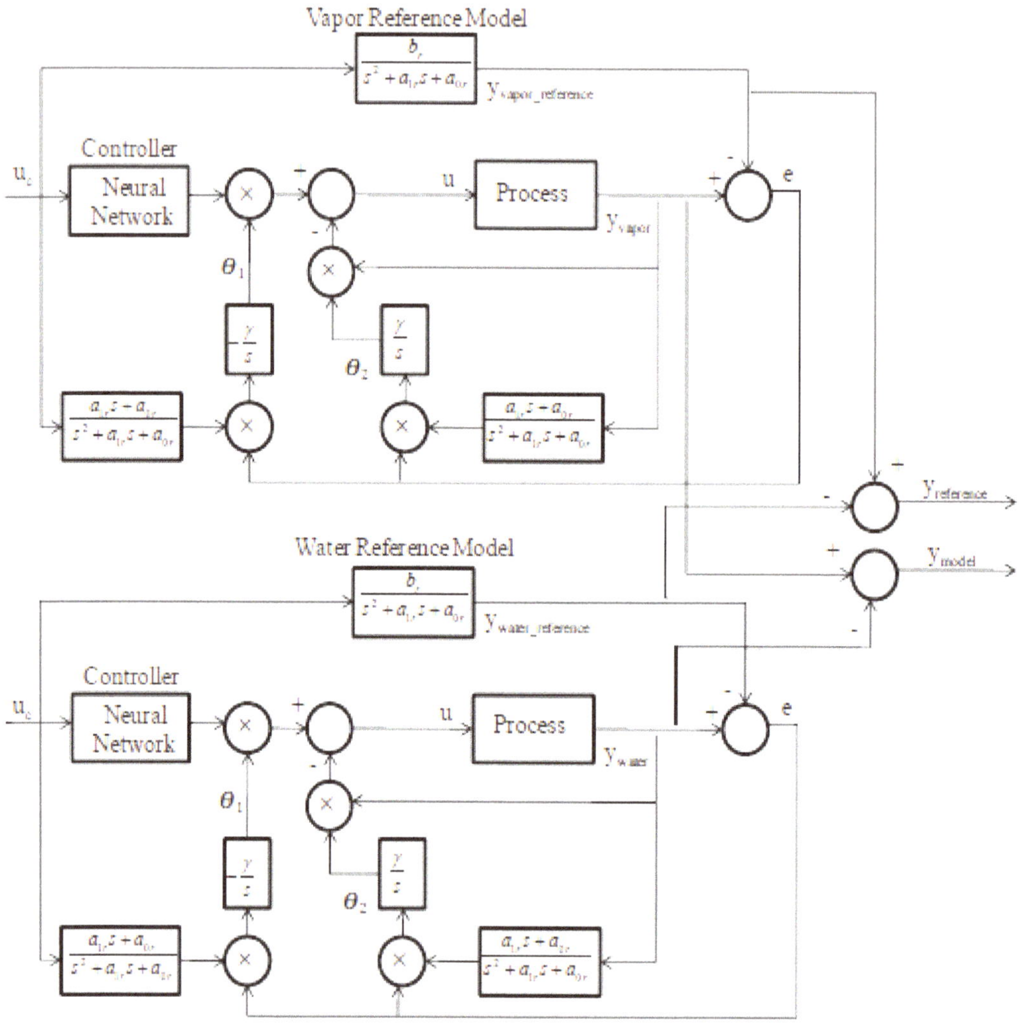


**Fig. 16. Fault tolerant MRAC-Neural Network Controller structure.**

With the data obtained in the PRBS test, the identification was achieved using software developed in Matlab®.

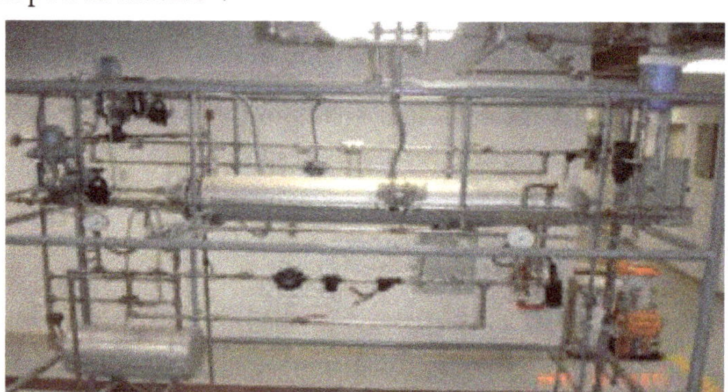

**Fig. 17. Industrial heat exchanger used in the experiments.**

The following model was obtained:

$$G_p = G_{vapor} - G_{water}$$

$$G_p = \frac{0.00002}{s^2 + 0.004299s + 0.00002} - \frac{0.000013}{s^2 + 0.007815s + 0.00008} \quad (5)$$

This model was used to implement the 3 different FTC Schemes mentioned above.

## 5.4 Results

A total of six different experiments were developed in Simulink®. Table 1 explains the results of each simulated experiment and shows the numerical performance of every method by using the Mean Square Error (MSE) obtained during the application of the fault.

**Table 1. Results of experiments with abrupt and gradual faults simulated in the 3 different fault tolerant MRAC schemes.**

| ***Method & Fault Type*** | ***Results when a the Fault was applied In Sensor*** | ***Results when an Fault was applied In Actuator*** |
|---|---|---|
| MRAC-<br>Abrupt Fault | If the fault magnitude is < 0.44, the system is robust against the fault.<br>If the fault magnitude is between [0.44, 1.52] the system accommodates the fault.<br>-If the fault magnitude is > 1.52, the system becomes unstable.<br>- **MSE= 0.501236189** | - If the fault magnitude is 1 the system response varies around +/- 3%. This means that the system is degraded but still works. This degradation becomes smaller over time, because the system continues accommodating the fault.<br>- **MSE=0.10521016** |
| MRAC-<br>Gradual Fault | If the fault has saturation < +/- 0.44, the system is robust against the fault.<br>If the fault has a saturation between +/- [0.44, 1.52] the system accommodate the fault.<br>-If the fault has saturation > 1.52, the system becomes unstable.<br>**MSE=0.50777113** | - If the fault saturation is +/- 1 the system response varies around +/- 4%. This means that the system is degraded but still works. This degradation becomes smaller over time, because the system continues accommodating the fault.<br>- **MSE=0.09163081** |
| MRAC- PID-<br>Abrupt Fault | If the fault magnitude is < 1.75, the system is robust against the fault.<br>If the fault magnitude is between [1.75, 1.97] the system accommodates the fault.<br>-If the fault magnitude is > 1.97, the system becomes unstable.<br>- **MSE=0.00036942** | - If the fault magnitude is 1 the system response varies around +/- 3.5%. This means that the system is degraded but still works. This degradation becomes smaller over time, because the system continues accommodating the fault.<br>- **MSE=0.13283874** |

| *Method & Fault Type* | *Results when a the Fault was applied In Sensor* | *Results when an Fault was applied In Actuator* |
|---|---|---|
| MRAC- PID- Gradual Fault | If the fault has saturation < +/- 1.75, the system is robust against the fault.<br>If the fault has a saturation between +/- [1.75, 1.97] the system accommodates the fault.<br>-If the fault has saturation > 1.97, the system becomes unstable.<br>- **MSE=0.0003694** | - If the fault saturation is +/- 1 the system response varies around +/- 4%. This means that the system is degraded but still works. This degradation becomes smaller over time, because the system continues accommodating the fault.<br>- **MSE=0.13508005** |
| MRAC- Neural Network- Abrupt Fault | - The system is robust against sensor faults<br>- **MSE=0.00030043** | - If the fault magnitude is 1 the system response varies around +/- 3%. This means that the system is degraded but still works. This degradation becomes smaller over time, because the system continues accommodating the fault.<br>- **MSE=0.13154736** |
| MRAC- Neural Network- Gradual Fault | - The system is robust against sensor faults.<br>- **MSE=0.00030043** | - If the fault saturation is +/- 1 the system response varies around +3% and - 4%. This means that the system is degraded but still works. This degradation becomes smaller over time, because the system continues accommodating the fault.<br>- **MSE=0.13149647** |

The graphics provided illustrate a comparison among many simulated experiments. Figure 18 depicts the behavior of the system when sudden faults are simulated. The three graphs in the left column represent sensor faults, while the graphs in the right column depict actuator problems. The sensor faults exhibit a magnitude of 1.8, while the actuator faults have a value of 1. The MRAC-Neural Network technique is considered the most effective due to its ability to handle sudden sensor faults and its strong performance in the presence of abrupt actuator faults.

The graphs in Figure 19 depict the behavior of the system in the presence of progressive defects. The sensor fault has a magnitude of 1.8, while the actuator fault has a magnitude of 1. Furthermore, it is evident that the MRAC-Neural Networks Controller scheme is superior due to its resilience to sensor defects and its minimal decline in performance during actuator breakdowns. Overall, the MRAC-Neural Network strategy suggested in this study is the most effective fault tolerant control scheme developed.

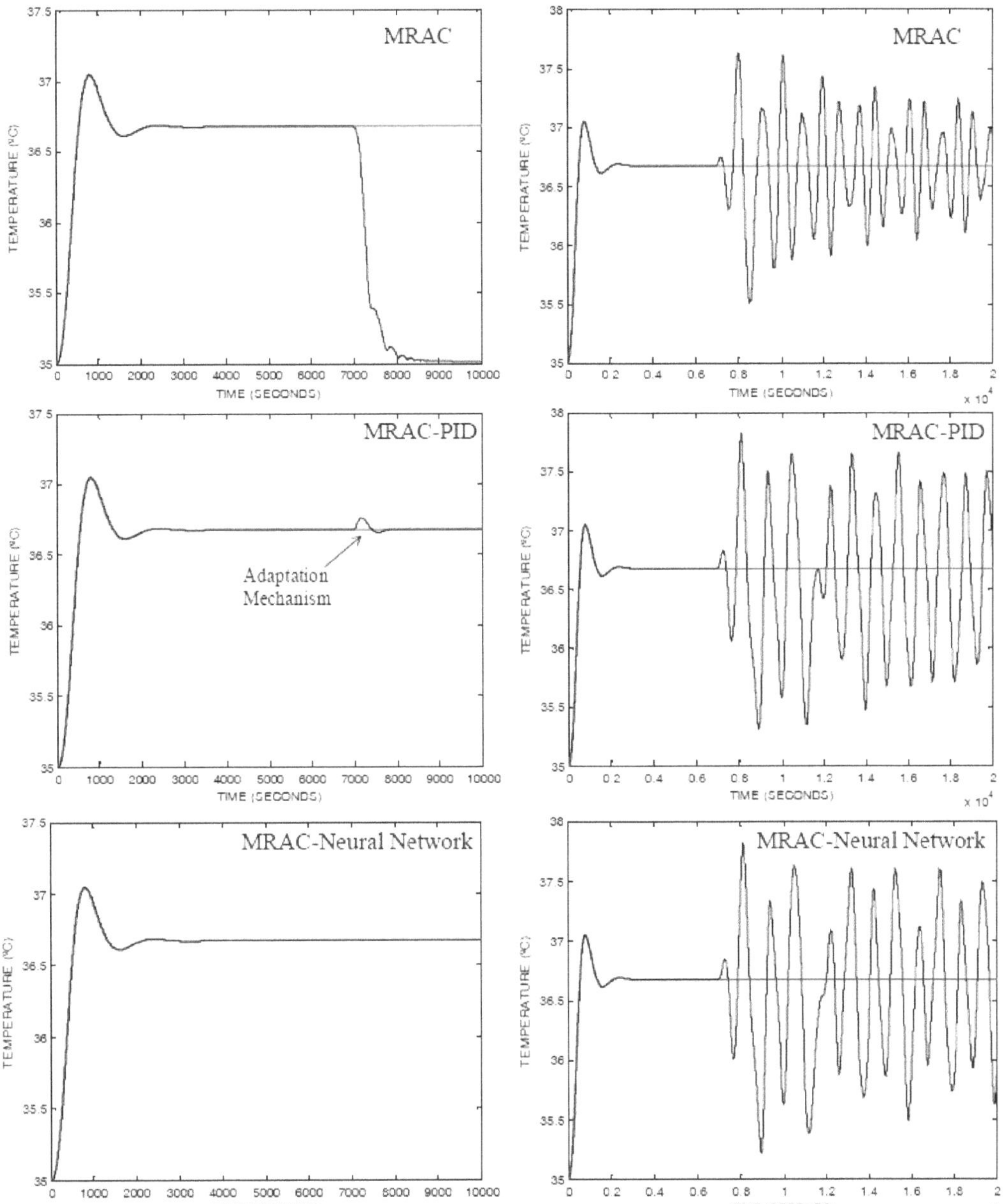


**Fig. 18. Abrupt-Sensor Faults (left column) and Abrupt-Actuator Faults (Right column) of the three different proposed schemes, the fault started at time 7000 secs.**

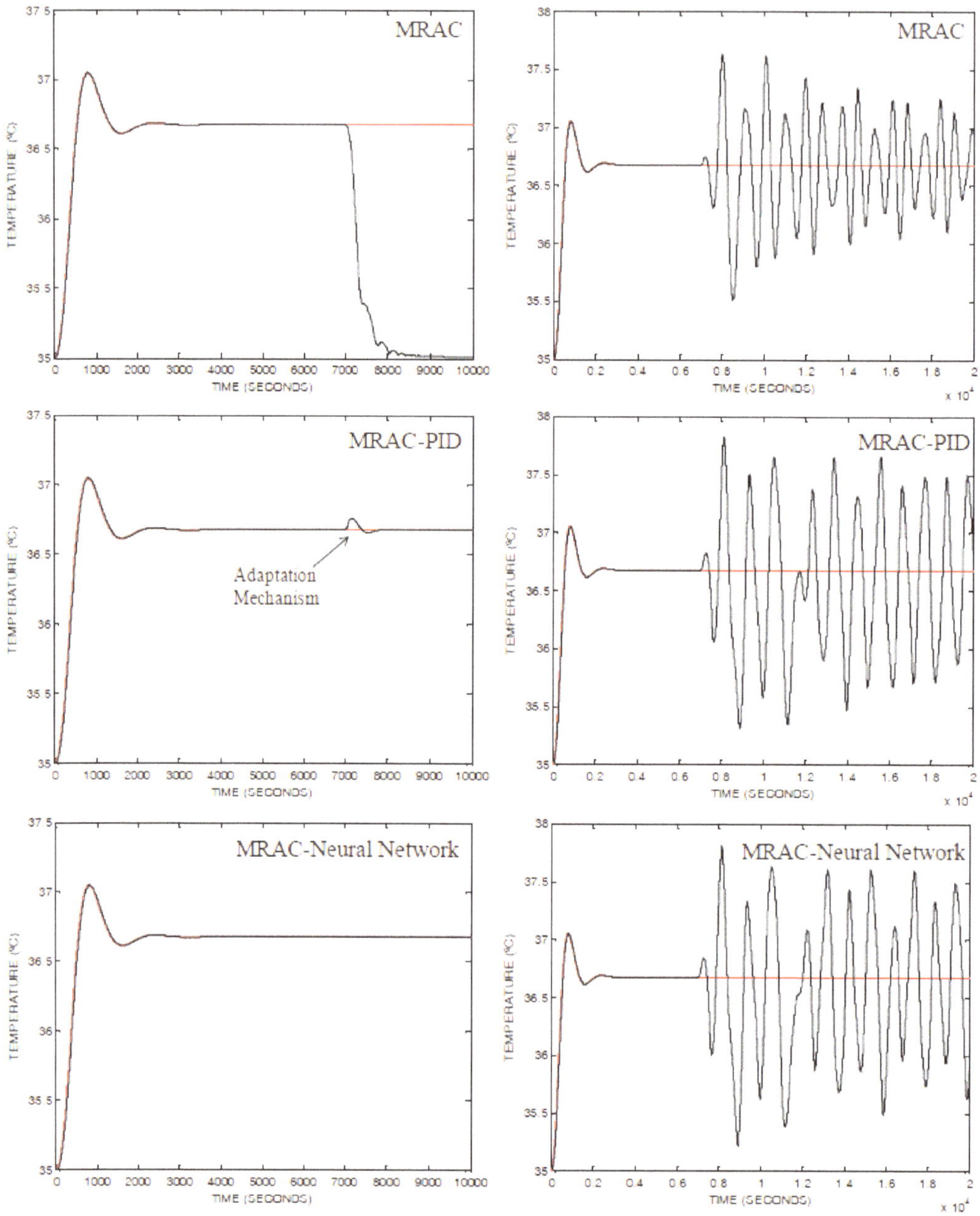


**Fig. 19. Gradual-Sensor Faults (left column) and Gradual-Actuator Faults (Right column) of the three different proposed schemes, the fault started at time 7000 secs.**

## REFERENCES

1. Ballé, P.; Fischera, M.; Fussel, D.; Nells, O. & Isermann, R. (1998). Integrated control, diagnosis and reconfiguration of a heat exchanger. *IEEE Control Systems Magazine*, Vol. 18, No. 3, (June 1998) 52–63, ISSN: 0272-1708.

2. Bastani, F., & Chen, I. (1988). The role of artificial intelligence in fault-tolerant process- control systems. *Proceedings of the 1st international conference on Industrial and engineering applications of artificial intelligence and expert systems*, pp. 1049-1058, ISBN:0-89791-271-3, June 1988, ACM, Tullahoma, Tennessee, United States.
3. Blanke, M.; Izadi-Zamanabadi, R.; Bogh, R. & Lunau, Z. P. (1997). Fault tolerant control systems—A holistic view. *Control Engineering Practice*, Vol. 5, No. 5, (May 1997) 693–702, ISSN: S0967-0661(97)00051-8.
4. Blanke, M., Staroswiecki, M., & Wu, N. E. (2001). Concepts and methods in fault-tolerant control. *In Proceedings of the 2001 American Control Conference*, pp. 2606–2620, Arlington, Virginia, ISBN: 0-7803-6495-3, June 2001, IEEE, United States.
5. Blanke, M.; Kinnaert, M.; Lunze, J. & Staroswiecki, M. (2003). *Diagnosis and Fault-Tolerant Control.* Springer-Verlag, ISBN: 3540010564 , Berlin, Germany.
6. Blondel, V. (1994). *Simultaneous Stabilization of Linear Systems.* Springer Verlag, ISBN: 3540198628, Heidelberg, Germany.
7. Caglayan, A.; Allen, S. & Wehmuller, K. (1988). Evaluation of a second generation reconfiguration strategy for aircraft flight control systems subjected to actuator failure/surface damage. *Proceedings of the 1988 National Aerospace and Electronics Conference*, pp. 520–529, May 1988, IEEE, Dayton , Ohio, United States.
8. Diao, Y. & Passino, K. (2001). Stable fault-tolerant adaptive fuzzy/neural control for turbine engine. *IEEE Transactions on Control Systems Technology*, Vol. 9, No. 3, (May 2001) 494–509, ISSN: 1063-6536.
9. Diao,Y. & Passino, K. (2002). Intelligent fault-tolerant control using adaptive and learning methods. *Control Engineering Practice*, Vol. 10, N. 8, (August 2002) 801–817, ISSN: 0967-0661.
10. Eterno, J.; Looze, D; Weiss, J. & Willsky, A. (1985). Design Issues for Fault-Tolerant Restructurable Aircraft Control, *Proceedings of 24th Conference on Decision and Control*, pp. 900-905, December 1985, IEEE, Fort Lauderdale, Florida, United States.
11. Farrell, J.; Berger, T. & Appleby, B. (1993). Using learning techniques to accommodate unanticipated faults. *IEEE Control Systems Magazine,* Vol. 13, No. 3, (June 1993) 40– 49, ISSN: 0272-1708.
12. Gao, Z. & Antsaklis, P. (1991). Stability of the pseudo-inverse method for reconfigurable control systems. *International Journal of Control,* Vol. 53, No. 3, (March 1991) 717–729. Goldberg, D. (1989). *Genetic algorithms in search, optimization, and machine learning*, Addison-Wesley, ISBN: 0201157675, Reading, Massachusetts, United States.

13. Gomaa, M. (2004). Fault tolerant control scheme based on multi-ann faulty models. Electrical, Electronic and Computer Engineering. *ICEEC International Conference*, Vol. , No. , (September 2004) 329 – 332, ISBN: 0-7803-8575-6.
14. Gurney, K. (1997). *An Introduction to Neural Networks*, CRC Press Company, ISBN: 1857285034, London, United Kingdom.
15. Holmes, M. & Ray, A. (2001). Fuzzy damage-mitigating control of a fossil power plant. *IEEE Transactions on Control Systems Technology*, Vol. 9, No. 1, (January 2001) 140– 147, ISSN: 1558-0865.
16. Isermann, R.; Schwarz, R. & Stölzl, S. (2002). Fault-tolerant drive-by-wire systems. *IEEE Control Systems Magazine*, Vol. 22, No. 5, (October 2002) 64-81, ISSN: 0272-1708.
17. Jaimoukha, I.; Li, Z. & Papakos, V. (2006). A matrix factorization solution to the H-/H infinity fault detection problem. *Automatica*, Vol. 42, No. 11, 1907 – 1912, ISSN: 000- 1098.
18. Jiang, J. (1994). Design of reconfigurable control systems using eigenstructure assignments. *International Journal of Control,* Vol. 59, No. 2, 395–410, ISNN 00-7179.
19. Karsai, G.; Biswas, G.;Narasimhan, S.; Szemethy, T.; Peceli, G.; Simon, G. & Kovacshazy, T. (2002). Towards Fault-Adaptive Control of Complex Dynamic Systems, In: *Software- Enabled Control*, Tariq Samad and Gary Balas, Wiley-IEEE press, 347-368, ISBN: 9780471234364, United States.
20. Kwong,W.; Passino, K.; Laukonen, E. & Yurkovich, S. (1995). Expert supervision of fuzzy learning systems for fault tolerant aircraft control. *Proceedings of the IEEE*, Vol. 83, No. 3, (March 1995) 466–483, ISSN: 0018-9219.
21. Liang, B. & Duan, G. (2004). Robust H-infinity fault-tolerant control for uncertain descriptor systems by dynamical compensators. *Journal of Control Theory and Applications*, Vol. 2, No. 3, (August 2004) 288-292, ISSN: 1672-6340.
22. Lunze, J. & J. H. Richter. (2006). *Control reconfiguration: Survey of methods and open problems.* , ATP, Bochum, Germany.
23. Mahmoud, M.; Jiang, J. & Zhang, Y. (2003). Active fault tolerant control systems: Stochastic analysis and synthesis, Springer, ISBN: 2540003185, Berlin, Germany.
24. Mitchell, M. (1996). *An introduction to genetic algorithms*, MIT Press, ISBN: 0262631857, Cambridge, Massachusetts, United States.
25. Nagrath, .J (2006). *Control Systems Engineering*, Anshan Ltd, ISBN: 1848290039, Indian Institute of Technology, Delhi, India.
26. Neimann, H. & Stoustrup, J. (2005), Passive fault tolerant control of a double inverted pendulum - a case study. *Control Engineering Practice*, Vol. 13, No 8, 1047-1059, ISNN: 0967-0661.

27. Nguyen, H.; Nadipuren, P.; Walker, C. & Walker, E. (2002). *A First Course in Fuzzy and Neural Control*, CRC Press Company, ISBN: 158488241, United States.

28. Oudghiri, M.; Chadli, M. & El Hajjaji, A. (2008). Sensors Active Fault Tolerant Control For Vehicle Via Bank of Robust H∞ Observers. *17th International Federation of Automatic Control (IFAC) World Congress*, July 2008, IFAC, Seoul, Korea.

29. Passino, K. and Yurkovich, S. (1997). *Fuzzy Control*, Addison-Wesley Longman, ISBN: 020118074, United States.

30. Pashilkar,A.; Sundararajan, N.; Saratchandran, P. (2006). A Fault-tolerant Neural Aided Controller for Aircraft Auto-landing. *Aerospace Science and Technology*, Vol. 10, pp. 49-61.

31. Patton, R. J. (1997). Fault-tolerant control: The 1997 situation. *Proceedings of the 3rd IFAC symposium on fault detection, supervision and safety for technical processes*, pp. 1033– 1055, Hull, United Kingdom.

32. Patton, R.; Lopez-Toribio, C. & Uppal, F. (1999). Artificial intelligence approaches to fault diagnosis. *IEEE Condition Monitoring: Machinery, External Structures and Health, I*, pp. 5/1 – 518, April 1999, IEEE, Birmingham, United Kingdom.

33. Perhinschi, M.; Napolitano, M.; Campa, G., Fravolini, M.; & Seanor, B. (2007). Integration of Sensor and Actuator Failure Detection, Identification, and Accommodation Schemes within Fault Tolerant Control Laws. *Control and Intelligent Systems*, Vol. 35, No. 4, 309-318, ISSN: 1480-1752.

34. Polycarpou, M. & Helmicki, A. (1995). Automated fault detection and accommodation: A learning systems approach. *IEEE Transactions on Systems*, Vol. 25, No. 11, (November 1995) 1447–1458.

35. Polycarpou, M. & Vemuri, A. (1995). Learning methodology for failure detection and accommodation. *IEEE Control Systems Magazine*, Vol. 15, No. 3, (June 1995) 16–24, ISSN: 0272-1708.

36. Polycarpou, M. (2001). Fault accommodation of a class of multivariable nonlinear dynamical systems using a learning approach. *IEEE Transactions on Automatic Control*, Vol. 46, No.5, (May 2001) 736–742, ISSN: 0018-9286.

37. Rumerhart, D.; McClelland, J.; & the PDP Research Group. (1986). *Parallel distributed processing: explorations in the microstructure of cognition*, MIT Press, ISBN: 0262631105, Cambridge, Massachusetts, United States.

38. Ruan, D. (1997). *Intelligent Hybrid Systems: Fuzzy Logic, Neural Networks, and Genetic Algorithms*, Kluwer Academic Publishers, ISBN: 0792399994, United States.

39. Schroder, P.; Chipperfield, A.; Fleming, P. & Grum, N. (1998). Fault tolerant control of active magnetic bearings. IEEE International Symposium on Industrial Electronics, pp. 573-578, ISBN: 0-7803-4756-0, July 1998, IEEE, Pretoria, South Africa.

40. Skogestad, S., & Postlethwaite I. (2005). *Multivariable Feedback Control-Analysis and Design*, John Wiley & Sons, ISBN: 9780470011676, United States.
41. Staroswiecki, M. (2005). Fault tolerant control: The pseudo-inverse method revisited. *Proceedings 16th IFAC World Congress*, pp. Th-E05-TO/2, IFAC, Prague, Czech Republic.
42. Steffen, T. (2005). *Control reconfiguration of dynamic systems: Linear approaches and structural tests*, Springer, ISBN: 3540257306, Berlin, Germany.
43. Stengel, R. (1991). Intelligent Failure-Tolerant Control. *IEEE Control Systems Magazine*, Vol. 11, No. 4, (June 1991) 14-23, ISSN: 0272-1708.
44. Sugawara, E.; Fukushi, M. & Horiguchi, S. (2003). Fault Tolerant Multi-layer Neural Networks with GA Training. *The 18th IEEE International Symposium on Defect and Fault Tolerance in VLSI systems*□pp. 328-335, ISBN: 0-7695-2042-1, IEEE, November 2003 Boston, Massachusetts, United States.
45. Venkatasubramanian, V.; Rengaswamy, R.; Yin, K. & Kavuri, S. (2003a). A review of process fault detection and diagnosis. Part I. Quantitative modelbased methods. *Computers and Chemical Engineering*, Vol. 27, No. 3, 293–311, ISSN-0098-1354.
46. Venkatasubramanian, V.; Rengaswamy, R. & Kavuri, S. (2003b). A review of process fault detection and diagnosis. Part II. Qualitative models and search strategies. *Computers and Chemical Engineering*, Vol. 27, No. 3, 313–326, ISSN: 0098-1354.
47. Venkatasubramanian, V.; Rengaswamy, R.; Kavuri, S. & Yin, K. (2003c). A review of process fault detection and diagnosis. Part III. Process history based methods. *Computers and Chemical Engineering*, Vol. 27, No. 3, 327–346, ISSN: 0098-1354.
48. Wang, H. & Wang, Y. (1999). Neural-network-based fault-tolerant control of unknown nonlinear systems. *IEE Proceedings—Control Theory and Applications*, Vol. 46, No. 5, (September 1999) 389–398, ISSN; 1350-2379.
49. Yang, G. & Ye, D. (2006). Adaptive fault-tolerant Hinf control via state feedback for linear systems against actuator faults, *Conference on Decision and Control*, pp. 3530-3535, December 2006, San Diego, California, United States.
50. Yen, G. & DeLima, P. (2005). An Integrated Fault Tolerant Control Framework Using Adaptive Critic Design. *International Joint Conference on Neural Networks*, Vol. 5, pp. 2983-2988, ISBN: 0-7803-9048-2.
51. Zhang, D.; Wang Z. & Hu, S. (2007). Robust satisfactory fault-tolerant control of uncertain linear discrete-time systems: an LMI approach. *International Journal of Systems Science*, Vol. 38, No. 2, (February 2007) 151-165, ISSN: 0020-7721.
52. Zhang, Y., & Jiang, J. (2008). Bibliographical review on reconfigurable fault-tolerant control systems. *Elsevier Annual Reviews in Control*, Vol. 32, (March 2008) 229-252.

# 16 AN EXPERT SYSTEM THAT OPERATES IN REAL TIME TO FACILITATE DECISION MAKING IN ROTARY RAILCAR DUMPERS

## 1. INTRODUCTION: AN OVERVIEW

A variety of approaches have been employed in industrial production to automate a wide range of procedures. These applications require a high level of control, the ability to tolerate failures, and the ability to make decisions, among other critical factors that contribute to the reliability of large-scale systems (Su et al., 2005), (Su et al., 2000).

Artificial Intelligence (AI) has a broad range of applications, especially when it comes to implementing its concepts through Expert Systems (Fonseca Neto et al., 2003). Applications utilizing this architecture expand knowledge-based systems and enable the machine to be organized into a model capable of acting and behaving in a manner that closely resembles how a human specialist utilizes their thinking when confronted with a decision-making dilemma (Feigenbaum, 1992).

The VALE production system includes multiple mining complexes, one of which is particularly well-known, the Ponta da Madeira Dock Terminal (PMDT). The Unloading, Storing, and Minerals Shipping procedures are carried out at a macro level, under the supervision of a highly dependable Operational Control Center (OCC).

This article explores the creation of an online expert system designed to make decisions on errors that occur in the VV311-K01, a machine used for unloading minerals at VALE's PMDT. This project focuses on managing a substantial amount of operational data generated during production and is responsible for organizing, interpreting, and comprehending this data.

In addition to automation technologies, we utilize other information technologies to support our concept. These include JESS, the JAVA programming language, and XML (eXtensible Markup Language), which enable the Expert System to run in real time.

The structure of this article is as follows: Section 2 provides an overview of the Expert System proposal. In Section 3, the specific details and functioning of the rotary railcar dumper system, the real-time hardware, and the monitoring conducted by the supervisor system are explained. The occurrence of faults is also explained based on the behavior of the VV311-K01 rotary railcar dumper. The precise processes for developing an Expert System using Knowledge Engineering approaches inside the CommonKADS methodology are described in Section 4. Furthermore, this section also includes the resources of the JESS environment, which serves as the inference engine for the system's decision module. It also covers the system's application and implementation global architecture, as well as the closing remarks.

## 2. EXPERT SYSTEM PROPOSAL

The proposed system aims to improve and expedite the decision-making process while dealing with defects in the minerals unloading system. It does this by taking into account the faults detected by the components of the VV311-K01 rotary railcar dumper system.

The identification of flaws is achieved by analyzing Microsoft electronic spreadsheets and Access databases. This refers to a significant amount of operational data and potential information that is not integrated with VALE's Plant Information Management System (PIMS). In order to identify possible remedies for a fault in the VV311-K01 positioner car, the engineers and technical team must analyze several devices that are relevant to the fault. They will trace the fault mode, effects, and related causes of these devices. In other words, this is constructed based on a specific model.

$$H_{fault} : x_{devices} \rightarrow y_{relevant}$$

Being $x$ the set of VV311-K01 devices or subsystems. The Expert System propose consider the plant devices mapping dealing and inferring the functional relationship (i.e. fault- device) between the set of plant devices and faults mode. By example:

$$x = \{x_i \mid devices\}$$

Being $x_i$ , **shaft**, **engine**, **sensors**, coupling, shock absorbers and furthermore VV311-K01 car positioner devices. Associated to this propose, these sets are inputs to begin the system modelling and discovery in which conditions the decision making procedure is sustained. In addition, the Expert System is built by using the AI symbolic reasoning paradigms (Luger and Stablefield, 2008) to be modelled for the industrial sector.

Notice that the Expert System considers the VV311-K01 significant characteristics based upon the knowledge of experts and the domain agents (i.e. engineers, operation analysts and operators), during positioner car operation in order to improve the unloading system's productivity along the execution of the involved tasks at the VALE industrial complex.

## 3. THE ROTARY RAILCAR DUMPER SYSTEM

The process of unloading minerals begins when a locomotive arrives to the rotary railcar dumper, pulling 102 to 104 rail-wagons. These rail-wagons are then positioned in the dumper, and the objective of each rotary is to empty 2 rail-wagons during each repetition. The iteration refers to the duration required for the positioner car to secure the rail-wagons during the dumping operation.

In order to achieve rotation, a positioner car secures the rail-wagons in the rotary mechanism. As a result, the material is unloaded by undergoing a rotation of 160°, and it can even be programmed to rotate up to 180°. This rotation occurs in the carrier-belts (Fonseca Neto et al., 2003). During the unloading process of the rail-wagons material, the positioner car simultaneously returns to prepare for the next rail-wagons. To ensure that the railroad-cargo remains stationary, latches are used until the rail-wagons in rotation are released.

### 3.1 Real Time Hardware

The devices that control the dumper typically consist of physical components, such as inductive and photoelectric sensors, load cells, pressure switches, and thermostats, as well as limit electromechanical switches and cam switches.

The peripherals of a dumper have a significant impact on several operations, such as stopping or interrupting displacement, signaling location, monitoring pressure and temperature, and other relevant elements in this context.

The hardware of rail-wagon dumpers can be seen as an intermediate layer, or middleware, that plays a crucial role in facilitating communication between the Expert System and the VV311-K01 hydraulic and mechanical components during operation.

### 3.2 Supervision Control System

The supervision process is carried out using programmable logic controllers (PLCs) that take input from the dumper hardware through input cards. The PLCs also operate the Motors operate Centre (MCC) using output cards. The programmable logic controllers in the dumper control the actuators and action drivers (converters).

The programming, implemented in LADDER, is organized in a manner where the initial section is dedicated to handling faults, followed by the control section, and finally the output section. The software is structured using subroutines, each dedicated to a certain component (such as the positioner car, rotation, latches, etc.) found in the dumper. The mesh command was designed to rely solely on the supervisory command in order to be closed.

The Operational Process is overseen by the Supervisory Control and Data Acquisition (SCADA), which consists of two servers running the InTouch software from Wonderware, and four clients that gather data for the SCADA system using the Dynamic Data Exchange (DDE) from Microsoft.

### 3.3 Faults occurrence

The occurrence of faults in the production process and the prolonged system shutdown due to equipment overload, sensor malfunctions, and issues with other component sets of the rotary railcar dumper have resulted in significant financial losses for the VALE industrial pole. This is based on the monthly average unloading of the VV311-K01, which amounts to approximately 16,120 rail-wagon cycles (equivalent to 155 trains, each consisting of 208 rail-wagons).

According to statistics reports from VALE, most of the defects in the dumper occur at the positioner car. This component is responsible for reducing the monthly average by 1095 cycles of rail-wagons.

Based on the provided information, the VV311-K01 positioner car was chosen as one of the key components to be examined in the previously indicated industrial area.

## 4. THE EXPERT SYSTEM DEVELOPMENT

Before initiating the Expert System developing stages, it is necessary to select some important characteristics that will be used to build the system, such as the JESS and the CommonKADS methodology.

### 4.1 JESS

The JESS is a tool for constructing the Expert System developed by Friedman Hill at Sandia National Laboratories. The JESS is totally developed in JAVA, and is characterized as an API for creating the expert Systems based on production rules. Its architecture involves cognition components defined like: Inference Engine, Agenda and Execution Engine. All these structures catch assertions or domain facts and also create new assertions.

The inference JESS engine is constituted by the Pattern-Matching mechanism (i.e. patterns joining) that decides which rules will be activated. The Agenda programs the order in which the activated rules will be fired, and the Execution Engine is in charge of the triggering shot (Friedman-Hill, 2006). Besides that, such rules can contain function callings that care of code statements in JAVA.

In JESS the facts have attributes or fields called slots, which must be grouped in templates in order to keep common feature assertions, and have some of their properties grouped in classes like Object-Oriented.

The reasoning formalism used by the JESS presents rules composed by if...then patterns, represented by the LHS (Left-Hand Side) and RHS (Right-Hand Side), respectively.The inference process is given by the Rete algorithm (Forgy, 1982) that combines the facts according to the rules and selects the one that will be shot to execute their corresponding actions.

Having JESS as decision core, the Expert System will operate by matching the facts, which are the right statements on the attributes contained in the VV311-K01 knowledge base, with the rules that translate the domain of the agent's explicit knowledge of the VALE unloading system's.

## 4.2 CommonKADS

The historical scope of the CommonKADS methodology was confirmed by the results of several projects of the ESPRIT program for building knowledge based systems. Even though it was conceived at the Amsterdam University, initially under the name KADS (Knowledge Acquisition Design System), it referred to a method for knowledge acquisition; later some contributions papers and European Science Societies developed various knowledge systems through it. As a consequence of the good results obtained with the KADS technique, they decided to expand it towards a set of techniques or methods applied to all development phases of systems based upon knowledge, creating the CommonKADS methodology, becoming acknowledged by several companies as a full pattern for knowledge engineering (Labidi,1997).

Products arisen from Expert Systems development that use this methodology are the result of the performed phases modelling activities, and characterize the input artifacts for the successive refinements undergone in the next steps of the CommonKADS life cycling.

Having in hands the particularities that will be used in the Expert System building, the steps of the system with actions such as Acquisition and Knowledge representation are organized- also including the analysis phase - Rules representation - ruling the Design phase - and the System's Settling- satisfying the settling phase.

## 4.3 Acquisition and Knowledge representation

Knowledge acquisition is the most important step when developing Expert Systems, and aims at the detailed attainment of the knowledge used by the expert to relate problems. All the knowledge elicitation was done by means of interviews with the expert through information kept in the operational reports, spredsheets and off-line database. The method used to the knowledge representation was built based upon production rules. These rules map the knowledge of the VV311-K01 operation expert onto computing artefacts take into consideration the set of relevants faults (i.e. $y$) instance and its generator sources, modeling the conditions in which the faults deduction can points out the diagnosis or support the expert's decision making. Highlighting the relevant fauls, they establish a vector in which the positoiner car devices are relevant faults attributes according to following set.

$$y = [y_1 \, y_2 \, y_3]$$

In this set, $y_1$ is the kind of generator source, $y_2$ is the priority and the $y_3$ is the historic, reminding that only generator source is treated in this chapter. In order to undestand the relevant faults model instance, was considered the car positioner in agreement with the following set.

$$y(x) = \begin{bmatrix} y_1 \\ y_2 \\ y_3 \\ y_4 \end{bmatrix}$$

Being x the VV311-K01 positioner car and y, the set of faults belongs to it, y1 is the engines situation, y2 positioner arm situation, y3 latches situation and y4 coupling situation. These mathematical elucidation are early requirements to understand the amount of situations that enginers and technicians team have to deal during the productivy system. The following model represents the possibilites during a decision making scenario.

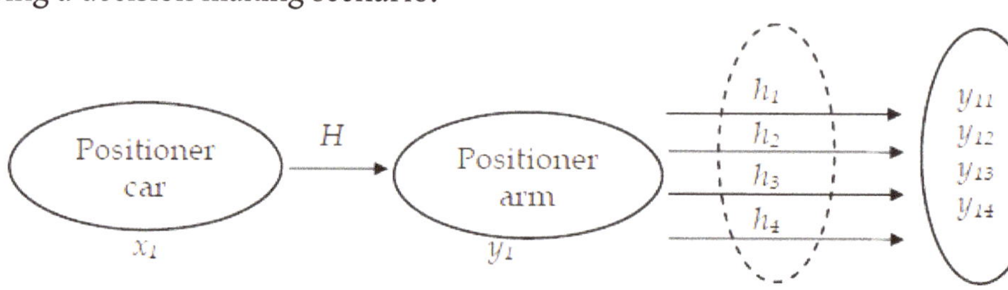


In fact, all this situations and conditions will be handled by JESS inference engine. The JESS handles the knowledge representation as production systems and rules them like condition- action pairs. A rule condition is treated as a pattern that decides when it can be applied, while the action will define the associated problem solution step (Friedman-Hill, 2003). In this way, there were defined the sort of problems presented by the positioner car, along the mineral unloading process, for the elaboration of production rules.

There were observed the main concepts related with the dumper's positioner car along activities in the operational productive system, aiming at getting knowledge elements description by elaborating the organizational model that complements the CommonKADS (Breuker et al., 1994) Analysis phase.

**Table 1. Organization Model.**

| SLOT | OPPORTUNITIES | PROBLEMS |
|---|---|---|
| Situation | Engine | Vibration |
| | | Broken Rollers |
| | | locked |
| | | Lack of voltage |
| | Positioner arm | Short-circuit |
| | | Broken fixing screws |
| | | Broken Counter-bolts |
| | Latches | Disruption |
| | | Infiltration |
| | | Internal Fugue |
| | | Insufficient outflow |
| | Coupling | High oil level |
| | | Low oil level |

The domain facts and experiences deal with the equipment situation and the potential causes that promote the main system stopping or reduce its productivity. Therefore, in Table 1 is presented the organizational model, correlating problems and opportunities that can be solved or enhanced by the Expert System from which extracted the identified slots for building the VV311-K01 templates. The slot called 'Situation' is one of the units that comprise the templates for representing the knowledge in the JESS inference engine (Friedman-Hill, 2006).

**Table 2. Knowledge Model.**

| SLOT | INFERENCE LEVEL | TASK LEVEL |
|---|---|---|
| Cause | Vibration | Motor basement snap |
| | | Resonance |
| | | Bend axle |
| | Short-circuit | Terminal out of order |
| | | Low isolation |
| | | Falling's wire material |
| | Insufficient outflow | Worn Pump |
| | | Obstructed Tabulation |
| | | Safety Valve with insufficiently fixed |
| | Obstruction | Dirt |

It was observed that the causes that lead the dumper to reach certain circumstances are pointers for guiding what must be done as to specify derivations that constitute a method for the VV311-K01 positioner car problem resolution, and the strategies to attain this solution. The efforts spent in this stage are described through the knowledge model of the CommonKADS methodology, as shown in Table 2.

According with (Labidi, 1997), the inference and task levels are layers that describe the expert Knowledge; thus, the model in Table 2 constitutes a set of knowledge instances on the VV311-K01 positioner car component. Starting from the model in Table 2, in order to better characterize the system's knowledge mechanism in agreement with the Common KADS methodology, the activities organized in the inference model presented in Figures 1, were decomposed.

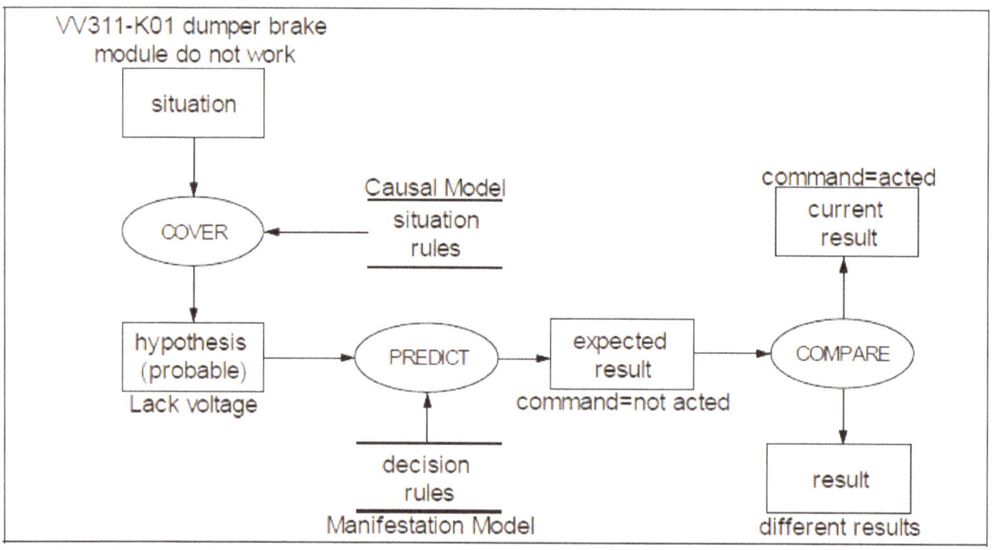


**Fig. 1. Inference Model.**

This model aims at elaborating a declarative specification of input and output properties, and the inference actions used in the Expert System reasoning. The Inference Model in Figure 1 describes a deduction example done for the VV311-K01 positioner car component, and can be explained as follows: the knowledge's roles are functional names that capture the elements participant in the reasoning process as diagnostic the positioner car state, and can present variable actual status (i.e. temporal stopping, overheating, etc). Inference actions assume as inputs static roles, represented by the manifestation and causal models.

Within the causal model, the rules relate the positioner car fault modes taking into account their attribute's values, while in the manifestation model are reunited the production rules that express their responsibilities through the attributes' values, which satisfy some given conditions (Labidi, 1997).

The COVER, PREDICT and COMPARE inference concepts, represent reasoning axioms that will be mapped by the JESS inference engine used in the Expert System. Basically, is done a transition from abstract concepts as input artefacts to synthesized concrete concepts in a set of assertions as output artefacts (Friedman-Hill, 2003).

Once we have in hands the analysis of the main elements that model the general goal of the knowledge specification stage, according to the CommonKADS and taking into account the granularity of the acquired information for the VV311-K01, significant levels of detail were obtained for representing the knowledge, under the form of production rules.

## 4.4 Rules Representation

After establishing the knowledge sources, the effort used in this step are linked to the structuring of rules of the type if...then. Firstly, the selection of this type of representation was justified in function of the shell JESS inference engine promotes the building of rules in production systems, and also as a consequence of the attained results in a system built by (Fonseca Neto et al., 2003), which used production rules to detect faults in the mineral belts conveyors at the VALE PMDT in off-line mode.

The example below shows the 'dumper' template that will treat the attributes 'situation' and 'cause' at the Expert System knowledge base.

```
 (deftemplate dumper
(slot situation (default NF))
(slot cause (default NF))
)
```

The 'dumper' template gathers the characteristics considered in the interviews for the possible situations (new status) that the VV311-K01 can generate and the pre-attributed values – NF: Not Found – in events within which there is no chance of pattern unification (i.e. pattern-matching).

In the 'decision' template are grouped the possible causes as particularities that can correspond to 'dumper' template situations (current status) expressed by means of the sensors. The next template was also defined with slots and predefined default valued.

```
(deftemplate decision
   (slot dec-sit (default NF))
   (slot dec-cause (default NF))
   (slot dec-decision(default NF))
```

After the templates specifications, the rules were built taking as basement their slots definition. These slots in the JESS rules structure are part of the so called LHS patterns (i.e. premises) and will be unified (pattern-matching) by the inference engine through the Rete algorithm. The templates and the rules were built with the 7.0p1 version of shell JESS, licensed for academic use.

## 4.5 System Implementation

An executable version of the Expert System was developed to encode the system and identify any functionalities that were not covered during the interviews, in order to capture any remaining knowledge. The tool selection was facilitated by the linguistic resources provided by JESS, which enhance inter-operability and mobility inside the VALE operational producing system.

The Expert System decision module scans the faults reported by the sensors in the VV311-K01 instrumentation to facilitate the operation handling of the system. The Expert System rules that activate a given situation are triggered in the working memory of the JESS inference engine by checking the fault addresses generated by the programmable logical controllers and mapped into fault tagnames stored in the relational database. Tagnames, such as AFF_CEP_F01@VV311K01, are included in the historical register of PIMS. The Expert System utilizes input data to deduce the condition of the VV311-K01 components. This is achieved by forming the rules' LHS pattern in the JESS language syntax, as demonstrated in the extract below:

```
                    (defrule rule198
(test
(eq TRUE(actedTag "AFF_CEP_F01@VV311K01")))
=>
(store RESULT198 "Loose-Wire-Connection")
(assert (decision(decCausa Loose-Wire-Connection)
(decPrint "Loose-Wire-Connection"))))
```

The above rule presents in its LHS pattern, a actedTag function, which returns true for 1 and false for 0, according to the result read in the XML file generated by stored procedure into the PIMS (Plant Information Management System) oracle database server (see Figure 2), location at which are stores the faults tagnames. The XML document is shown as following structure:

```
<?xml version="1.0" encoding="UTF-8" ?>
- <tags>
 - <tagname id="ASC_B11@VV311K01">
       <value>0</value>
   </tagname>
 - <tagname id="AIN_ALI_001@VV311K01">
       <value>1</value>
   </tagname>
 - <tagname id="AFL_BEP_001@VV311K01">
       <value>0</value>
   </tagname>
 - <tagname id="ATA_BRT_M01@VV311K01">
       <value>0</value>
  </tagname>
 ...
  </tags>
```

The function 'actedTag' assumes as input the description of a tagname to check whether its value acted or not using set and get methods upon JAVA call functions.

The 'actedTag' function can be seen in JESS language statement as following:

```
(deffunction actedTag (?tagName)
   (bind ?sb (new BeansTag))
   (definstance beans ?sb)
   (call ?sb setTgName ?tagName)
   (if (eq (call ?sb getTgValue)1)
        then return TRUE else
        (if (eq (call ?sb getTgValue)0)then return FALSE))
)
```

Starting from an acted tagname, the rule's LHS pattern is evaluated by the conditional function test (Friedman-Hill, 2006) – which is part of the own JESS language – responsible for determining that this pattern will only be unified if the result of the actedTag function returns true. Taking as example the following parcel of the system's decision rule, is possible to observe that the cause 'Loose-Wire-Connection' deduces that the ultrasound on VV311-K01 must be made. The RHS pattern of the rule's parcel stores the result through the command store, and next inserts in the 'decision' template, the action to be done.

```
 (defrule rule204
(decision (decCause Loose-Wire-Connection))
 =>
(store RESULT204 "Do ultrasound on VV311-K01")
(assert (decision
          (decDecision  Do-ultrasound-on-VV311-K01)
          (decPrint  "Do ultrasound on VV311-K01")
)))
```

The architecture components collaboration is achieved by integration of JESS inference engine and PIMS database server, which is supplied by historical database connected to VV311-01 throughout PLC device. The key point is the usage of the XML pattern aiming convert the SQL queries from PIMS database to knowledge base.

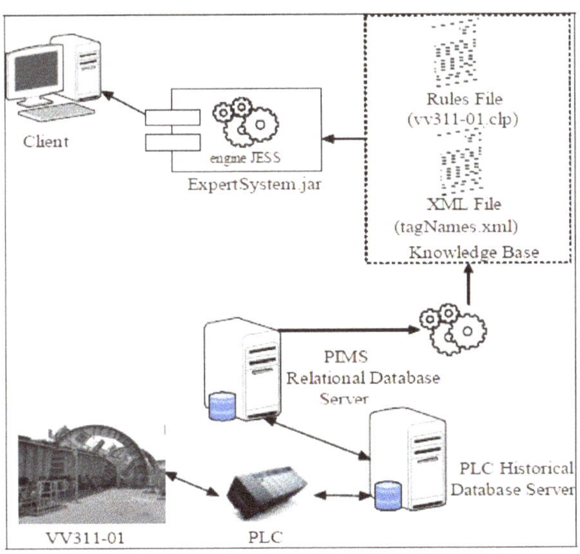


**Fig. 2. Expert System Architecture.**

Aiming to highlight the global and external view of the system, in Figure 2 was described the Expert System architecture and the nodes distribution that form its structure in order to summarize the Expert System solution and its knowledge base composition.

At the end of the performed deductions, in the shell JESS working memory is shown a window with the recommendations that were stored by the command store (Friedman-Hill, 2006), based in the causes that led to the inference process pointed out by the system, along the responses given by the user. Figure 3 shows the decision making delivered by the system.

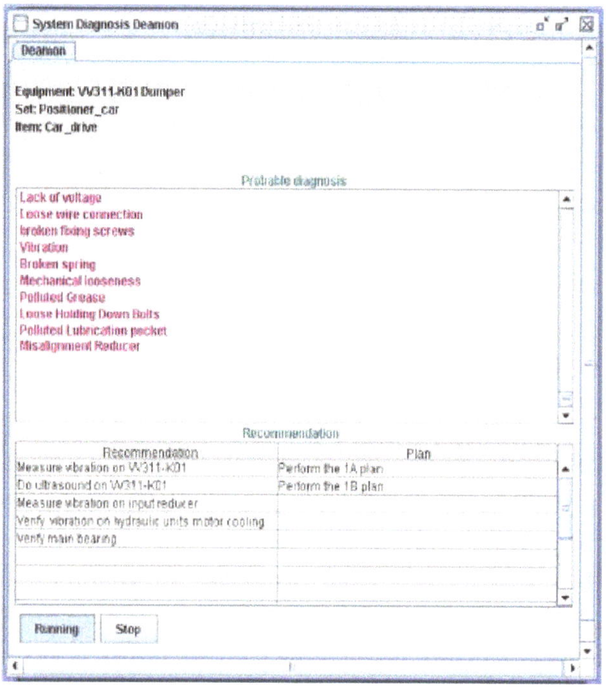


**Fig. 3. Expert System Recommendations**

Foreach plan recommended, the Expert System detail the needful activities in order to its running. Besides, the amount of people that will perform the tasks related to the faults. By example, in Figure 3 the necessary information to measure vibration onVV311-K01 is listed by class of maintain plans.

Starting from recommendation, the user (technician or engineer) can access its details by double clicking over plans line, according to Figure 4. Through this screen is possible to perform the activities related with decisions and maintain plans lunched by Expert System. Notice that these plans comprise maintain methods developed by VALE's especialists and include all background acumulated by themself too. In fact, this feature can promote the activity decomposition by steps, the labor envolved with each recommendation.

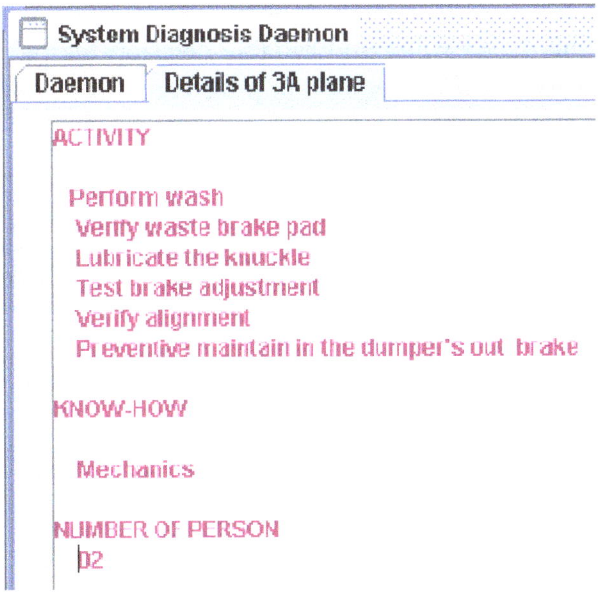


**Fig. 4. Expert System Plans Details.**

The recommendations viewed in Figure, represent the rules that were activated in the working memory of the JESS engine were triggered (shot) only because they got the unification of patterns present in its structure, deduced by the Rete algorithm. The control structure used for the Expert System rules chaining was based on backward chaining (Friedman-Hill, 2003).

## 5. EXPERIMENTS AND RESULTS

The developed expert system approach is focused in the information and automation architecture and infrastructure. For the first results obtained through its putting into operation, the system was set on the corporative mineral unloading mesh of the VALE Company at the OCC installments.

The system's performance while processing operative data in the JESS inference engine was considered satisfactory once the search frequency of such data elements was tested at 5 s. interval for the deduction of 250 rules, although its standard configuration requires 5 minutes of response time, in order to update in the working memory of the expert system the events occurred at the PIMS server.

The speed and proper timing obtained in terms of the updating processes of the expert system rules base was due to the use of the XML technology as to feed the systems' knowledge base. In addition, the expert system is nor directly connected to the PIMS server, thus avoiding the expenditure of computational resources due to the flux of information between the application and the VALE's data manager, once all the programming necessary for generating the file XML was implemented in the own database.

The comparison of the results between the system and the human expert was characterized by the need of fast and reliable decision taking, fact that allowed operators and maintenance teams to minimize the time spent from the fault's detection to its correction.

**Table 3. Comparison between Expert System and Technician.**

| Set | Fault | Is it correctly detected by technician? | Is it correctly detected by Expert System? |
|---|---|---|---|
| Arm command | Damaged contact block | Yes | Yes |
| Car command | Lack of Measuring | No | Yes |
| Drive command | Pulley out of place | Yes | No |
| Drive command | Defective switching | Yes | Yes |
| Drive command | Vibration | No | Yes |
| Drive command | Heat of rollers | Yes | Yes |
| Arm command | Misalignment | No | No |
| Drive command | Burn | No | Yes |

As a critical point observed during the results comparison are highlighted atypical events such as a slight misalignment and pulley out of place which was noticed by the system, but did not work in the next cycle. These experiments are very important because the results can contribute to modify the knowledge base inference logic in order to improve these decisions which human expert still have some advantage over the system.

## 6. CONCISE SUMMARY

This effort built a system that automated the strategic knowledge needed for VALE's mineral unloading system activities. Given the complexity of mineral exploitation, a comprehensive understanding of the domain is essential. To facilitate this, a knowledge representation system was developed utilizing the JESS inference engine, which is capable of applying production rules.

The implementation of the Expert System in JESS made it possible to utilize established and effective techniques for designing knowledge-based systems, such as CommonKADS. Utilizing JAVA technology objects and XML directly facilitated the implementation of the system's architecture in real time, ensuring that operations are displayed immediately as they occur. Additionally, this approach enables seamless interchange with any database.

Ultimately, this system provided improvement and a suitable level of preparedness for the processing of knowledge, serving as a valuable tool for the decision-making process of specialists in VALE's rail-wagon unloading system.

## REFERENCES

1. Breuker, J. Velde, Van de. CommonKADS Library for Expertise Modeling: Reusable problem solving components. IOS Press, Amsterdam, 1994.
2. Schreiber, Guus. Akkermans, Hans. Anjewierden, Anjo. Hoog, Robert de. Knowledge engineering and management: The CommonKADS methodology. 1st ed. The MIT Press. 2000.
3. Feigenbaum, A.E. Expert Systems: principles and practice. In: Encyclopedia of computer science and engineering, 1992.
4. Forgy, C.L. Rete: A fast algorithm for the many pattern/many object pattern match problem, Artificial Intelligence 1982.
5. Friedman-Hill, E. J. Jess in Action: Rule-based systems in java. USA: Manning Press, 2003.

# 17 A MODULAR AND HYBRID EXPERT SYSTEM DESIGNED FOR THE MANAGEMENT OF PLANT ASSETS

## 1. INTRODUCTION: AN OVERVIEW

Industrial firms must enhance their manufacturing process as a vital prerequisite for their existence in a highly competitive national and international market. This encompasses safeguarding the accessibility and functionality of the manufacturing equipment. This equipment comprises basic components like as sensors and actuators, as well as intricate machinery and even more sophisticated production lines and factories. The term "improvement" refers to the optimization of production performance and the reduction of unplanned production downtime.

This chapter presents a method for facilitating the management of plant assets, which is a crucial aspect of company business management. Plant asset management specifically pertains to the maintenance and optimization of the production equipment within an industrial organization. On the other hand, general asset management encompasses the administration of all tangible and intangible assets of a company, such as buildings and patents. Plant asset management primarily focuses on the operational phase of manufacturing equipment. In the following, we will use the term "asset" as a synonym for "production equipment."

The primary objectives of plant asset management are to enhance the dependability and effectiveness of the assets while minimizing the resources required to maintain them in a production-ready condition. Equipment maintenance is a crucial component of plant asset management.

Important strategies of maintenance are:

- **Preventive maintenance**: The fitness to work of the assets is preserved by maintenance actions before any failure occurs. The action planning is based on production time and/or load cycles.

- **Predictive maintenance:** Maintenance actions are executed before any damages. The abrasion state of the assets is continuously monitored and logged here. The remaining useful life is predicted. This information builds the base for the planning of maintenance actions. This strategy is also called condition based maintenance. It leads to longer useful life time of assets compared to the preventive maintenance strategy.
- **Reactive maintenance:** Maintenance actions are executed after emerging failures of the assets. Repair actions or exchange of assets are necessary actions to take in most cases. This strategy has been used widely in former times when technical preconditions like computer aided planning systems and measurement systems where not available.

Applying preventative or predictive maintenance strategies can greatly enhance the reliability of manufacturing systems compared to using reactive maintenance. Nevertheless, technical systems are prone to failure. Therefore, it is impossible to fully implement the reactive maintenance plan.

The implementation of predictive and preventive maintenance procedures, along with economic limitations, results in a significant reduction of maintenance personnel. The complexity of technical systems increases in tandem with the use of advanced automation systems and specialized technical components aimed at conserving energy and materials throughout the production process. Therefore, in several instances, a limited number of maintenance workers are faced with very intricate systems in infrequent but extremely crucial scenarios, where equipment malfunctions result in output loss.

Computerized assistance systems can aid maintenance staff in completing their jobs. This chapter presents a methodology for diagnosing the reasons of failures in intricate production systems. Merely employing discrete theories is inadequate for providing best assistance in diagnostic activities. The approach incorporates a synthesis of different theories from expert system technology and probability theory. In this context, the optimality criteria are the accuracy of identifying the correct failure reasons and the level of work put into formalizing the required diagnostic information.

## 2. SYSTEM REQUIREMENTS

Expert systems for diagnosing technological systems have been known since the 1960s and early 1970s. They have been effectively utilized for specialized systems. However, these methods have not been utilized for large-scale, diverse, and intricate systems. Conducting an analysis of end user requirements is essential to identify solutions that may be applied more widely to these systems.

Figure 1 presents a concise overview of the discovered use cases in the WISA research project (see to acknowledgements in section 5). It is important to highlight that the diagnosing procedure is just one of many different use cases. The successful use of expert system technology relies heavily on the ease with which the knowledge base may be created. Therefore, it is essential to take into account the requirements of other groups of individuals in addition to prioritizing the maintenance people. Significant user roles include:

- **Diagnostics user**. This is the user of the expert system. Mostly this will be a member of the maintenance staff. She is responsible for the availability of the assets.
- **Asset manager**. The asset manager is responsible for the efficient production. She controls the processes on business level. In most cases she is supervisor of the diagnostic user.
- **Asset expert**. The asset expert has special knowledge about the disturbance behaviour of the assets. This knowledge contains relations between failure causes and their symptoms. In many cases the expert is the constructor of the asset, but it is also likely that she is the asset user with long term experiences.

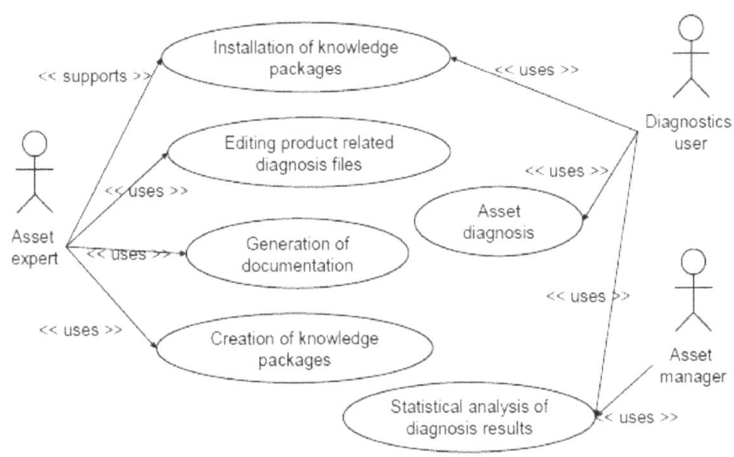


**Fig. 1. Summary on use cases for a diagnosis expert system framework**

The following list gives some details about the requirements resulting from the use case analysis:

- **Asset diagnosis:** This is the main use case. A failure event occurs and becomes visible for the diagnostic user who searches for the failure cause. The information about the failure may be distributed via industrial networks or may be a perception of the operation staff or the diagnosis user itself. Requirements are:

- The input of the diagnosis user into diagnosis dialogs should be minimized.
- If there are multiple failure causes possible then the list of causes should be sorted according to their probability.
- The diagnosis system should ask back for additional symptoms, which make sense.

- **Editing product related diagnosis files:** The asset expert has to formalize his diagnosis knowledge. In most cases the asset expert is not an IT expert. Thus the input of the knowledge should be done not directly by using a programming language. Instead a forms based editor will be used. Requirements regarding such an editor are:
  - The graphical user interface of the editor should use maintenance terminology.
  - The editor should recognize and inform about syntactic and semantic errors within the knowledge base as far as possible.
- **Creation of knowledge packages:** The diagnosis knowledge of a complex asset includes the knowledge about the asset elements and the causal relations between them. A system is needed, which gathers all the knowledge, which contain the necessary information in order to create a knowledge package for the considered asset. Requirements regarding such a packaging system are:
  - The packaging system should gather all necessary knowledge in a file container.
  - The file container should be compressed for hard drive space efficiency.
- **Generation of documentation:** The equipment manufacturers should act as asset experts. Thus they investigate more time into the development of the products. The generation of a part of the necessary product documentation is a pay-back possibility. Requirements on a generator system are:
  - The documentation generator system should be included into the editor framework.
  - The generator system should only be parameterized by the asset main context.
  - Necessary descriptions of asset elements will be included automatically.
  - The documentation result should be linked HTML pages, which may be transformed into other text processing formats.
- **Installation of knowledge packages:** The diagnosis expert system to be used by the diagnostics user will include the knowledge of a variety of

products, which aggregate other components. Thus there is the need for installation of asset descriptions. Some requirements are:

- The installation system should inform about type and version conflicts regarding descriptions of assets.
- The installation system should contain features to gather knowledge packages via the Internet and from local resources.

- **Statistical analysis of diagnosis results:** If the failure behaviour of assets is tracked over a long time period, then there is the possibility to identify the assets, which are responsible for the highest production loss. Failure frequency and amount of production loss has to be tracked for this purpose. Requirements for this analysis are:
  - The statistical analysis system should identify the assets, which are responsible for the highest production loss.
  - The result presentation may be done in text form on-screen.
  - These system requirements should be satisfied by an intelligent language construction for the formalized asset descriptions and by a related software tool chain.

## 3. DESIGN OF THE KNOWLEDGE BASE

### 3.1 Special Criteria for the Knowledge Base

The introduction and the system requirements state, that the knowledge base will be constructed by multiple person groups. Thus there must be a way of collaboration between them. In the WISA project the following approach has been used:

- A language has been designed for the formal description of diagnosis knowledge. It has been named **Hybrid Logic Description Language (HLD)** for purposes described below.
- Tools have been designed for creation, transportation and installation of HLD files.
- An HLD interpreter has been designed for the diagnostic process.

A variety of approaches are known to work fine for special diagnostic purposes. The problem is to select the right one(s) to meet industrial needs and strength. Thus a selection had to be made by matching against a catalogue of criteria for such a language. Some of these criteria are:

- **Processing of discrete causal relations should be possible.** That means that a specific symptomatic is uniquely related to a distinct failure cause. Such relations are often described in tabular form in conventional machinery and plant documentation.

- **Processing of vague diagnosis information should be possible.** Vague information here means such information, which don't may be acquired by measurement or only with inadequately hard effort. This kind of information is estimated by the diagnostics user mostly based on his/her visual, acoustical and tactile perceptions.
- **Processing of uncertain causal relations should be possible.** In opposite to the discrete causal relations it is often necessary to formalize relations between symptomatic and a variety of causes. These relations are weighted by different probabilities. Thus the language should provide constructs to formalize these kinds of relations.
- **The information processing concepts should be kept simple.** The concepts should be easy to understand for asset experts.
- **There should be the possibility to validate language instances.** The process of editing HLD files should be supported by checking them for syntactic and semantic errors.
- **There should be the possibility to re-use asset type based diagnosis knowledge.** The asset experts know the scope of the assets they describe, which includes all aggregates of the considered asset. The diagnosis knowledge about the aggregates should simply be referenced by the type and version of the aggregate in the description of the asset.
- **There should be efficient algorithms for the information processing concepts.** The knowledge base for a complex technical system will itself become very complex. The algorithm implementations should provide answers on the failure causes in an appropriate time (some seconds).

The following section describes the design of the HLD language, which has been developed to satisfy this list of criteria.

## 3.2 Hybrid Logic Description Language (HLD)

### 3.2.1 Propositional Logic and Its Representation as HLD Rules

The first means of expression is the propositional logic. Later on we want to introduce certainty factors. In order to motivate the introduction of these concepts let us have a look on the following example dialog between a machinery operator and an expert for the machinery:

*Operator*: "Production line 3 is currently out of order" *Expert*: "How do the last products look like?" *Operator*: "The surfaces are scratched."

*Expert*: "Did the forming machine made noise?"

*Operator*: "No."

*Expert*: "Check if there are crumbs in the raw material."

*Operator*: "Yes, there are crumbs."

*Expert*: "Then we have got low quality of the component X within the raw material.

This example shows a systematically diagnosis controlled by the expert. This expert has to be replaced by the expert system. Thus there is the need to formalize the shown diagnosis knowledge.

The kind of knowledge as mentioned in the example may be expressed by a rule base like the following:

- Products_scratched if line_3_not_working.
- Tool_loose if products_scratched and forming_machine_noisy.
- Crumbs_in_material if products_scratched.
- Component_X_bad_quality IF crumbs_in_material

These rules constitute only a small section of a diagnosis knowledge base for a real world application. The causes of symptoms are situated on the left side of the IF statement, while the symptoms itself are positioned on the right side. This is in opposite direction of the causal direction. The results of the application of the reasoning algorithms are the conclusions of the rules on the left-hand side of the IF statement and the result should be by definition the cause of symptoms.

The syntax of the propositional logic has been defined in various books like (Kreuzer and Kühling, 2006), (Russel and Norvig, 2003) or (Poole et al., 1998). Propositional formulae deal only with the truth values {TRUE, FALSE} and a small set of operations is defined including negation, conjunction, disjunction, implication and bi-conditional relations. The possibility to nest formulae enables arbitrary large formulae.

The HLD restricts the propositional logic to Horn-logic, which is not a big limitation. A Horn formula is a propositional formula in conjunctive normal form (a conjunction of disjunctions) in which each disjunction contains in maximum one positive literal. The set of elements of these disjunctions is also called a Horn clause. A set of Horn clauses build a logic program. If a Horn clause contains exactly one positive and at least one negative literal, then it is called a rule. The positive literal is the conclusion (or head) of the rule, while the negative literals constitute the condition part (or the body) of the rule. If a rule is part of a HLD file, then we call it a HLD rule.

The form of horn clauses is chosen for the HLD, since there exist an efficient reasoning algorithm for this kind of logic - namely the SLD resolution. This resolution algorithm may be combined with the breadth-first search (BFS) or with the depth-first search (DFS) strategy.

- **Breadth-first search:** The algorithm proves for each rule whether the conclusion is a consequence of the values of the conditions. Each condition value is either looked up in a variable value mapping table or it will be determined by consideration of rules, which have the same literal as conclusion. If there exist such rules, but a direct evaluation of their conclusion is not possible, then a reference to this rule is stored, but the algorithm proceeds with the next condition of the original rule. If there is no condition left in the original rule, then references are restored and the same algorithm as for the original rule is applied to the referenced rules. This approach needs a huge amount of memory.
- **Depth-first search:** This algorithm proves for each rule whether the conclusion is a consequence of the values of the conditions. Each condition value is looked up in a variable value mapping table or it will be determined by consideration of rules, which have the same literal as conclusion. If there exist such rules, but a direct evaluation of the conclusion is not possible then the first of these rules is evaluated directly. Therefore this algorithm does not need the references and saves a lot of memory compared to BFS.
- It may be shown that the SLD resolution with BFS strategy is complete for Horn logic while the combination with DFS is incomplete. However, DFS is much more memory efficient than BFS and in practise it leads often very quickly to the result values. Thus both resolution algorithms have been prototypically implemented for evaluation of HLD files. The syntax of the HLD does not depend on the selection of search algorithms.
- The propositional variables of HLD rules have special meanings for the diagnosis purposes. Following has been defined:
- **Symptoms** are propositional variables, which appear only as conditions within HLD rules.
- **Indirect failure cause**s are propositional variables, which appear as conclusion in some HLD rules and in other HLD rules condition part.
- **Direct failure causes** are propositional variables, which appear only as conclusions of HLD rules.

Thus simple propositional logic is modelled in the HLD by direct and indirect failure causes as conclusion of rules and by symptoms and indirect failure causes as conditions of rules.

### 3.2.2 HLD Rules with Empirical Uncertainty Factors

The application of HLD rules is not always applicable or at least not very comfortable because of the following reasons:

- A huge amount of rules and symptoms have to be defined in order to find failure causes in complex technical systems. This is accompanied by very large condition parts of the rules. The establishment of the knowledge base becomes too expensive.
- A diagnosis expert system, which has a high complex knowledge base, has to ask the users for a lot of symptoms in order to find a failure cause. Guided diagnosis becomes too time-consuming.
- Complex knowledge bases lead to long-term reasoning.

All these effects should be avoided according to the defined requirements. The mapping of simple cause-effect relations with simple HLD rules continues to be applicable. But complex circumstances need other kinds of expressivity.

A simple extension of HLD rules is the introduction of certainty factors (CF). Therein the conclusion of a rule is weighted with a certainty factor. Such systems are described for example in (Bratko, 2000), (Norvig, 1992) and (Janson, 1989). In these resources the value range for the certainty factors is the interval [-1, +1]. For a better comparability of the CFs with probabilities the interval [0, 1] has been chosen for certainty factors of HLD rules.

All propositions, which are evaluated by application of an algorithm on a HLD knowledge base, are weighted by a CF, since the conclusion parts of the rules are weighted by certainty factors. Certainty factors of propositions have the following semantic within HLD files:

| | |
|---|---|
| CF = 0.0 | The proposition is false. |
| CF = 0.5 | It is unknown if the proposition is true or false. |
| CF = 1.0 | The proposition is true. |

CF values between 0.5 and 1.0 have the meaning that the related propositions are more likely true than false, while CF values between 0.5 and 0.0 mean, that the related propositions are more likely false than true.

Two algorithms for the evaluation of HLD rules with certainty factors have been tested. These are the simple evaluation algorithm according to (Janson, 1989) and the EMYCIN algorithm as shown in (Norvig, 1992). The simple algorithm is based on the following instructions:

- The CF of the condition part of a rule is the minimum CF of all the conditions.
- The CF of the conclusion of a rule is the CF of the condition part of this rule multiplied with the CF value for this rule.
- If the knowledge base contains multiple rules with the same conclusion, then the CF of this conclusion is the maximum of the related CF values.

The algorithms for certainty factors are proved to provide incorrect results in some situations. On the other hand for MYCIN it has been shown that such systems may provide better results than human experts. In addition the rule CFs may be empirically determined and thus the creation of a knowledge base is very easy. For these reasons the concept of certainty factors has been included into the HLD language.

### 3.2.3 Fuzzy Logic as Part of the HLD

The rule sets discussed in the previous sections utilize mappings of physically relevant diagnostic values to discrete values represented as propositions. Therefore, it is necessary to define rules for each individual value period. This results in a significant endeavor to establish the knowledge foundation. In this section, we present Fuzzy Logic as a potential method to enhance the accuracy of reasoning and minimize the need for highly detailed discretization of physical data. An illustration of a high-level design (HLD) fuzzy logic rule is as follows:

motor_defect WITH 0.9 IF motor_windings_hot AND load_low.

The way of diagnosis is different from that of the propositional logic. The diagnosis user inputs values of continuous value spaces (in the example for motor winding temperature and mechanical load), instead of providing discrete symptoms and binary answering of questions. The result is again a value out of a continuous value space (in the example an estimation of the degree of abrasion of the motor). Special diagnosis relevant output variables have been defined for the HLD language.

The use of Fuzzy Logic for diagnosis purposes works in following steps:

- Definition of the knowledge base: (A) Fuzzy variables have to be defined and (B) a Fuzzy rule set has to be integrated into the knowledge base.
- Evaluation of the knowledge base: (C) The user inputs variable values and (D) the implementation of a Fuzzy Logic interpreter provides results by fuzzyfication of input variables, applying of inferences and by defuzzyfication of output variables.

Fuzzy variables may be defined by mapping of triangles, trapezoids or more round function shapes to terms of natural language. Input variables within the HLD fuzzy logic may be defined by piecewise linear membership functions, while output variables are defined by singletons (see figure 2).

This specification conforms to the standard (IEC61131-7, 1997). PLC programming languages adhere to this established norm. PLCs are the predominant automation systems utilized for the control of machinery and plants. Therefore, if the maintenance employees possess knowledge of Fuzzy Logic, it is quite probable that they are familiar with the nomenclature and semantics of this standard.

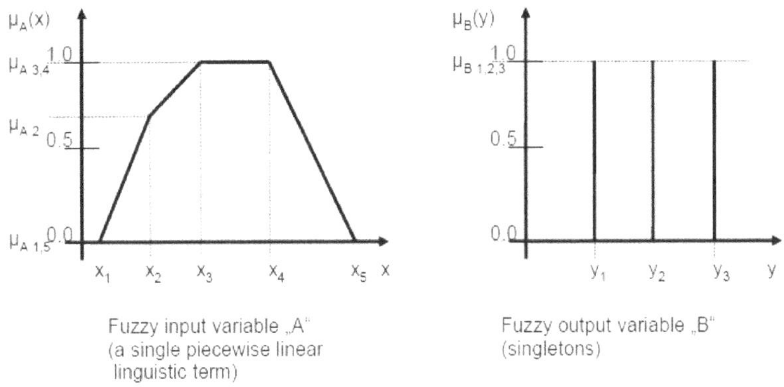


**Fig. 2. HLD Fuzzy Logic input and output variables**

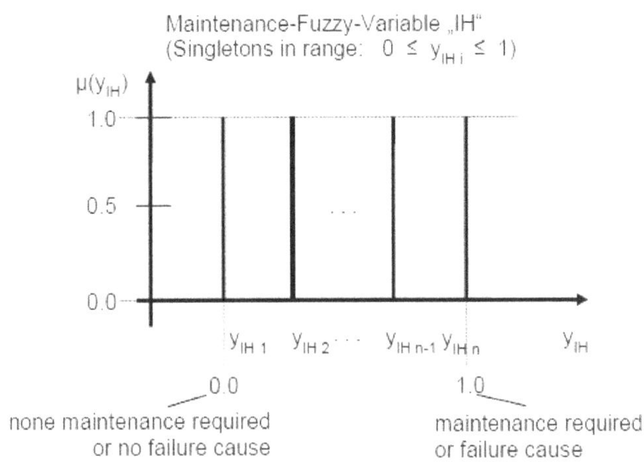


**Fig. 3. HLD maintenance fuzzy variables maintenance required or failure cause**

Beside the common semantics of Fuzzy output variables there are special definitions for maintenance variables in the HLD specification. This is illustrated in figure 3. The values of such variables yIH are defined only within a range of [0, 1.0]. Only within this value range singletons may be defined. Similar to the definitions for certainty factors following conventions have been defined for these maintenance variables:

| | |
|---|---|
| yIH = 0.0 | Maintenance is not necessary or this is not a failure cause. |
| yIH = 0.5 | It is not decidable if this is a failure cause. |
| yIH = 1.0 | Maintenance is necessary since this is a failure cause. |

As mentioned above the processing of the maintenance knowledge base is done in three steps:

- **Fuzzyfication**: Memberships are computed for each linguistic term of the input variables if there are numerical values available for the physical input variables.

- **Inference:** The inference is done very similar to the approach used for rule sets with certainty factors:
  - a. The membership of the condition part of a fuzzy rule is the minimum of all the memberships of the condition variables.
  - b. The membership of the conclusion of a fuzzy rule is the membership of the condition part of this rule multiplied with the weighting factor for this rule.
  - c. If the knowledge base contains multiple fuzzy rules with the same conclusion, then the membership of this conclusion is the maximum of the membership values of the conclusion variables.

***Defuzzyfication:*** Within the basic level of conformance of the standard (IEC61131-7, 1997) the method "Center of Gravity for Singletons" (COGS) has been defined as defuzzyfication method. This has been taken over for the HLD specification. The result value of the fuzzy logic output variable is computed by evaluation of following formula:

$$y = \frac{\sum_{i=1}^{p} (\mu^*_{Bi} \cdot y_i)}{\sum_{i=1}^{p} \mu^*_{Bi}}$$

This formula uses the terminology as presented in figure 2. The μ*Bi are the membership values computed in the inference process for the *p* singletons at the values *yi*. The result value *y* is the value of the output variable. Thus it is not a membership but a value of the value range defined for this output variable. Especially for the maintenance output variables the value range is [0, 1].

The approach of using singletons fits the need of fast computations as specified in the requirements analysis, since only multiplication and addition operations are used.

### 3.2.4 Bayesian Networks

Bayesian Networks have been introduced into the HLD, since the handling of uncertainty with certainty factors is not as mathematically correct as the probability theory does. The example introduced in the propositional logic section could be extended by probabilities as follows

Component_X_bad_quality (p=0.9) IF crumbs_in_material.

Component_X_bad_quality (p=0.5) IF product_color_grey.

This example has the meaning that if there are crumbs in the raw material then the probability are very high (90%) that the material component X has not a good quality. In other words there are not many other reasons for crumbs than a bad material X. But there is another phenomenon in that approach: the variables crumbs_in_material and product_color_grey are not independent from each other. If there are crumbs in the material, then it is likely that the component X has a bad quality, but then there is also a good chance that the product looks a little bit grey.

Bayesian Networks are graphical representations (directed acyclic graphs) of such rules as shown in the example. (Ertel, 2008) gives a good introduction to Bayesian Networks based on (Jensen, 2001). One of the earlier literature references is (Pearl, 1988). There are following principles of reasoning in Bayesian Networks:

- **Naive computations of Bayesian Networks.** This algorithm computes the probabilities for every node of the network. The computation is simple but very inefficient. (Bratko, 2000) presents an implementation of this algorithm for illustration of the principles.
- **Clustering algorithms for Bayesian Networks.** This approach uses special properties of Bayesian Networks (d-Separation) for dividing the network into smaller pieces (clusters). Each of the clusters may be separately computed. For each cluster it is decided if it is influenced by evident variables. The computation of probabilities is done only for these clusters. The approach is much more efficient than the naive approach.
- **Approximation of Bayesian Networks.** Algorithms of this concept estimate the probability of variables. Such algorithms may be used even in cases where clustering algorithms need too much time.

The naive algorithm has been implemented for the evaluation of the usability of Bayesian Networks for the HLD. Further evaluation has been done by using the SMILE reasoning engine for graphical probabilistic models contributed by the Decision Systems Laboratory of the University Pittsburgh (http://dsl.sis.pitt.edu).

### 3.2.5 Summary and the HLD Language Schema

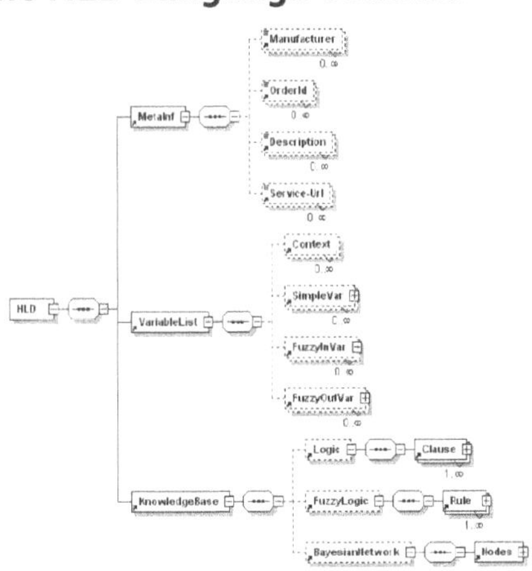


**Fig. 4. HLD schema overview.**

XML has been chosen as basic format of the HLD. Thus an XML schema according to W3C standards has been developed, which contains language constructs for the methodologies described in the previous sections. The structure of this schema is shown in figure 4.

The HLD schema contains following top level information:

- Meta Information. The element MetaInf contains various common information about the asset described by the HLD file. This includes for example the manufacturer name, an ordering number, a short description and a service URL for getting further information from the manufacturer.
- Variable Declarations. The element VariableList contains lists of variables. Propositional variables (with and without certainty factors) are separated from Fuzzy Logic input and output variables due to their different representation models.
- Knowledge Base. This element contains the following sub elements:
    - Logic: This element contains rules with and without the use of certainty factors.
    - Fuzzy Logic: This element contains fuzzy logic rules and it references the Fuzzy Logic input and output variables.
    - Bayesian Network: This element contains the definition of a Bayesian Network for discrete variables. It contains conditional probability tables and references to the declarations of propositional variables.

The other attributes and elements define the semantics as specified in the sections above. The full HLD scheme may be downloaded at "http://i2service.ifak.eu/wisa/".

## 4. FRAMEWORK FOR THE HANDLING OF THE KNOWLEDGE BASE

The central application of the HLD framework is the diagnosis system. It is implemented as a web application. This provides the possibilities to:

- Maintain the knowledge base on one place,
- Enable the access to the diagnosis system from any place,
- Reduce the necessity of installation of special software (a Web browser is expected to be installed on any modern operating system by default).

Figure 5 gives an overview of this application. On the left side the expert system provides a list of possible symptoms. The diagnosis user marks, which symptoms he has percepted. The diagnosis results are listed on the right side, sorted by their probability or their membership to a maintenance fuzzy variable.

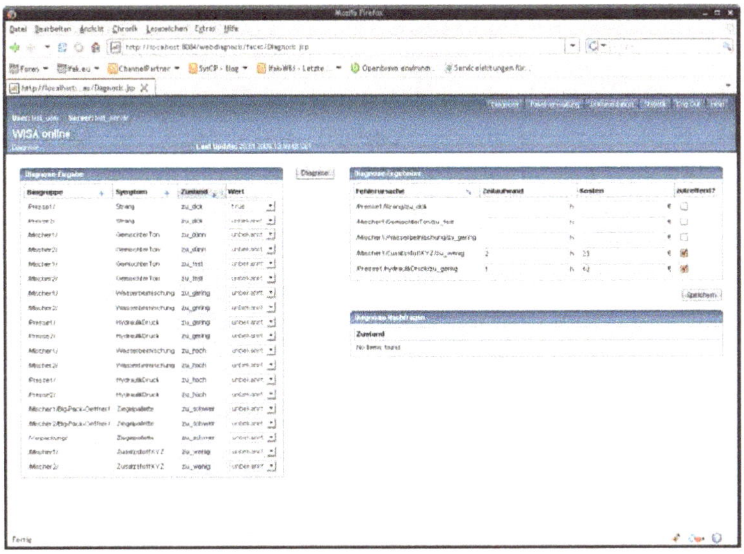


**Fig. 5. HLD interpreter as web application**

The expert has another application for the creation of the knowledge for a specific asset type. This is an editor for HLD files. A screenshot of a prototype application is shown in figure 6. On the left side there is a tree representing the asset hierarchy. Elements of this tree are set into relations by definition of rules. This is done by entering some input into the forms of the right side. The screenshot shows the forms for input of logic rules. The entry fields are labelled by using the maintenance terminology. Thus a transformation of the terminology of artificial intelligence terminology to the application domain is done by this user frontend for the asset experts. The HLD editor uses for example the term "failure cause" ('Schadensursache') instead of the term "conclusion" or "clause head".

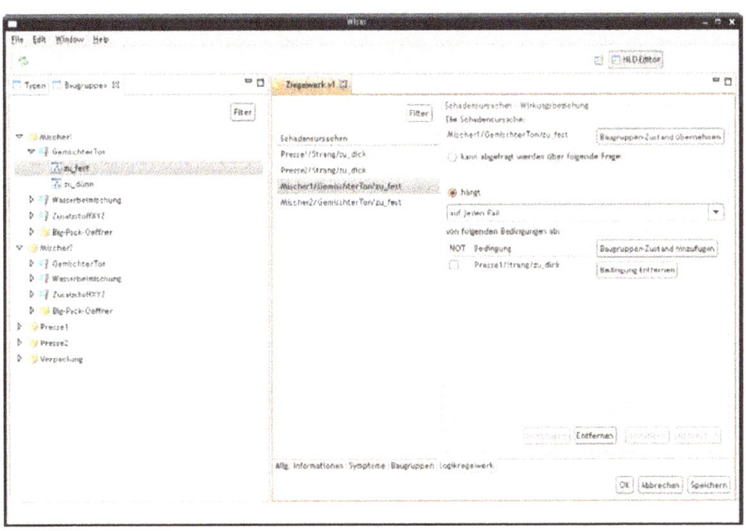


**Fig. 6. HLD editor.**

Assets like machine and plants are recursively nested when considering the aggregation relation. This is illustrated in fig. 7. If we consider a plant as asset, then the machines are the asset elements. If we further consider the machines as assets, then the tools, HMI elements and the control system are the asset elements. The HLD language introduces elements with the name "Context" in order to reference aggregated asset elements (see also fig. 4).

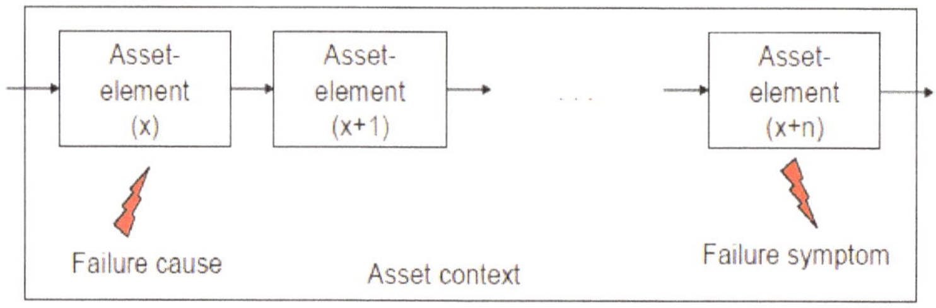


**Fig. 7. Failure cause and symptom relation within an asset**

Failures in one asset element often result in symptoms in another asset element. The relationships can be described in HLD files that are specifically dedicated to the higher asset context. These files contain the asset pieces that are directly or indirectly related.

The knowledge base of the diagnosis expert system is constructed by incorporating all high-level design (HLD) descriptions of the assets and their recursively nested aggregates. The files are arranged in a horizontal line next to each other in an HLD file repository. Each High-Level Design (HLD) description is specifically assigned to a certain type of asset and its corresponding version, which are represented by structural elements within the repository. Therefore, the repository does not have a free form. From figure 7, it is evident that an asset description is comprised of the asset context description and the descriptions of all asset elements. A High-Level Design (HLD) description consists of a collection of HLD files that have the same structure as an HLD repository. The tool set includes a packaging mechanism that gathers the required high-level design (HLD) descriptions from the asset expert's repository and compresses them. In addition, the tool set includes a package installation mechanism that unpacks the packages and installs them in an HLD repository, taking into account the asset type and version information. Furthermore, a documentation generating system has been established. This system creates HTML pages from a repository based on a specified asset context.

## 5. SUMMARY AND PROSPECTS FOR FUTURE RESEARCH

A hybrid knowledge base has been included into an expert system, which utilizes many paradigms from the field of artificial intelligence research. There existed a disparity between the achievement of the theoretical research and its recognition inside the industry. An essential issue is the level of effort required to create the knowledge base. This challenge can be addressed by implementing a collaborative approach, where makers of production equipment contribute to the establishment of the knowledge base.

Additional research efforts will be dedicated to developing and refining methods for the structure and parameter learning of Bayesian Networks. The results need to be included into the HLD editor. In addition, an online data gathering will be incorporated into the diagnosis system, which is particularly essential for a proficient implementation of the Fuzzy Logic thinking.

The majority of the work has been completed as a component of the WISA research project. The funding for this project has been provided by the German Ministry of Economy and Employment. The registration number is IW06215. The authors express their gratitude for the help provided by the German government.

## REFERENCES

1. Bratko, Ivan (2000). PROLOG - Programming for Artificial Intelligence, 3.Ed., Addison- Wesley
2. Bronstein, Semendjajew, Musiol, Mühlig (1997). Taschenbuch der Mathematik, 3. Ed., Verlag Harri Deutsch
3. Ertel, W. (2008). Grundkurs künstliche Intelligenz, 1. Ed., Vieweg Verlag
4. IEC61131-7 (1997). IEC 61131 - Programmable Logic Controllers, Part 7 - Fuzzy Control Programming, Committee Draft 1.0, International Electrotechnical Commission (IEC)
5. Janson, Alexander (1989). Expertensysteme und Turbo-Prolog, 1. Ed., Franzis Verlag GmbH München Jensen, Finn V. (2001). Bayesian networks and decision graphs, Springer Verlag
6. Kreuzer, M.; Kühling, S. (2006). Logik für Informatiker, 1. Ed, Pearson Education Deutschland GmbH
7. Norvig, Peter (1992). Paradigms of Artificial Intelligence Programming - Case Studies in Lisp, 1. Ed., Morgan Kaufman Publishers, Inc.
8. Pearl, J. (1988). Probabilistic Reasoning in Intelligent Systems, Morgan Kaufmann Publishers, Inc.
9. Poole, D.; Mackworth, A.; Goebel, R. (1998). Computational Intelligence - A Logical Approach, 1. Ed., Oxford University Press, Inc.
10. Russell, S. and Norvig, P. (2003). Artificial Intelligence - A Modern Approach , 2. Ed., Pearson Education, Inc.

# 18 AN IMAGE RETRIEVAL SYSTEM DESIGNED TO OPERATE IN A HETEROGENEOUS DATABASE

## 1. INTRODUCTION: AN OVERVIEW

This chapter aims to provide readers with theoretical and practical strategies for describing and recognizing images in huge databases across many applications. There is a growing demand for searching and browsing visual information due to the emergence of various computer vision system applications, including computer-human interaction, regulating processes such as autonomous vehicles and industrial robots. Moreover, as a result of the rapid advancement of internet technologies, multimedia archives are expanding at a fast pace. This is particularly true for digital image libraries, which are becoming increasingly significant repositories of knowledge. Therefore, it is prudent to create robust browsing computer systems capable of efficiently managing, organizing, categorizing, and identifying photos inside these extensive databases. An image retrieval system typically consists of several processes, with the most crucial ones being feature extraction and classification.

Image material is analyzed by feature extraction and description. Each feature captures specific aspects of the image, such as shape, color, or texture. However, no one feature can encompass all visual features. Hence, the utilization of diverse and varied characteristics to depict an image is deemed acceptable. This chapter presents and discusses the extraction of various features and the description of a heterogeneous image database. Testing and selecting appropriate classifiers for effective clustering and picture recognition: The selection of the utilized classifier in a CBIR system (Content Based Image Retrieval) is of utmost significance. This chapter introduces Support Vector Machines (SVMs) as a supervised classification technique that consists of two primary stages: training and generalization. Based on these two crucial factors, we want to determine the relevance of the extracted features in a broad and diverse database, as well as ensuring that the response to a query image is satisfactory. Given these circumstances, it is necessary to strike a balance between the accuracy of image identification and the amount of time the CBIR system can allocate for computation. This study aims to evaluate the performance of a heterogeneous image retrieval system by comparing statistical and hierarchical feature type models. The results are analyzed and discussed in connection to the pictures database, the chosen features, and classification approaches. This chapter highlights the significance and impact of the features relevance and categorization in image recognition and retrieval systems. In part 3, various methods for extracting features are provided and discussed. In section 4, alternative approaches for classification are also presented and explored. In sections 5, we present the principle and the outcomes gained from the optimization approaches.

## 2. RELATED WORKS

The image recognition system involves retrieving from a Database all the images that are comparable to the query images selected by the user, as shown in figure 1. This topic has garnered significant research attention in recent years. The main challenges were the ability to extract visual features from the image, the ability to handle geometric distortions, and the quantification of the similarity between images. Indexing and recognition are achieved through the application of classification methods to image features. In order to comprehend the impact of the picture database on the description method and the suitable classification tool, it is preferable to divide the image databases into two distinct types.

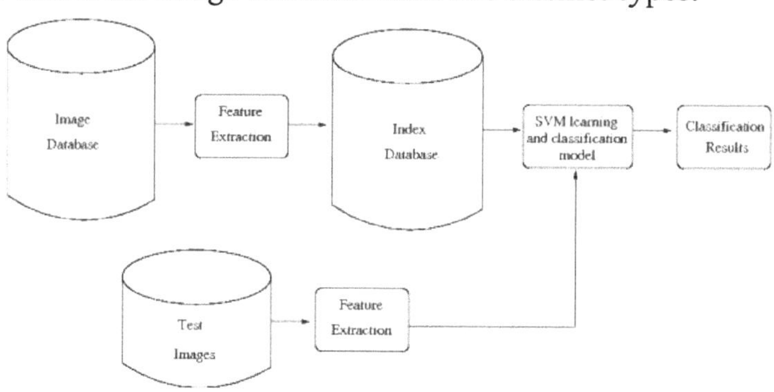


**Fig. 1. Content Based Image Recognition system.**

The initial category comprises picture databases that are typically heterogeneous in nature. The purpose of the image recognition system in this case is to aid users in conducting intelligent searches within an image library for specific subjects, while adapting to the individual subjective preferences of each user. The second category pertains to particular picture databases. In this context, the visuals often consist of a consistent semantic content. The applications in question are typically intended for professional use. In order to index these picture databases, the user needs to incorporate additional information provided by the expert to create a tailored algorithm. The goal is to enhance the efficiency of the system and its responsiveness, matching that of the expert. It is important to consider this classification while creating a content image retrieval system. In order to achieve satisfying outcomes, it is crucial to select or create description methods that are suitable for the specific type of picture database being evaluated. This is mostly attributed to the immense challenge of acquiring a comprehensive algorithm for universal description.

The content-based picture retrieval, as depicted in Figure 1, has garnered the interest of numerous experts across various fields and has been the subject of significant research efforts for some years. There has been a noticeable surge in interest in categorizing huge biological photographs based on their content. This is mostly due to the advancements in digital imaging technology and the developing field of computer vision. Image identification in biological databases is crucial for medical diagnosis. Various techniques have been devised and suggested to streamline the process of automatically categorizing and identifying image databases. The image retrieval system consists of two crucial steps: feature extraction and classification. In this part, we provide an overview of significant prior research that is relevant to these stages. We have a variety of feature options available, ranging from basic to advanced levels. These include shape, geometry, and symbolic features, among others. The primary objective of content-based picture retrieval and recognition is to overcome the disparity between the fundamental image characteristics. As a result, we can immediately get the things that people typically seek in image databases. Color histograms, such as those proposed by Stricker and Swain in 1994, and by Swain and Ballard in 1991, are frequently employed for image description and have demonstrated their utility. However, this method of global categorization does not provide information regarding the spatial distribution of colors. Various scholars have endeavored to address this constraint by integrating spatial information throughout the process of extracting characteristics. Stricker and Dimai (1997) store the mean color and the color covariance matrix in each of the five fuzzy image regions.

(Huang et al., 1997) store a color correlogram that represents the spatial correlation between pairs of color bins. In (Smith & Chang, 1996), the locations of colors that are present in significant quantities are stored in areas that are generated using histogram back-projection. (Lipson et al., 1997) extract images by analyzing spatial and photometric connections inside and between basic image sections. Minimal or negligible segmentation is performed; the regions are obtained from low-resolution images. The authors in (Jacobs et al., 1995) utilize multiresolution wavelet decompositions to conduct queries that rely on iconic matching. Some of these systems incorporate data on the spatial arrangement of color characteristics, while others carry out basic automated or manually-assisted segmentation. However, none of them offer the degree of automated segmentation and user control required to facilitate object queries in extremely large image databases. (Carson et al., 2002) view picture retrieval as an object recognition problem and approach it in three sequential steps.

Initially, pixels were classified into areas that typically relate to things or specific sections of objects. The user is provided with descriptions of these regions that are relevant and significant. The suggested system enables the retrieval of required photos by providing access to region descriptions, either automatically or with human participation. This approach fails to capture all the crucial information in images through its features, resulting in image retrieval without categorization. This might lead to recognition issues when dealing with huge image databases. The study conducted by Antani et al. (2002) provides a thorough examination of the application of pattern recognition techniques for content-based picture and video retrieval, with a specific focus on classification approaches. Research endeavors have resulted in the creation of techniques that enable the retrieval of picture and video data. These techniques are derived from pattern recognition. The methods are employed to ascertain the similarity in the visual information content derived from several low-level features.

In (Lin et al., 2006), a method is proposed that combines Support Vector Machines (SVM) with hybrid Bayesian networks (HBN) to provide a multi-level semantic modeling approach. SVM categorizes the continuous variables of medical image data by assigning them to discrete states, which represent middle-level semantics. The HBN can serve as the basis for designing a semantic model for medical picture retrieval that incorporates multi-level semantics. In order to verify the effectiveness of their approach, a model is constructed to automatically annotate images at the content level using a limited number of astrocytoma MRI (magnetic resonance imaging) samples. Multi-level annotation is a viable technique for enabling image retrieval at several semantic levels. The experiment findings demonstrate that this approach is highly efficient in facilitating the multi-level interpretation of astrocytoma MRI scans. This study presents an innovative approach to connecting the disparity between the abstract meaning and the specific visual characteristics of an image.

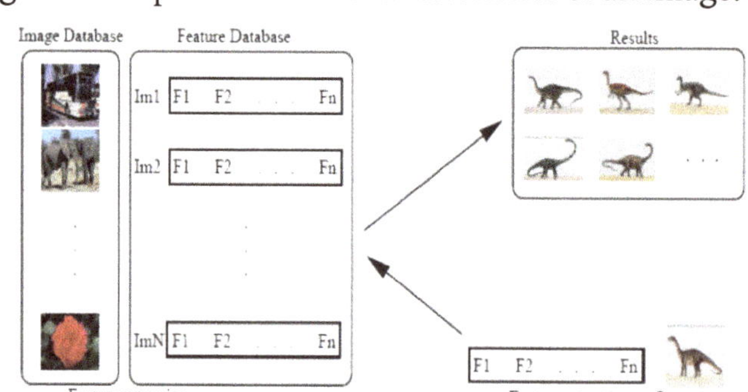


**Fig. 2. Features extraction and image recognition: form a query image, a relevant features extraction allows to obtain all similar images in database.**

Several indexation and recognition methods have been created using image content description and categorization to facilitate image identification in extensive datasets (figure 2). These systems utilize basic characteristics such as the histograms of colors and orientations, as well as Fourier and wavelet transformations. Although the findings are adequate, the classification method based on computing similarity distances is not sufficiently robust to handle large dimensions of the extracted feature vectors. In order to address this issue, other techniques suggest computing a feature vector for each pixel in every image. This vector comprises the elements pertaining to the color, texture, and position descriptors, so providing a more comprehensive depiction of the image. However, there is still room for improvement in the quality of the image description performances. Furthermore, numerous studies utilized wavelet technologies as their foundation. In their study, Serrano et al. (2004) improved their image representation by using texture data obtained by wavelet transform. A novel method for extracting rotation invariants is proposed in the study conducted by Sastry et al. in 2004. This strategy yields satisfactory outcomes by considering the rotational characteristics. In order to enhance the accuracy and comprehensiveness of images, a novel technique known as the Trace transform is introduced in the study conducted by Kadyrov and Petrou in 2001. This transformation simultaneously provides a comprehensive visual description while remaining unaffected by rotation, translation, and scaling. Following the extraction of features, classification is next performed using a classifier, such as the KNN classifier. However, this classifier's performance is hindered by its inability to handle large feature vector dimensions, resulting in delayed processing. A more efficient technique utilizing a Bayesian approach (Sclaroff & Pentland, 1995), involving the merging of feature blocks, yielded superior outcomes. Artificial training approaches, such as artificial neural networks (ANN) and evolutionary algorithms, were also employed in the domain of picture categorization. Their widespread use is attributed to the capabilities they provide in terms of computer time and their performance in classification. Undoubtedly, there are several works that are founded on the Artificial Neural Network (ANN). The results obtained demonstrate the high efficiency of the Artificial Neural Network (ANN) due to its fast processing capabilities and easy implementation, as mentioned by Takashi & Masafumi (2000) and Egmont-Petersen et al. (2002). The authors in (Djouak et al., 2007) propose a fusion of feature vectors. The vectors are derived by a visual search process that relies on color histograms, geometric characteristics, and texture elements. The wavelet transform is used to provide rotation invariants and texture information. Furthermore, the Trace transform is implemented to achieve greater invariance. The RBF classification technique achieves a favorable balance between efficacy and computational simplicity. Every content heterogeneous image recognition system consists of two stages: feature extraction and classification (figure 1). Feature extraction is the crucial stage, as the quality of image classification depends largely on the accuracy of image description.

**Fig. 3. Images database sample.**

Typically, the selection of a classification algorithm is determined by the content of the data sets being analyzed. Furthermore, while dealing with heterogeneous image databases, it is advisable to do appropriate feature selection. Section 3 focuses on the process of feature extraction and selection. We assessed our modular statistical optimization and hierarchical type features model, as described in section 5, using a widely recognized heterogeneous pictures database. The database was selected from the Wang image database, which is accessible on the following website: http://wang.ist.psu.edu/docs/home.shtml/

The image collection comprises 1000 photographs of several genres, exhibiting significant variations in colors, forms, and textures. Several examples are depicted in figure 3.

This chapter focuses on the two crucial stages of an image recognition and retrieval system, which are presented in sections 3 and 4. Section 5 introduces two CBIR systems: one based on modular statistical approaches and the other on hierarchical best features type selection.

## 3. FEATURES EXTRACTION

This section provides a comprehensive explanation of the feature extraction and evaluation methods used for recognizing a picture database with diverse characteristics. The primary goal of feature extraction is to identify a compact and easily compared representation (signature) for each image. This representation should be readily available and provide a thorough characterization of the image.

The most frequently utilized qualities mostly encompass the fundamental attributes present in an image, including color, texture, and/or shape (Bimbo 2001). Color features are the primary components utilized in CBIR systems, and they continue to be the most widely employed because to their straightforward extraction process, comprehensive depiction, and efficient recognition capabilities. The utilization of average color (Faloutsos et al. 1994) and color histograms (Hafner et al. 1995) is highly advantageous due to their insensitivity to variations un position and size. The correlogram feature was introduced by Huang et al. in 1997 as an enhancement to the color histogram. The matrix represents the spatial relationship between colors in a picture based on distance and orientation. The auto-correlogram, a derivative of the correlogram, was introduced by Huang et al. in 1999. It specifically measures the spatial auto-correlation of identical color intensities inside a picture. (Bimbo 2001) conducted a comprehensive analysis of several color indexing techniques. In addition, a group of color characteristics was examined to determine their suitability for incorporation into the established MPEG-7 standard (Manjunath et al., 2001). Texture is being utilized more frequently in image indexing due to its ability to address specific challenges that arise from color indexing, particularly when the color distributions are extremely similar. There are two main categories of existing texture descriptors: The first one is deterministic and involves the spatial repetition of a fundamental pattern in various orientations. This structural approach aligns with large-scale patterns. First order statistics, as described by Press in 1987, is a basic and straightforward method for computing features of this nature. The data is obtained from the histogram of the normalized image. The co-occurrence matrix, introduced by Haralick and Dinstein in 1973, is utilized for texture analysis in the spatial domain. This matrix enables the calculation of pixel numbers in the image that have the same gray level, are separated by a specific distance, and are positioned according to a particular direction. Thirteen texture features are calculated from this matrix, as described by Haralick and Dinstein in 1973. The second category of texture is probabilistic, aiming to describe the chaotic element that lacks identifiable patterns or dominant repetition frequencies. This strategy is referred to as microscopic or multi-resolution. The paper by Serrano et al. (2004) provides examples of texture features obtained from the wavelet treatment.

In addition to color features, a set of texture features was also examined for inclusion in the standard MPEG-7 (Manjunath et al., 2001). Shape attributes, in addition to color and texture, provide a means of characterizing the various objects within a picture. Typically, this type of feature signifies the outline of an item. Therefore, image preprocessing, namely segmentation, is often required. Additionally, ensuring accurate segmentation is particularly crucial for accurately characterizing shape features. Image segmentation is the process of dividing an image into smaller parts that are of interest and can be used for analysis or further manipulation. Once the region segmentation is completed, two types of form features can be recovered, as described by Chen et al. in 2004 and 2006. The first method relies on the geometric properties of image regions. The second method relies on analyzing the statistical distribution of pixel grey levels in various sections of the image. One of the most well-liked characteristics of this kind is the use of invariant moments of HU (Hu 1962).

These moments remain unchanged under many geometric transformations, such as translation, rotation, and scale variation. Due to the intricate and diverse nature of the image contents, it is necessary to have a comprehensive description that encompasses many visual attributes (Manjunath et al., 2001). In this chapter, we utilize a collection of features that fall under three distinct visual categories: The color features we utilize are the four average color feature vectors (Faloutsos et al. 1994) computed in various color spaces, histograms, and correlogram (Huang et al. 1997), (Huang et al. 1999). The texture features utilized in this study include first order statistics (Press, 1987), cooccurrence matrix (Haralick & Dinstein, 1973), and Daubechies wavelet transformation. The shape features included in this study include the gradient norm (Delingette & Montagnat, 2001) and the invariant moments of Hu (Hu, 1962). Figures 4, 5, and 6 display a selection of extracted feature samples.

## 3.1 Used and improved features

In this area, we present the various aspects that have been utilized and enhanced. The discussion revolves around three distinct sorts of features: color, textures, and shape aspects. The enhancement is achieved by the utilization of gradient norm projection and Hu moments, which are derived from the gradient norm image after eliminating local maxima.

### 3.1.1 Color features

Fist we compute the average color in the RGB, HSV, HMMD and YCbCr spaces.

$$R_{avg} = \frac{1}{N}\sum_{p=1}^{N} R(p), \quad G_{avg} = \frac{1}{N}\sum_{p=1}^{N} G(p), \quad B_{avg} = \frac{1}{N}\sum_{p=1}^{N} B(p)$$

Where T

$X=(R_{avg,} G_{avg,} B_{avg})T$ and N the image pixel number. The second color

$$histR(x) = \frac{1}{N}\sum_{p=1}^{N} R(x_p = x) \quad histG(x) = \frac{1}{N}\sum_{p=1}^{N} G(x_p = x)$$

$$histB(x) = \frac{1}{N}\sum_{p=1}^{N} B(x_p = x)$$

With $x \in [0,1,2,.......,255]$.

We use the color autocorrelogram, defined in (Huang et al. 1999), captures the spatial correlation between identical colors only, it is a subset of the correlogram (Huang et al. 1997) which expresses how the spatial correlation of pairs of colors changes with distance.

### 3.1.2 Textures features

As texture features we use the First Order Statistics, the Daubeshies Wavelett transform and the Spatial Gray Level dependence matrix.

The first order statistics are considered by the mean, the standard deviation and the coefficient of variation.

$$Mean = \frac{1}{255}\sum_{x=0}^{255} x.hist(x), \quad StandardDeviation = \frac{1}{255}\sum_{x=0}^{255} x - Mean^2.hist(x)$$

$$CoefficientOfVariation = \frac{StandardDeviation}{Mean}$$

The Daubeshies Wavelett transform in the fourier space is given by the following equations:

$$\hat{\varphi}(\omega) = \Pi_{k=1}^{\infty} \frac{h(2^{-k}\omega)}{\sqrt{2}}$$

$$\hat{\psi}(\omega) = \frac{1}{\sqrt{2}} g\left(\frac{\omega}{2}\right)\hat{\varphi}\left(\frac{\omega}{2}\right) \qquad h(\omega) = \sum_{k\in Z} h_k e^{-ik\omega} \qquad g(\omega) = \sum_{k\in Z} g_k e^{-ik\omega}$$

And the decomposition is given as follow:

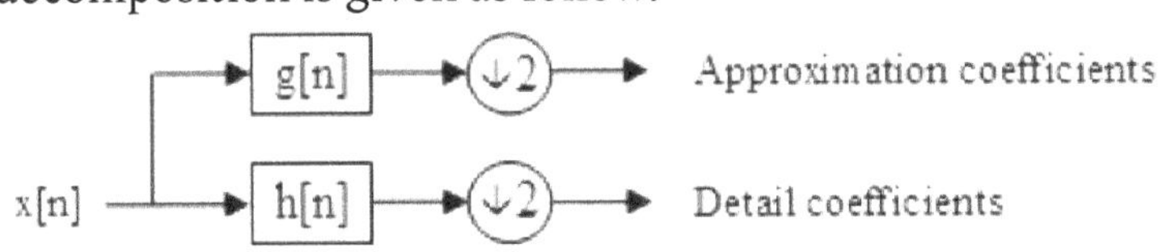


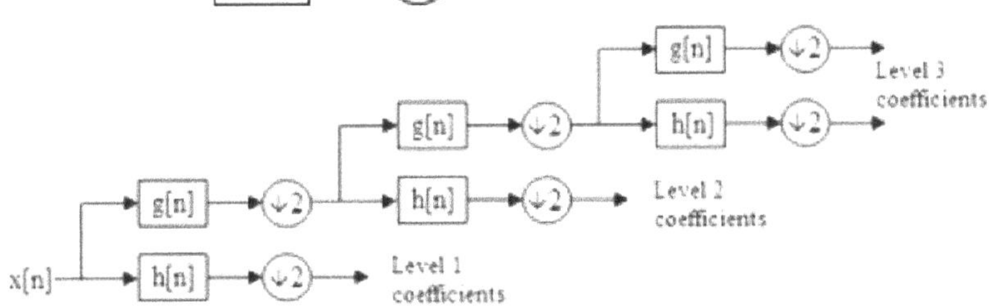


The Wavelet coefficients are *clij(x,y)*, where *l* is the decomposition level.

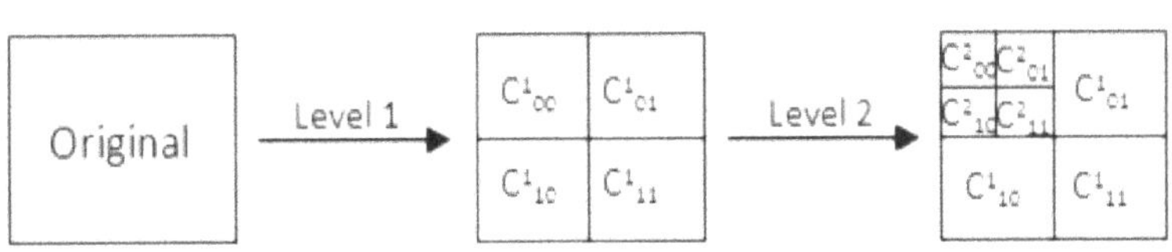

The lowest frequency coefficients *c200(x,y)* are not inherently useful for texture analysis. Therefore, a direction-independent measure of the high-frequency signal information is obtained by filtering the raw coefficients *c200(x,y)* with the Laplacian. The texture features are obtained by computing the subband energy of all wavelet coefficients (including the Laplacian filtered *c200(x,y)* coefficients) :

$$c_{ij}^{l} = \frac{1}{MN}\sum_{m=1}^{M}\sum_{n=1}^{N}\left|C_{ij}^{l}(mn)\right|^{2}$$

For the Spatial Gray Level dependence matrix (SGLD Matrix), the co-occurrence matrix (Julezs, 1975), called also the spatial gray level dependence (SGLD) matrices, counts how often pairs of gray level of pixels, separated by a certain distance and lying along certain direction, occur in an image. From these SGLD matrices, thirteen features related to the image texture, could be calculated, (Haralick & Destein, 1973).

### 3.1.3 Shape features

The invariant moments of Hu and the Gradient norm projection are used as shape features. The Hu invariants (Hu, 1962) are obtained from quotients or powers of moments. A moment is a sum over all pixels of the image model by weighted polynomials related to the positions of pixels. Let *I(x,y)* the grayscale of a pixel in the image *I*, we define the moment of order *(p+q)(p,q>0)* of an image *I* by:

$$m_{p,q} = \int_{R^2} x^p y^q I(x,y)\,dxdy$$

Let *(x0,y0)* the centroid of the function *I* and the centered image *IT(x,y)=I(x+x0,y+y0)* is invariant by translation. The central moment of order *(p+q)* of the function *I* is written as follows:

$$v_{p,q} = \int_{R^2} x^p y^q I(x+x_0, y+y_0)\,dxdy$$

The central moments are invariant by translation. We introduce the normalized moments as follows:

$$\mu_{p,q} = \frac{v_{p,q}}{v_{0,0}^{(1+(p+q)/2)}}$$

These moments are invariant to translation and scale changes. Hu moments are calculated from normalized moments and are invariant by translation, rotation and change of scale:

$$\phi_1 = \mu_{2,0} + \mu_{0,2}$$
$$\phi_2 = (\mu_{2,0} - \mu_{0,2})^2 + 4\mu_{1,1}^2$$
$$\phi_3 = (\mu_{3,0} - 3\mu_{1,2}^2 + (3\mu_{2,1} - \mu_{0,3})^2$$
$$\phi_4 = (\mu_{3,0} + \mu_{1,2})^2 + (\mu_{2,1} + \mu_{0,3})^2$$
$$\phi_5 = (\mu_{3,0} - 3\mu_{1,2})(\mu_{3,0} + \mu_{1,2})[(\mu_{3,0} + \mu_{1,2})^2 - 3(\mu_{2,1} + \mu_{0,3})^2]$$
$$+ (3\mu_{2,1} - \mu_{0,3})(\mu_{2,1} + \mu_{0,3})[3(\mu_{3,0} + \mu_{1,2})^2 - (\mu_{2,1} + \mu_{0,3})^2]$$
$$\phi_6 = (\mu_{2,0} - \mu_{0,2})[(\mu_{3,0} + \mu_{1,2})^2 - (\mu_{2,1} + \mu_{0,3})^2] + 4\mu_{1,1}(\mu_{3,0} + \mu_{1,2})(\mu_{2,1} + \mu_{0,3})$$
$$\phi_7 = (3\mu_{2,1} - \mu_{0,3})(\mu_{3,0} + \mu_{1,2})[(\mu_{3,0} + \mu_{1,2})^2 - 3(\mu_{2,1} + \mu_{0,3})^2]$$
$$- (\mu_{3,0} - 3\mu_{1,2})(\mu_{2,1} + \mu_{0,3})[3(\mu_{3,0} + \mu_{1,2})^2 - (\mu_{2,1} + \mu_{0,3})^2]$$

The Gradient Norm projection feature is obtained after some processing steps. First, we apply Sobel masks to obtain the directional gradients according to x and y:

$$G_x(i,j) = h_x(i,j) \otimes I(i,j) \qquad G_y(i,j) = h_y(i,j) \otimes I(i,j)$$

where *I(i, j)* is the image gray level information and *hx(i, j), hy(i, j)* are Sobel masks:

$$h_x(i,j) = \begin{bmatrix} -1 & -2 & -1 \\ 0 & 0 & 0 \\ 1 & 2 & 1 \end{bmatrix} \qquad h_y(i,j) = \begin{bmatrix} -1 & 0 & -1 \\ -2 & 0 & -2 \\ -1 & 0 & -1 \end{bmatrix}$$

then, gradient norm is computed as follow:

$$G(i,j) = \sqrt{G_x(i,j)^2 + G_y(i,j)^2}$$

As the coefficient number in the gradient norm is the same as the pixel number in the image, we compute the gradient norm projection according to x and y, in order to reduce this feature size:

$$P_{Xi} = \frac{1}{max\, G_{i,j}} \sum_j G(i,j) \qquad P_{Yi} = \frac{1}{max\, G_{i,j}} \sum_j G(i,j)$$

## 3.2 Examples of description results and discussion

From the presented variety feature vector extraction, an improvement of certain vectors (Kachouri et al., 2008b) is ensured, in order to reduce their size while reserving their relevance description and even improve it.

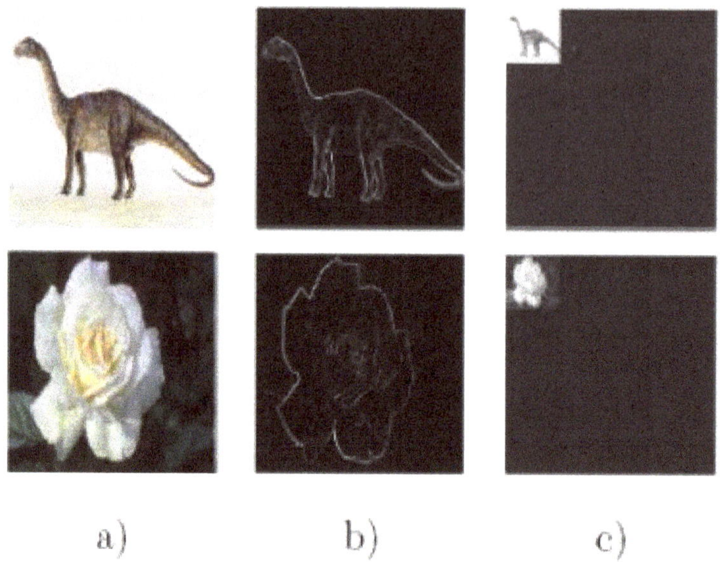


**Fig. 4. Extracted feature samples: a) Dinosaur and Flower images , b) Dinosaur and Flower image respectively gradient norm and c) Dinosaur and Flower image respectively two level decomposition Daubechies wavelet coefficients.**

For this, we employ the auto-correlogram, computed from correlogram feature (Huang et al. 1999), we retain only the eight most commonly used features since thirteen extracted from image co-occurrence matrix. Also a condensation to compute single value resulting from different level decomposition detail coefficients of the Daubeshies wavelet transformation and the Laplacian approximation filtering, is made, figure 4.

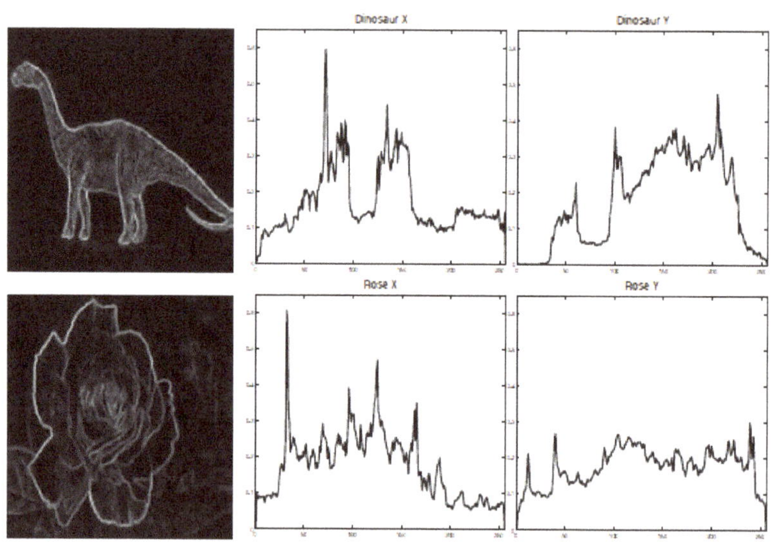


**Fig. 5. Extracted features samples: Gradient Norm projection (/x and /y), obtained from Dinosaur and Rose images.**

Furthermore, we replace the original gradient norm matrix with a gradient norm projection along the x and y axes. This approach effectively reduces the dimension of this feature while maintaining its descriptive quality. Figure 5 illustrates the discriminatory capacity of this novel characteristic in distinguishing between images belonging to distinct classes.

Ultimately, we select only the initial four unchanging Hu moments from the seven that were recovered. Furthermore, these instances are calculated based on the magnitude of the gradient rather than the original pixel values. Considering the contour of the image is more important than other information, as it further increases the description of the shape of this feature (figure 6).

These properties will facilitate the construction of models of objects for the purpose of recognition and indexing in a database. In order to effectively do this work, we require a classification system that is capable of categorizing all objects and images in a database based on their description. In the following chapter, we observe various average color vectors represented by the acronyms RGB, HSV, HMMD, and YCrCb. We also refer to histograms as Hist, auto-correlogram as Auto-Corr, first order statics as FOS, co-occurrence matrix as SGLD, Daubeshies wavelet transformation as Daub, gradient norm as G-norm, and the invariant moments of Hu as Hu.

**Fig. 6. Contour extraction and Histograms examples: From real images, we compute first the gradient norm, the contours extraction, the histograms and the Hu moments.**

## 4. CLASSIFICATION

We call automatic classification, the algorithmic categorization of the objects and images. This one is to determine in which class or category you can “store” each object, based on similarities criteria. Many classification or learning methods have been already developed, and used in a wide spectrum of applications: diagnosis, bioinformatics, fraud detection, speech recognition, etc.

Once training image feature set is built, a classifier is used to ensure image recognition. There are two classification kinds in the literature. The first is unsupervised classification where the classifier deals directly feature dataset without any prior knowledge (Kachouri et al. 2008c). Classification here is based, generally, on clustering approach. So, CBIR systems allow image access according to their visual characteristics by means of similarity measures. The smaller the similarity distance is, the closer the two images are. The most used distance in literature is the Euclidian one which is really a particular case of the Minkowski metric (Kachouri et al. 2008c), various similarity measures are discussed in (Ramesh et al. 1995). The second type is supervised classification (Kachouri et al. 2008a), where the classifier assumes that there is already image classification, a training stage is required, it is the case of a library or a search engine. Since the Nineties, Support vector machines (SVMs) did not cease arousing the interest of several researcher communities of various expertise fields. Nowadays, SVM performance, in several applications (Bi et al. 2003), (Chen et al. 2006), exceeds those of already established traditional models. In fact, SVM originally formulated for twoclass classification problems, have been successfully extended for multiclass classification problems, and become in a very short period of time the standard state of-the- art tool. Also in the SVM-multiclass, original input space is mapped into a higher dimensional feature space in which an optimal separating hyper-plane is constructed on the basis of Structural Risk Minimization (SRM) to maximize the margin between the different clusters, i.e., generalization ability.

## 4. 1 Support Vector Machines (SVM)

This method of supervised classification was published by (Vapnik, 1998). An SVM is a supervised learning algorithm allows to learning a separator. We have a finite set of separated vectors of *Rn*, as part of a binary classification into two groups, or two classes. Learning to a group is defined by an associated label to each vector, which is inscribed group 1 or group 2. Find a separator is equivalent to reconstruct a function that takes a vector from a set of samples and can say what group it belongs. However, it expects the SVM good generalization properties, ie if a new vector is introduced and it was not in the set of database, the SVM will say which group it is probably belongs.

The linear separator is the lower part of the SVM, that is a relatively simple problem, and that allows later on to the SVM providing some well more powerful separators. We saw that we started from a finite set of labelled vectors. We say that all the labelled vectors that we give are the set of samples noted *S*, which contains *p* elements.

$$S = (\bar{x}_l, y_l)_{1 \le l \le p} \, with \, \forall l, y_l \in \{-1,1\}$$

The scalar product of two vectors is noted where is $\langle x, y \rangle$ , we can then define the linear separator

$$f_{\vec{w}.b}(\vec{x}) = \langle \vec{w}.\vec{x} \rangle + b$$

This separato r does not provide valid values only -1 or 1, but we assume that the result when $f_{\vec{w}.b}(\vec{x})$ is positive, then the vector is of the same class as samples of label 1, and the result when $f_{\vec{w}.b}(\vec{x})$ is negative, the vector belongs to the class of examples labelled -1. The equation $f_{\vec{w}.b}(\vec{x}) = 0$ defines the border between the two classes. This boundary is an affine hyper-plane in the case of linear separator.

### 4.4.1 Separability and margin

We suppose that S is linearly separable. The basic idea of SVM is to separate samples of each class, but it is also necessary that the hyper plane passes as possible between the two classes. It is to define this notion that we introduce the margin. The margin is the minimum distances of each sample to the hyper plane (figure 7).

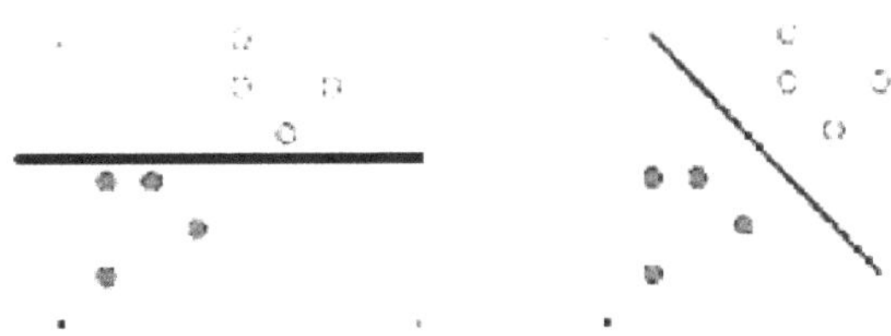

**Fig. 7. A separator that the margin is not maximized, have less good generalization properties than that which the margin is maximized.**

The aim of the SVM, in case where S is separable, is to give separator S whose margin is maximal, while ensuring that properly separates the samples with label -1 and the samples with label 1.

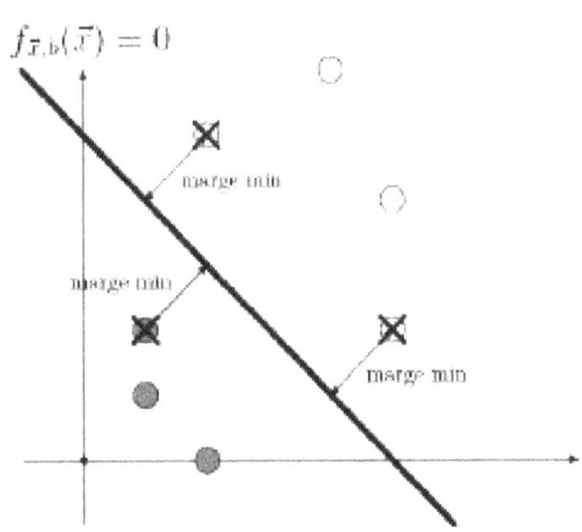


**Fig. 8. The separator which should maximize the margin.**

The maximum margin separator is such that, the smaller sample has a margin wider than the sample of the smallest margin of the other possible separators. In fact, there are really at least two samples of smaller margin, a class 1 and class -1. They force this margin, and the border of separation passes between them (figure 8). These are the only samples that force the margin, and remove all other samples of the database does not change the separator. These samples are called *support vectors*, hence the name of the method.

### 4.1.2 General case

For the general case where S is not separable, the solution is to allow some samples to have a lower margin than the margin chosen as the smallest margin or even negative. However, the solution of the problem may be very bad if too many samples are allowed to have a small margin. The idea is to add value margins lower than the maximum margin in the expression to minimize. This avoids that the margins are too low, which limits the samples that do not respect the separability through a separator solution of optimization problem. This is a problem of quadratic convex optimization, i.e. an optimization problem that admits no local optimum, but only one optimum, thus overall. This is crucial because the convexity of the problem is a guarantee of convergence to the SVM solution.

The interest of the kernel functions is that they allow using what we just presented on the linear separation to the non-linear separations. Let *S* a set of samples labelled by 1 or -1 depending on the class to which they belong, which is not at all linearly separable. The method we have seen works in this case but may give poor results, and many samples became support vectors. The idea of using kernels comes from the assumption that if a set is not linearly separable in the descriptors space, it can be in a space of higher dimension. A better way to separate the samples is to project them into a different space, and perform a linear separation in this space, where this time it should be more adapted. The kernel functions can achieve this projection, and must check a number of properties to ensure the effectiveness of this technique, so you do not have to make calculations in very large dimensions. With the kernel functions, we can work in very large dimensions. However, a linear separation, and a linear regression is facilitated by the projection of data in a space of high dimension. Projecting in the space of descriptors and using an algorithm to maximize the margin, SVM managed to get a severability retaining good generalization capacity, is the central idea of SVM.

For more details on SVMs, we refer interested readers to (Cristianini & Taylor, 2000). A comparison between SVM-multiclass, as supervised classification and Euclidian distance based k-means, as unsupervised classification, is presented in (Kachouri et al., 2008b). The obtained results prove that SVM classifier outperforms the use of similarity measures, chiefly to classify heterogeneous image database. Therefore, we integrate SVM classifier in our proposed image retrieval systems in this chapter.

## 5. SELECTION OF APPROPRIATE FEATURES ENABLES THE ACHIEVEMENT OF IMAGE RECOGNITION AND RETRIEVAL RESULTS.

In order to assure optimal feature selection for image retrieval, we describe and analyze the efficacy of various types of features and their aggregation methods. Due to the presence of diverse photos, a heterogeneous image database exhibits significant variations in content. The decision to implement a system optimization tool became necessary when it was discovered during the conducted testing that the utilization of all extracted features could be burdensome to handle. Indeed, as the number of feature vector dimensions increases, the classifier encounters greater difficulty in classifying them. The conventional approach described in (Djouak et al., 2005a) and commonly employed in numerous CBIR systems involves utilizing all extracted features during the classification phase. Regrettably, this strategy has a significant drawback as it utilizes all the characteristics, resulting in a high number of dimensions managed by the classifier. This results in a significant computational time, which poses a substantial challenge for large image databases. Indeed, this issue, which arises directly from the problem of high dimensionality, has been extensively studied and several approaches have been developed to address it.

Content extraction is the fundamental process in CBIR. Modern Content-Based picture Retrieval (CBIR) systems employ visual features as the basis for picture retrieval. Since we utilize a diverse image library, the photographs belong to different categories and exhibit significant variations in their visual qualities. A singular or distinctive characteristic, whether it be a unique feature or a unique type of feature, is not sufficient to accurately characterize the entirety of the picture database.

Furthermore, although SVM is a robust classifier, it faces limitations when dealing with heterogeneous images due to their complex content. One limitation is the presence of redundant or irrelevant features, which may not contribute to the observed image classification or may be similar to each other. Moreover, the presence of numerous irrelevant features in the index dataset can negatively impact the generalization performance. Therefore, it is crucial and necessary to choose a subset of features that are highly relevant to the problem of interest categorization. Therefore, a new problem arises, which is not related to describing images, but rather focuses on selecting significant features. In order to provide optimal classification performance, a high-quality content image recognition system must primarily possess the capability to identify the most significant collection of features and effectively discretize the corresponding spaces. The topic of feature selection for classification purposes has been extensively researched, as seen by the work of Blum and Langley in 1997. More recent studies have focused on feature selection specifically for Support Vector Machines (SVMs). There is a significant amount of research available on proposed algorithms in this area, which will continue to be relevant for many years (Guyon & Elisseeff 2003). There are two primary branches of selection methods: wrappers and filters (John et al., 1994; Yu & Liu, 2004). Filters are highly efficient and depend on theoretical principles, enabling a more comprehensive knowledge of variable interactions in most cases. Linear filters such as PCA (Principal Component Analysis) or FLD (Fisher's Linear Discriminant) are commonly employed. However, these methods are effective only when there is an initial redundancy in the data. (Daphne & Mehran 1996) present Markov blanket techniques, which can identify a set of variables that do not include a specific variable xi but yet make xi unnecessary. Once a Markov blanket is identified, xi may be confidently removed. However, it is important to note that this is merely a summary approximation, as the practical implementation of this idea is not feasible. However, as it does not consider the classifier used in the generalization stage, all approaches for selecting filters for kind classification are often unable to ensure a high recognition rate. Wrappers, which were developed by (John et al. 1994), are a relatively recent idea that is conceptually simpler than filters. This type of selection utilizes the classifier as an essential component of the selection process. The notion of feature subset selection is based on its efficacy in classifying test images. Hence, the chosen subset of features is highly compatible with the classification algorithm, resulting in elevated recognition rates due to the consideration of the inherent bias of the algorithm. Several relevant studies on feature selection using SVM classifier have been documented in the literature, including works by Guyon et al. (2002), Zhu & Hastie (2003), Bi et al. (2003), and Chen et al. (2006). The primary drawback of this selection strategy is the requirement for costly computation, particularly as the number of variables increases. Additional information can be found in the work of Guyon and Elisseeff (2003) and the references cited therein. In order to optimize the benefits of both selection types, namely filters for speed and wrappers for adaptability of the selected feature subset with the classifier being used, new selection methods have been developed to ensure a constant search for a compromise. In a recent study by Bi et al. (2003), they suggested using 1-norm SVM as a linear classifier for feature selection to avoid processing costs. After feature selection, a non-linear SVM is employed for generalization. Additional techniques that integrate filters and wrappers are discussed in the publication by Guyon and Elisseeff (2003). In this section, we provide the modular statistical optimization approach (section 5.1) and the best features type selection method (section 5.2) inside the given framework.

## 5.1 Modular statistical optimization

The modular statistical architecture depicted in figure 9 is constructed around a feedback loop technique. The main concept (Djouak et al., 2006) behind this design is to categorize features into many blocks or modules instead of using all of them in the classification stage. The goal is to get the highest possible precision with the fewest number of blocks. The modular features database contains all the features that were given in section 3.

By utilizing all of these features, four distinct feature modules were created. These modules can be described as follows: The initial module (b1) collects all the characteristics related to shape, the second module (b2) collects the characteristics related to color, the third module (b3) collects the characteristics related to texture, and lastly the fourth module (b4) collects the characteristics related to Daubeshies.

**Table. 1. Used features blocks table**

| Features Blocs | B1 | B2 | B3 | B4 | B5 | B6 |
|---|---|---|---|---|---|---|
| Concerned Modules | b1 | b1 b2 | b3 | b1 b2 b3 | b4 | b1 b2 b3 b4 |

The table (table.1) summarizes the obtained features blocks (B1 to B6) by combining the exposed features modules (b1 to b4).

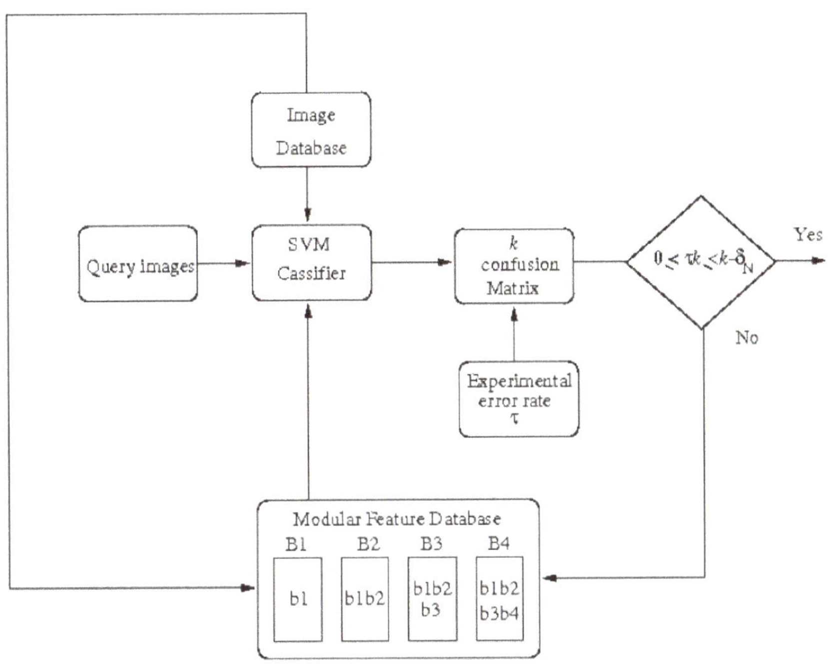


**Fig. 9. Modular Statistical optimization architecture.**

We can remark in figure 10, for the request (query) image number 4 that the classification rate error is very important for bloc B1. However, the rate error decrease progressively when the others features bloc are used. The presented modular architecture presents some disadvantages, the query images must be included in the database, the experimental rate error is used as prior information. To solve this problem, we propose in the next section the classification procedure based on hierarchical method using the best feature type selection.

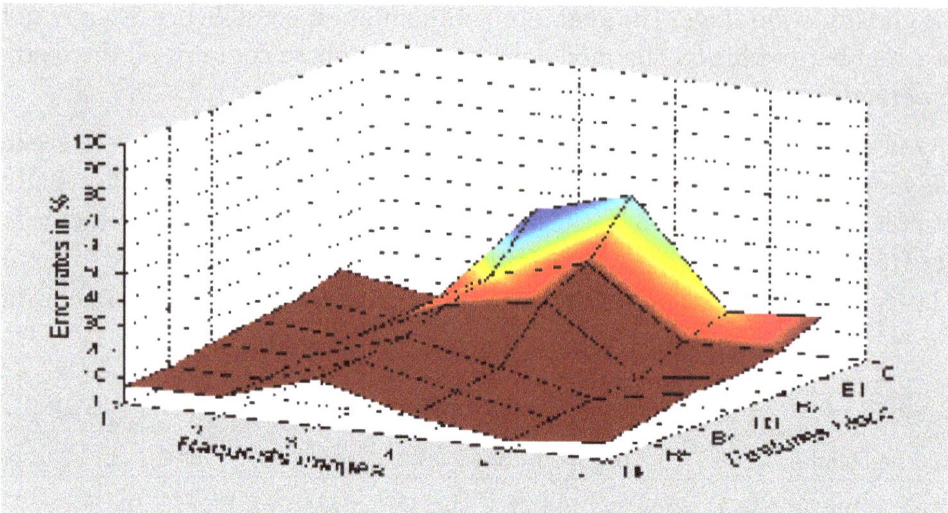

**Fig. 10. Average of the classification rate error obtained for different feature blocs.**

## 5.2 Best feature type selection method

The hierarchical feature model is suggested as a substitute for the traditional use of combined features (Djouak et al., 2005a), (Djouak et al., 2005b). This approach has the capability to automatically choose and organize features based on their types and the contents of the image database. During the offline stage, the feature extraction phase allows us to retrieve a corresponding feature dataset from an image database. Firstly, we begin by conducting a training step where we analyze each group of the same type of features individually. We next use a classifier to compute a training rate criterion, which helps us determine the best group of the same feature kind in a hierarchical manner. During the online stage, we individually categorize each image from the test image database using distinct feature types. Each image will yield distinct clusters as a retrieval outcome. Next In order to make a decision among these different outputs, we group together outputs with the same type of features and compare them using the hierarchical feature selection model, as outlined in figure 11. We initiate the procedure using the two most recent outputs with identical features, continuing until we reach the optimal outcome. After analyzing two groups with similar characteristics, a comparison block determines whether to utilize the Nearest Cluster Center (NCC) block or not. The NCC block calculates the Euclidean distance between the candidate picture and the two cluster centers. The clusters utilized are the two groups of outputs with the same kind of features.

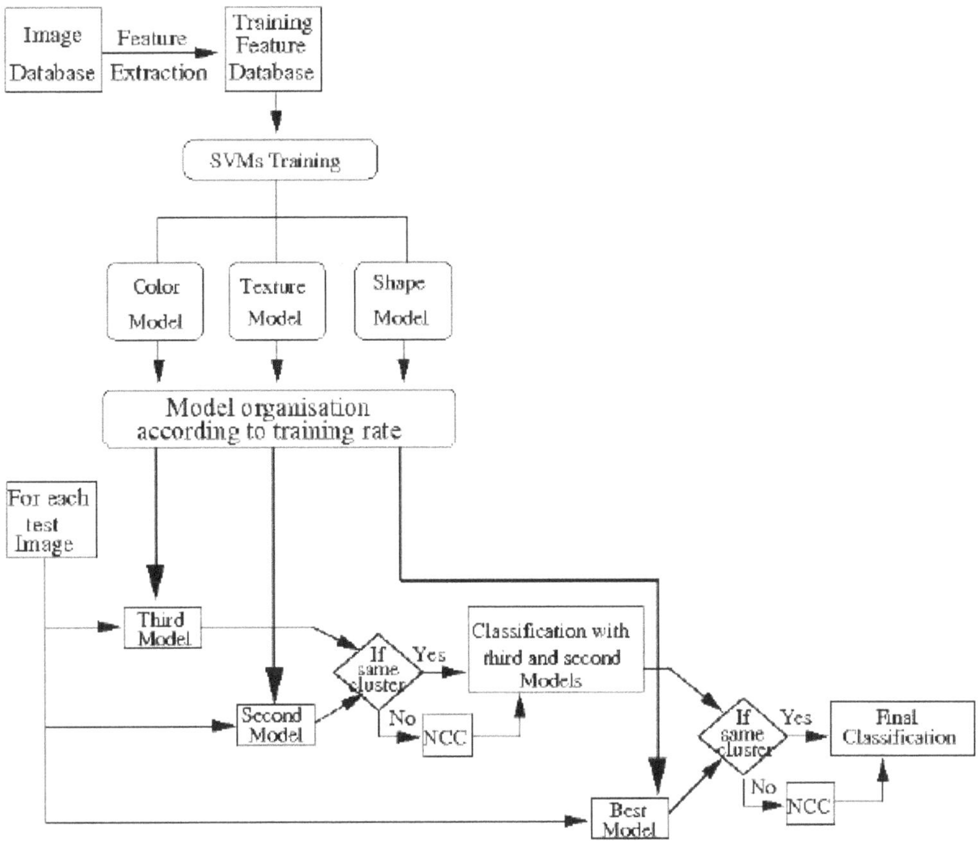


**Fig. 11. Hierarchical best features type selection and organization architecture using different SVM models.**

An analysis is conducted to compare traditional mixed features with the proposed hierarchical feature model. The hierarchical feature model, as shown in figure 11, surpasses the utilization of aggregated features, which involves combining multiple types of features such as color, texture, and shape, without any distinction. In Figure 12 and Figure 13, we display the initial 15 photos that were obtained for a query image. The retrieval was performed using aggregated features and hierarchical features, respectively. Among these two figures, the initial image represents the requested image. It is evident that the hierarchical feature model has a higher retrieval accuracy compared to the use of aggregated features. However, we will show in this section that feature aggregation is not sufficiently efficient when we just combine different types of features. Each descriptor category has a distinct range that differs from the ranges of other descriptor categories.

**Fig. 12. Retrieval examples, using classical aggregated features.**

So, each feature vector extracts signature which is not uniform with the other feature signature extracted from images.

**Fig. 13. Retrieval examples, using hierarchical feature model.**

Consequently, we prove that using proposed hierarchical feature model is more efficient than using aggregated features in an heterogeneous image retrieval system.

Figure 14 proves that using the hierarchical feature model is more efficient than using aggregated features in an image retrieval system. Indeed, we obtain with hierarchical features model 0,815 % representing the good classification results and 0,68 % with aggregated features method.

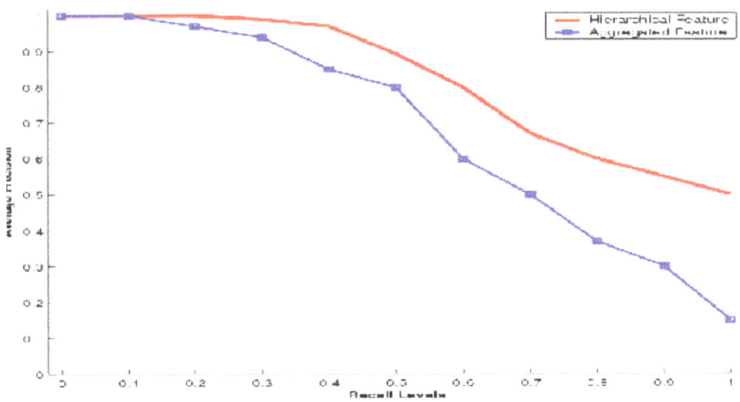

**Fig. 14. Precision-recall graph comparing hierarchical features and Aggregated Features.**

## 6. CONCISE SUMMARY

This chapter has outlined the various phases of an image identification and retrieval system that is specifically designed for various applications within the field of computer vision. The picture description and classification are two crucial stages of an image recognition system in extensive heterogeneous databases. We have elucidated the fundamental concepts of feature extraction, picture description within extensive databases, and the significance of robustness. We have discussed the process of extracting features and making improvements. Additionally, we have emphasized the significance of the classification problem and introduced the supervised SVM classifier.

In order to guarantee optimal feature selection in image retrieval, we have provided and analyzed the efficacy of various types of features and their aggregation methods. We have elucidated the necessity of optimization approaches in CBIR systems and put forth two architectures: the modular statistical optimization and the hierarchical features model. The good results achieved demonstrate the significance of optimization and feature selection in this field.

Searching for Content-Based Image Retrieval (CBIR) systems continues to be a challenging subject. Despite the recognized significance of image retrieval and identification methods and systems in the face of increasing use of picture databases in mobile robotics, research, and education, the various fields have been unable to utilize them effectively. The complex nature of the images requiring treatment and the absence of appropriate technologies have impeded their uptake. Although creating a single all-encompassing system is challenging, it would be feasible to capitalize on the increasing research focus and the existence of other effective systems that have already established strategies for recognizing images in extensive datasets.

## REFERENCES


1. Antania, S., Kasturi, R. & Jain, R. (2002). A survey on the use of pattern recognition methods for abstraction, indexing and retrieval of images and video, *Pattern recognition,* 35(4), pages: 945-965.
2. Bi J., Bennett K., Embrechts M., Breneman C., & Song M., (2003), Dimensionality reduction via sparse support vector machines, *J. Machine Learning Research (JMLR)*, 3, 1229– 1243.
3. Bimbo A. D., Visual Information Retrieval, (2001), *Morgan Kaufmann Publishers, San Francisco, USA*.
4. Blum A.L. & Langley P., (1997), Selection of Relevant Features and Examples *in Machine Learning, Artificial Intelligence*, 97(1- 2), 245–271.
5. Carson, C., Belongie, Se., Greenpan, H. & Jitendra, M. (2002). Blobworld: Image segmentation using Expectation-Maximization and its Application to Image Querying, *IEEE Transaction on Pattern Analysis and Machine Intelligence,* Vol. 24, NO. 8.
6. Chen Y. & Wang J.Z., (2004), Image Categorization by Learning and Reasoning with Regions, J. Machine Learning Research, vol. 5, pp. 913–939.
7. Chen Y., Bi J. & Wang J.Z., (2006), MILES: Multiple-Instance Learning via Embedded Instance Selection, *IEEE Transactions on Pattern Analysis and Machine Intelligence*, vol. 28, no. 12, pp. 1931–1947.
8. Cristianini N. & Taylor J. S., (2000), An Introduction to Support Vector Machines and Other Kernel-Based Learning Methods, *Cambridge University Press.*
9. Daphne K. and Mehran S., (1996), Toward optimal feature selection, *In International Conference on Machine Learning*, 284–292.
10. Delingette H. & Montagnat J., (2001), Shape and topology constraints on parametric active contours, *Computer Vision and Image Understanding*, vol. 83, no. 2, 140–171.
11. Djouak, A., Djemal, K. & Maaref, H. (2007). Image Recognition based on features extraction and RBF classifier, *Journal Transactions on Signals, Systems and Devices, Issues on Comminucation and Signal Processing,* Shaker Verlag, Vol. 2, N°. 3, pp: 235-253.
12. Djouak, A., Djemal, K. & Maaref, H., (2005a), Image retrieval based on features extraction and RBF classifiers, *IEEE International Conference on Signals Systems and Devices, SSD 05*, Sousse. Tunisia.
13. Djouak, A., Djemal, K. & Maaref, H.k, (2006). Modular statistical optimization and VQ method for images recognition, International Conference on Artificial

Neural Networks and Intelligent Information Processing, pp: 13-24, ISBN: 978-972-8865-68- 9, Setúbal, Portugal, August.

14. Djouak, A., Djemal K. & Maaref, H., (2005b), Features extraction and supervised classification intended to image retrieval. *In IMACS World Congress: Scientific Computation, Applied Mathematics and Simulation,* Paris, France.
15. Egmont-Petersen, M., de Ridder & Handels, D. H. (2002). Image processing with neural networks-a review. *Pattern Recognition,* 35(10):2279–2301.
16. Faloutsos C., Equitz W., Flickner M., Niblack W., Petkovic D. & Barber R., (1994), Efficient and Effective Querying by Image Content, *Journal of Intelligent Information Systems,* vol. 3, No 3/4, 231– 262.
17. Guyon I.,Weston J., Barnhill S., & Vapnik V., (2002), Gene selection for cancer classifcation using support vector machines, *Machine Learning* 46, 389–422.
18. Guyon I., & Elisseeff A., (2003), An introduction to feature and variable selection, *Journal of Machine Learning Research*, vol. 3, 1157–1182.
19. Hafner J., Sawhney H.S., Equitz W., Flickner M., & Niblack W., (1995), Efficient color histogram indexing for quadratic form distance function, IEEE Trans. *Pattern Anal. Mach. Intell.*, 17, 7, 729–736.
20. Haralick R.M.S.K., & Dinstein I., (1973), Textural Features for Image Classification, *IEEE Transactions on Systems, Man and Cybernetics*, 3(6), 610-621.
21. Hu M.K., (1962), Visual pattern recognition by moment invariants, *IEEE Transactions information Theory*, 8, 179–187.
22. Huang J., Kumar S.R., Mitra M., & Zhu W.J., Spatial color indexing and applications, (1999), *Intl. Conf. on Computer Vision*, 35, 3, 245–268.
23. Huang, J., Kumar, S. R., Mitra, M., Zhu, W.-J. & Zabih, R., (1997). Image indexing using color correlograms. *In Proc. IEEE Comp. Soc. Conf. Comp. Vis. and Patt. Rec.,* pages 762–768.
24. Jacobs, C., Finkelstein, A. & Salesin, D. (1995). Fast multiresolution image querying. *In Proc. SIGGRAPH.*
25. Julezs. B., (1975), Experiments in the visual perception of texture. *Scientific American*, 232(4):2–11.
26. John G.H., Kohavi R., & Pfleger K., Irrelevant features and the subset selection problem, 1994, *In International Conference on Machine Learning, pages 121–129. Journal version in AIJ*, available at http ://citeseer. nj.nec.com/13663.html.
27. Kachouri R., Djemal K., Maaref H., Sellami Masmoudi D., & Derbel N., (2008b), Heterogeneous Image Retrieval System Based On Features Extraction and SVM Classifier, *International Conference on Informatics in Control, Automation and Robotics, ICINCO 08*, Funchal, Madeira, Portugal.

28. Kachouri R., Djemal K., Sellami Masmoudi D., & Derbel N., (2008c), Content Based Image Recognition based on QUIP-tree Model, *IEEE International Conference on Signals Systems and Devices, SSD 08*, Amman, Jordan.
29. Kachouri R., Djemal K., Maaref H., Masmoudi, D.S., Derbel N., (2008), Content description and classification for Image recognition system, *3rd IEEE International Conference on Information and Communication Technologies: From Theory to Applications, ICTTA08* 1– 4.
30. Kadyrov, A. & Petrou, M. (2001). The Trace transform and its applications. *IEEE Transactions on Pattern Analysis and Machine Intelligence, PAMI*, Vol:811–828.
31. Lin, Chun-Yi., Yin, Jun-Xun., Gao, X., Chen, Jian-Y. & Qin, P. (2006). A Semantic Modeling Approach for Medical Image Semantic Retrieval Using Hybrid Bayesian Networks, *Sixth International Conference on Intelligent Systems Design and Applications (ISDA'06)*, vol. 2, pp.482-487.
32. Lipson, P., Grimson, E. & Sinha, P. (1997). Configuration based scene classification and image indexing. In *Proc. IEEE Comp. Soc. Conf. Comp. Vis. and Patt. Rec.*, pages 1007– 1013.
33. Manjunath B.S., Ohm J.-R., Vasudevan V.V., & Yamada A., (2001), Color and texture descriptors, *IEEE transaction on circuits and systems for video technology*, 11, 703–715.
34. Meng J.E., Shiqian W., Juwei L., & Hock L.T., (2002), Face Recognition With Radial Basis Function (RBF) Neural Network, *IEEE Transaction on Neural Networks*, 13, 3.
35. Press W.H., Flannery B.P., Teukolsky S.A., & Vetterling W.T., (1987), Numerical Recipes, The Art of Scientific Computing, *Cambrigde, U.K.:* Cambridge Univ.
36. Rezai-Rad G., & Aghababaie M., (2006), Comparison of SUSAN and Sobel Edge Detection in MRI Images for Feature Extraction, *Information and Communication Technologies, ICTTA 06*. 1, 1103–1107.
37. Ramesh J. R., Murthy S.N.J., Chen P.L.-J., & Chatterjee S., (1995), Similarity Measures for Image Databases, *Proceedings of IEEE International Joint Conference of the Fourth IEEE International Conference on Fuzzy Systems and The Second International Fuzzy Engineering Symposium*, 3, 1247-1254.
38. Sastry, Ch. S., Pujari, Arun K., Deekshatulu, B. L. & Bhagvati, C. (2004). A wavelet based multiresolution algorithm for rotation invariant feature extraction, *Pattern Recognition Letters*, 25:1845–1855.
39. Sclaroff, S. & Pentland, A. (1995). Modal Matching for Correspondence and Recognition, *IEEE Trans. Pattern Analysis and Machine Intelligence*, 17(6):545–561.
40. Serrano N., Savakisb A.E., & Luoc J., (2004), Improved scene classification using efficient low-level features and semantic cues, *Pattern Recognition*, 37 , 1773–1784.

41. Smith, J. R. & Chang, S.-F. (1996). Tools and techniques for color image retrieval. *In SPIE Proc. Storage and Retrieval for Image and Video Databases*, volume 2670, pages 426–437
42. Stricker, M. & Dimai, A. (1997). Spectral covariance and fuzzy regions for image indexing. *Machine Vision and Applications,* 10(2):66–73
43. Stricker, M., & Swain, M. (1994). Capacity and the sensitivity of color histogram indexing, *Technical Report*, 94-05, University of Chicago.
44. Swain, M. & Ballard, D. (1991). Color indexing, *International Journal of Computer Vision,* 7(1), pages:11–32.
45. Takashi, I. & Masafumi, H. (2000). Content-based image retrieval system using neural networks. *International Journal of Neural Systems,* 10(5):417–424.
46. Vapnik, V., (1998). Statistical learning theory, *Wiley-Interscience.*
47. Wu, J., (2003), Rotation Invariant Classification of 3D Surface Texture Using Photometric Stereo. *PhD Thesis, Heriot-Watt University*.
48. Yu L. & Liu H., (2004), Efficient Feature Selection via Analysis of Relevance and Redundancy, *J. Machine Learning Research*, vol. 5, pp. 1205–1224.
49. Zhu J., & Hastie T., (2003), Classication of gene microarrays by penalized logistic regression, *Biostatistics*.

# INDEX

www.ingramcontent.com/pod-product-compliance
Lightning Source LLC
LaVergne TN
LVHW062216200926
848484LV00006B/1407